URBAN AREAS AND GLOBAL CLIMATE CHANGE

RESEARCH IN URBAN SOCIOLOGY

Series Editor: Ray Hutchison

Recent Volumes:

Volume 1:	Race, Class and Urban Change, 1989
Volume 2:	Gentrification and Urban Change, 1992
Volume 3:	Urban Sociology in Transition, 1993
Volume 4:	New Directions of Urban Sociology, 1997
Volume 5:	Constructions of Urban Space, 2000
Volume 6:	Critical Perspectives on Urban Redevelopment, 2001
Volume 7:	Race and Ethnicity in New York City, 2004
Volume 8:	Ethnic Landscapes in an Urban World, 2006
Volume 9:	Gender in an Urban World, 2008
Volume 10:	Suburbanization in Global Society, 2010
Volume 11:	Everyday Life in the Segmented City, 2011

RESEARCH IN URBAN SOCIOLOGY VOLUME 12

URBAN AREAS AND GLOBAL CLIMATE CHANGE

EDITED BY

WILLIAM G. HOLT

Birmingham-Southern College, Birmingham, AL USA

United Kingdom – North America – Japan
India – Malaysia – China

Emerald Group Publishing Limited
Howard House, Wagon Lane, Bingley BD16 1WA, UK

First edition 2012

British Library Cataloguing in Publication Data
A catalogue record for this book is available from the British Library

ISBN: 978-1-78190-036-9
ISSN: 1047-0042 (Series)

CONTENTS

LIST OF CONTRIBUTORS *ix*

CLIMATE CHANGE IMPACTS ON URBAN AREAS
William G. Holt *xiii*

SECTION 1: THE GLOBAL NORTH

CHAPTER 1 INSTITUTIONAL AND SOCIAL CAPACITIES IN LEAD CITIES IN EUROPE AND THE UNITED STATES: SUCCESS FACTORS FOR URBAN SUSTAINABILITY?
Corinna Altenburg *3*

CHAPTER 2 GROWTH MACHINES AND CARBON EMISSIONS: A COUNTY-LEVEL ANALYSIS OF HOW U.S. PLACE-MAKING CONTRIBUTES TO GLOBAL CLIMATE CHANGE
Matthew Thomas Clement and James R. Elliott *29*

CHAPTER 3 TACKLING CLIMATE CHANGE ADAPTATION AT THE LOCAL LEVEL THROUGH COMMUNITY PARTICIPATION
Geneviève Cloutier and Florent Joerin *51*

CHAPTER 4 CLIMATE PROTECTION AND CIVIL SOCIETY: DOES EFFECTIVE LOCAL CLIMATE POLICY NEED THE PARTICIPATION AND ENGAGEMENT OF CITIZENS? A COMPARISON BETWEEN THE CITIES OF POTSDAM AND MUENSTER
Mirjam Neebe and Fritz Reusswig *75*

CHAPTER 5 MILAN'S ANSWER TO THE CLIMATE CHANGE PROBLEM
Ilaria Beretta *105*

CHAPTER 6 WIN, LOSE, OR DRAW? ASSESSING THE SUCCESS OF THE ENVIRONMENTAL JUSTICE MOVEMENT IN EMISSIONS TRADING SCHEMES
Krystal Tribbett *135*

SECTION 2: THE GLOBAL SOUTH

CHAPTER 7 CLIMATE ADAPTATION IN THE FACE OF RESOURCE CONSTRAINTS: LESSONS FROM A COASTAL SOUTH ASIAN MEGA-CITY
Madhu C. Dutta-Koehler *171*

CHAPTER 8 ADAPTING TO WHAT? CLIMATE CHANGE IMPACTS ON INDIAN MEGACITIES AND THE LOCAL INDIAN CLIMATE CHANGE DISCOURSE
Fritz Reusswig and Lutz Meyer-Ohlendorf *197*

CHAPTER 9 ENVIRONMENTAL STATE IN TRANSFORMATION: THE EMERGENCE OF LOW-CARBON DEVELOPMENT IN URBAN CHINA
Yifei Li *221*

CHAPTER 10 URBAN VULNERABILITY AND ADAPTATION TO THE HEALTH IMPACTS OF AIR POLLUTION AND CLIMATE EXTREMES IN LATIN AMERICAN CITIES
Patricia Romero-Lankao, Hua Qin, Sara Hughes, Melissa Haeffner and Mercy Borbor-Cordova *247*

CHAPTER 11 CITIES IN THE FLOOD: VULNERABILITY AND DISASTER RISK MANAGEMENT: EVIDENCE FROM IBADAN, NIGERIA
Andrew Onwuemele *277*

CHAPTER 12 GLOBAL ENVIRONMENTAL CHANGES AND IMPACTS ON FISHING ACTIVITIES IN THE NORTHERN COAST OF SÃO PAULO, BRAZIL
Sônia Regina da Cal Seixas, Michelle Renk, João Luiz de Moraes Hoeffel, André Luiz da Conceição and Gabriela Farias Asmus *301*

CHAPTER 13 CLIMATE CHANGE AND THE NORTH COAST OF JAKARTA: ENVIRONMENTAL JUSTICE AND THE SOCIAL CONSTRUCTION OF SPACE IN URBAN POOR COMMUNITIES
Rita Padawangi *321*

CHAPTER 14 THE GENDER DIMENSIONS OF CLIMATIC IMPACTS IN URBAN AREAS: EVIDENCE AND LESSONS FROM KAMPALA CITY, UGANDA
Buyana Kareem *341*

LIST OF CONTRIBUTORS

Corinna Altenburg	Potsdam Institute for Climate Impact Research, University of Potsdam, Potsdam, Germany
Gabriela Farias Asmus	Centre for Environmental Studies and Research-NEPAM, Cidade Universitaria Zeferino Vaz, São Paulo, Brazil
Ilaria Beretta	Postgraduate School for Environmental Studies, Catholic University, Brescia, Italy
Mercy Borbor-Cordova	Ministry of Environment, Quito, Ecuador
Matthew Thomas Clement	Department of Sociology, University of Oregon, Eugene, OR, USA
Geneviève Cloutier	School of Planning and Regional Development, Université Laval, Quebec, Canada
Sônia Regina da Cal Seixas	Centre for Environmental Studies and Research-NEPAM, Cidade Universitaria Zeferino Vaz, São Paulo, Brazil
André Luiz da Conceição	School of Mechanical Engineering, Cidade Universitaria Zeferino Vaz, São Paulo, Brazil
João Luiz de Moraes Hoeffel	Centre for Sustainability Studies, FAAT College, São Paulo, Brazil
Madhu C. Dutta-Koehler	Department of Urban Studies and Planning, Massachusetts Institute of Technology, Cambridge, MA, USA

James R. Elliott	Department of Sociology, University of Oregon, Eugene, OR, USA
Melissa Haeffner	Resilient and Sustainable Cities (RAL-ISP), National Center for Atmospheric Research, Boulder, CO, USA
William G. Holt	Urban Environmental Studies Program, Birmingham-Southern College, Birmingham, AL, USA
Sara Hughes	Resilient and Sustainable Cities (RAL-ISP), National Center for Atmospheric Research, Boulder, CO, USA
Florent Joerin	School of Planning and Regional Development, Université Laval, Quebec, Canada; University of Applied Sciences Western Switzerland, Vaud, Switzerland
Buyana Kareem	Department of Socioeconomic Sciences, Cavendish University, Kampala, Uganda
Yifei Li	Department of Sociology, University of Wisconsin, Madison, Madison, WI, USA
Lutz Meyer-Ohlendorf	Potsdam Institute for Climate Impact Research, University Potsdam, Potsdam, Germany
Mirjam Neebe	Potsdam Institute for Climate Impact Research, Potsdam, Germany
Andrew Onwuemele	Nigerian Institute of Social and Economic Research (NISER) and Governance Policy Research Development, Ibadan, Oyo State, Nigeria

Rita Padawangi	Institute of Water Policy, Lee Kuan Yew School of Public Policy, National University of Singapore, Singapore
Hua Qin	Resilient and Sustainable Cities (RAL-ISP), National Center for Atmospheric Research, Boulder, CO, USA; Department of Environmental Studies, University of Illinois at Springfield, Springfield, IL, USA
Michelle Renk	Centre for Environmental Studies and Research-NEPAM, Cidade Universitaria Zeferino Vaz, São Paulo, Brazil
Fritz Reusswig	Potsdam Institute for Climate Impact Research, University Potsdam, Potsdam, Germany
Patricia Romero-Lankao	Resilient and Sustainable Cities (RAL-ISP), National Center for Atmospheric Research, Boulder, CO, USA
Krystal Tribbett	History of Science and Science Studies Program, University of California, San Diego, La Jolla, CA, USA

CLIMATE CHANGE IMPACTS ON URBAN AREAS

On the eve of the UN Commission on Sustainable Development's (UNCSD) conference on sustainable development (Rio + 20) in June 2012, the United Nation's Environmental Program (UNEP)'s Global Outlook Report (GEO-5) provided sobering data on global environmental progress. The report indicated that world nations are making little headway on significant environmental targets set for themselves under the Millennium Development Goals. The UN indicated that world nations made progress on only 4 of the 90 most significant objectives from the Millennium Plan: reducing substances depleting the ozone layer, removing lead from fuel, increasing access to water supplies, and increasing research on ways to reduce pollution in marine environments. However, in some cases no progress or regression occurred in reaching goals on climate change issues including limiting increases in average global temperatures to less than two degrees above preindustrial levels nor advances in issues such as revitalization of depleted fish stocks, protection of biodiversity, and combating desertification (UNEP, 2012).

The UNEP report highlighted the different environmental threats among developed and developing nations. North America and Europe continue to maintain "unsustainable levels of consumption" with North America trailing the world in renewable energy. The report indicated that Asia, Africa, and the Pacific face environmental threats from rapid urbanization combined with increased consumption that stresses natural resources. The report showed similar issues for Latin America and the Caribbean. The hope of the Rio + 20 Conference was to jumpstart these environmental concerns (UNEP, 2012).

With a world population that is now majorly urban, the threat of global climate change to the world's urban regions is an international concern. Climate change will impact water cycles with wet areas becoming wetter and dry areas drier. Rising sea levels will change where people can live in urban areas. The purpose of this edited volume is to examine how urban areas address these environmental issues at the local level. Since environmental issues take different forms in developed and developing nations, this edition is divided into two sections focusing on the Global North and Global South.

Each section begins with a comparative look at urban areas. Then, there are chapters examining how specific urban areas address climate change. Finally, the final chapters discuss the individual impacts of climate change on residents, focusing on social stratification issues crossing class, race, and gender.

In Chapter 1, Corinna Altenburg compares the relationships between urban climate governance in European and American cities examining the idea of what makes a sustainable city work well. Altenburg's case studies indicate that efficient, creative, and participatory urban governance is key for the quest of sustainability. She argues that dynamic governance involving both high levels of institutional and social capital or capacity produces the most successful sustainability programs.

Drawing on growth machine theory in Chapter 2, Matthew Thomas Clement and James R. Elliott take an urban-level approach to examining carbon emissions in US cities. Their regression analysis indicates that local demographic, economic, and consumption activity all relate independently and positively to total carbon emissions at the local level. They argue that social systems that treat land as a commodity increase local carbon emissions over those that do not.

Moving from comparisons among cities to local initiatives, we see that North American and European cities are dealing with issues involved in environmental planning from development to citizen participation. In Chapter 3, Geneviève Cloutier and Florent Joerin look at Québec City, Canada's cross-sectorial environmental initiative adaption of planning processes that considers local issues in a global debate. Their research provides a framework from which other cities may examine the roles for territorial actors and community stakeholders in this local decision-making process. In Chapter 4, Mirjam Neebe and Fritz Reusswig compare German cities examining differences between Western capitalism and former-Soviet cities in addressing climate change issues. They explore two medium-sized German cities – Muenster in Rhineland Westphalia, and Potsdam, the capital of Brandenburg next to Berlin. While Muenster, a prospering university city in former West Germany, has a long-standing tradition in environmental policy and climate protection, Potsdam, located in the former German Democratic Republic (GDR), has started only recently to actively engage with local climate policy. Neebe and Reusswig explore the role of civil society as these two different cities develop environmental initiatives. In Chapter 5, Ilaria Beretta describes Milan, Italy's policies and plans to reduce greenhouse gas emissions under the European Union's policies. Beretta's work examines Milan's roles in national and international networks while developing its own initiatives demonstration cooperation among cities.

The Global North section concludes by examining issues of environmental injustice through regulations. In Chapter 6, Krystal Tribbett questions the implementation of emissions trading in Southern California's South Coast Air Basin. Showing the parallel development of emissions trading policies and the environmental justice movement during the 1960s through 1980s, Tribbett uses a historical perspective to explain how their different approaches to cleaning air created opposition to each other. Tribbett analyses the policy clashes calling simultaneously for economically efficient antiregulatory-ism and environmental equity. Tribbett's work shows how efforts to address climate change may lead to environmental injustices within local areas.

The Global South section begins with Madhu C. Dutta-Koehler's analysis of the planning and implementation of climate-adaptation and vulnerability-reduction strategies in coastal megacities of the Global South, utilizing Kolkata, India as a case study. In Chapter 7, Dutta-Koehler uses field interviews to identify factors that aid the implementation of climate-centered action in resource-constrained environments of developing countries while providing a set of policy guidelines reflecting best practices. In Chapter 8, Fritz Reusswig and Lutz Meyer-Ohlendorf discuss the role of the media in shaping public awareness and opinions. Reusswig and Meyer-Ohlendorf show how India is moving away from its "anticolonial climate discourse," and accepting the common, but differentiated climate responsibility of India as a nation.

The globalization of cities in developing nations is characterized by rapid urbanization coupled with rises in socioeconomic levels creating new energy demands. In Chapter 9, Yifei Li compares five Chinese cities. Li argues that as cities move into postindustrialization phases more support emerges for innovative low-carbon development strategies. Li shows that a new decentralized process involving local levels of governments as key players of low-carbon policy-making emerges to address these issues. In Chapter 10, Patricia Romero-Lankao, Hua Qin, Sara Hughes, Melissa Haeffner, and Mercy Borbor-Cordova examine the environmental vulnerability and health risks of four Latin American cities: Bogota, Columbia; Buenos Aires, Argentina; Mexico City, Mexico; and Santiago, Chile. Combining quantitative and qualitative data, they focus on the intersection of these environmental risks with socioeconomic issues. The research team found that while wealthier populations have access to education, good quality housing, and health services to mitigate some environmental risks, overall, the data show that health impacts from air pollution and temperature in the study cities do not necessarily depend on socioeconomic differentiations.

Continuing this examination of the intersection of urbanization and the environment, Andrew Onwuemele in Chapter 11 analyzes the impacts of reoccurring flooding in Ibadan, Nigeria. Onwuemele discusses the impacts, vulnerability factors, and disaster risk management framework based on the August 26, 2011 flood disaster in Ibadan that caused monumental destructions in the city. Onwuemele identifies the continuous construction on floodplains, indiscriminate dumping of refuse, excessive rainfall, and deforestation as the main vulnerability factors. In Chapter 12, Sônia Regina da Cal Seixas, Michelle Renk, João Luiz de Moraes Hoeffel, André Luiz da Conceição, and Gabriela Farias Asmus examine the impacts on the economy as well as daily lives in Cocanha beach (Caraguatatuba city, north coast of the state of São Paulo). Their qualitative research indicates a decrease in fish stocks linked to changes in fishing, in climate, and beach landscape. Moreover, the category related to increased water temperature was highly significant, since the fishing and mariculture activities are directly influenced by this factor.

Environmental justice issues emerge as urban officials are incorporating climate change initiatives into their long-term planning efforts. In Chapter 13, Rita Padawangi examines Jakarta, Indonesia's master plan 2010–2030. Padawangi argues that the top-down nature of planning would likely produce, reproduce, or reaffirm unjust urban geographies in the name of climate change adaptation. North Jakarta and its coastal area are areas prone to climate change risks and are home to around 48,000 poor households, most of which live in houses less than $50\,m^2$ in informal settlements with lack of basic needs infrastructures. Padawangi describes how these communities are unaware of city officials' plans but desire to be part of this planning process. In Chapter 14, Buyana Kareem explores the gender dimensions of climatic impacts in urban areas based on African cities. Kareem draws on focus group discussions with female and male residents of Kasubi-Kawaala neighborhood in Kampala city dealing with prolonged dry spells, erratic heavy rains, and seasonal floods, which destroy physical infrastructure, expose households to environmental health hazards, contaminate air and water sources, and lead to unprecedented spread of cholera and malaria. Kareem argues that these climatic impacts on one hand do worsen gender inequalities across different urban sectors while on the other such gender inequalities contribute to the intensity of climatic impacts. These are the gender dimensions of climatic impacts in urban area that require deep examination while planning to adapt or reduce emissions.

International environmental issues are viewed typically from a global rather than a local perspective. Since the 1945 Bretton Woods meetings held after World War II to regulate international financial order, the World Bank expanded developed nations' roles in developing countries' social and economic policies. The original Bretton Woods policies evolved over the 1970s and 1980s ending in 1995 with the establishment of the World Trade Organization. Growing out of the 1972 Stockholm Conference on the Human Environment and the 1980 World Conservation Strategy of the International Union for the Conservation of Nature, world leaders recognized the growing tensions between development, growth, and industrialization with environmental impacts. In 1983 the UN established the World Commission on Environment and Development (WCED), known as the Bruntdland Commission, to be independent of the UN and focus on tensions between development and the environment. In 1987, the UN adopted the Report of the Bruntdland Commission, *Our Common Future*, defining sustainable development and policies to achieve these goals (WCED, 1987).

Twenty years ago, the first Earth Summit in Rio drew more than 100 world leaders and international press attention. The 2012 Rio + 20 Conference was not a transforming moment in world environmental policy. The organizers pushed back the schedule not to conflict with Queen Elizabeth's Diamond Jubilee in London. The event was only three days compared to 14 at the first Rio summit. World leaders did not attend (Pearce, 2010). Global North nations facing economic crisis and potential election failures made no new proposals. Developing nations in the Global South are pressing to weaken international guidelines as they move into developed status. Since global leaders are ignoring these issues, local urban regions are developing their own strategies. While global climate change cannot be solved locally in a piecemeal process, the world's urban areas are implementing strategies that should be addressed by world leaders.

William G. Holt
Editor

REFERENCES

Pearce, F. (2010). On the road back to Rio: Green direction has been lost. Retrieved from http://www.e360.yale.edu. Accessed on May 2, 2012.

United Nations Environmental Programme (UNEP). (2012). *Global environmental outlook-(GEO5)*. Nairobi, Kenya: United Nations Environmental Programme.

World Commission on Environment and Development (WCED). (1987). *Our common future: Report of the world commission on environment and development*. Oxford: Oxford University Press.

SECTION 1
THE GLOBAL NORTH

CHAPTER 1

INSTITUTIONAL AND SOCIAL CAPACITIES IN LEAD CITIES IN EUROPE AND THE UNITED STATES: SUCCESS FACTORS FOR URBAN SUSTAINABILITY?

Corinna Altenburg

ABSTRACT

Purpose – *The aim of this chapter is to review the relationship of urban climate governance against the background of sustainability and to identify driving forces in a set of leading cities in the United States and Europe.*

Design/methodology/approach – *Case-study research indicates that efficient, creative and participatory urban governance is key for the quest for sustainability. It is theorized that dynamic governance is composed of high levels of both institutional and social capital or capacity. Each capacity category is composed of different indicators or 'success factors' taken from qualitative data in overall 84 cities in the United States and Europe. By triangulating desktop research, mail surveys and review of existing studies, the role and influence of these success factors is evaluated and compared.*

Urban Areas and Global Climate Change
Research in Urban Sociology, Volume 12, 3–28

ISSN: 1047-0042/doi:10.1108/S1047-0042(2012)0000012004

Findings – *In both the United States and in Europe, the key driving forces are those related to institutional capacity. Yet, cities tend to be more successful in sustainable development if levels of institutional* and *social capacity are high. Linkages between institutional and social dimensions can be successfully strengthened by capacity building measures.*

Research limitations/implications – *As the comparison is based on existing studies on leading cities in sustainability, driving factors likely to play a stronger role in the early phase of urban sustainability measures are underrepresented.*

Practical implications – *Local authorities are advised to encourage social capacities in order to help them pursue local sustainability over time.*

Originality/value – *This chapter compares and evaluates underlying success factors going beyond a case-study approach across the Atlantic in order to draw broader conclusions.*

Keywords: Urban governance; institutional capacity; social capital; sustainable development; climate change

INTRODUCTION

During the past two decades paradigms of urban development have gradually moved from economic growth towards a more sustainable approach. This change is well reflected by a growing number of city projects featuring sustainable development, rising membership numbers of municipal networks across the world and also by an increasing output of scientific reports and papers on urban sustainability and climate issues (e.g. Brotchie, 1999; Camaren & Swilling, 2011; Chance, 2009; Girardet, 1999; Ooi, 2005; Rosenzweig, Solecki, Hammer, & Mehrotra, 2011).

While the environmental dimension of sustainable urban development has been arguable addressed most often, practitioners and scientists have been also incorporating social and economic aspects. In spite of a large amount of research and diverse local action, it is yet unclear how a 'true' sustainable city would look like or if it will or can ever exist. Even though many cities are implementing actions *towards* sustainability, the often used term of '*the* sustainable city' describes rather a process and not a current status-quo (Newman, 2006; Sorensen, 2004).

Some cities around the world are often referred to as leaders in their fields. For instance, Freiburg, Curitiba, Portland, Copenhagen, Stockholm and Cape Town have developed progressive sustainability plans. Some of these conceptual ideas come pretty close to what city planners would call a sustainable city. However, at the current state the majority of cities improve just specific sections, leaving behind other domains (Beatley, 2000).

Beginning in the early 1990s, climate change issues have become increasingly important within the quest for sustainable development. On the international scale, this is best reflected by the foundation of the Intergovernmental Panel of Climate Change (IPCC) in 1988, the kick-off of the United Nations Framework Convention on Climate Change (UNFCCC) in 1992 and its various subsequent international conferences of its parties (COPs). The most recent ones in Copenhagen (2009), Cancun (2010) and Durban (2011) underlined the ongoing struggles of the international community to establish binding rules for a climate mitigation treaty. Reflecting international developments, climate change has similarly emerged on the local level as a critical topic – albeit with less pressure to reach global agreements. Frontrunner cities were among the first to integrate climate protection in their portfolio of sustainability action around the early-1990s (Holden, Roseland, Ferguson, & Perl, 2008). With the urgency to respond to rising global greenhouse gas emissions as well as to adapt to current and future changes, sub-national climate governance has since then become ever more important. Many more cities are nowadays committed to some sort of climate change policies – mostly on a voluntary basis. Depending on specific vulnerabilities, location and energy usage, these policies may sometimes focus on adaptation or mitigation measures. Framing in cities also differs. In some cities, climate change mitigation or adaptation is a way to achieve urban sustainability goals, in others local sustainable projects are implemented 'anyway' and climate change is a additional pressure next to other issues such as the economic crisis or peak-oil (Slavin, 2011). In either case, governance for sustainable development and climate change are closely interlinked. Depending on the government structure, both policy arenas might be attached to varying interest groups, technical and legal capacities and policy outcomes. As most of sustainability or climate-related urban issues are intertwined with existing city departments such as transport, buildings or education, many municipal decisions touch upon these policies. Currently, many local governments are putting more and more effort into adaptation measures. Therefore, it will become important to better understand potential trade-offs between mitigation, adaptation and sustainable development. With a few notable exceptions, there has been relatively little research on particular trade-offs on a sub-national level and how this is addressed in various municipal departments (Corfee-Morlot et al., 2009; Viguié & Hallegatte et al., 2012).

Mirroring increasing urban measures on climate change and sustainability and ongoing rapid urbanization, particularly in the Global South, the international and scientific communities have acknowledged the topic. Notably, many international organizations ranging from the World Bank (2010) to the United Nations Human Settlements Programme (2011) and the Organisation for Economic Co-operation and Development (OECD) (2011) have lately published assessment reports on the issues of cities, climate change and sustainability. Research has evolved through a number of different approaches stemming from diverse disciplines such as urban planning, geography, economics, political science and sociology. A large part of this research refers to valuable case-study approaches in which single urban sites were studied. To name a few prominent topics, some authors have addressed urban governance issues (e.g. Alber & Kern, 2008; Brandon & Lombardi, 2005; Crosbie & Baker, 2010; Hoff, 1998), planning approaches to eco-communities (e.g. Barton, 2000; Beatley, 2000, Brotchie, 1999; Girardet, 1999; Inoguchi, Newman, & Paoletto, 1999; Register, 2006; Wheeler, 2005), or the role of civil society and equity issues (Agyeman & Evans, 2006; Barnes, Sullivan, Knops, & Newman, 2004; Carlye, Jenkins, & Smith, 2001; Portney & Berry, 2010). In 2011, a comprehensive assessment report by the newly founded 'Urban Climate Change Research Network' was published which takes up the latest results and case studies on various research threads (Rosenzweig et al., 2011). Comprehensive overviews of the state of the current literature, especially focusing on governance in urban areas, can be found in Anguelovski and Carmin (2011) and Bulkeley (2010).

Only a few authors have so far dealt with a coherent comparison of multiple sustainable or low-carbon cities focusing on governance. If this was the case, comparison was then confined to a particular continent or a small number of case studies (e.g. Portney, 2003: USA (44 cities), Beatley, 2000: Europe (27 cities); Evans, Joas, Sundback, & Theobald, 2005: Europe (40 cities); Sorensen, 2004: United States and East Asia; Moore, 2007: Curitiba, Austin, and Frankfurt; Bulkeley & Betsill, 2005: UK, United States, Australia (6 cities); Dhakal, 2008: Canada (3 cities)).[1]

These and many other case studies are very valuable as they increased our understanding of sub-national characteristics in the pursuit of sustainable development and effective climate governance. Because of the variety of urban features, ranging from varying financial, political and human resources to differences in the size, location or climatic condition of municipalities, generalization of results is often difficult. In fact, this is one of the inherent problems in urban comparative studies. However, to reach broader conclusion about the whereabouts of sustainable cities on different continents, it is useful to study multiple cities on a meta-level. The purpose of this study is to go beyond the single case-study approach. Hence, this chapter draws on two other studies which have done so in the United States and in Europe and compare these results with each other. Therefore it is possible to detect underlying common enablers which seem to be crucial for sustainable local policy outcomes. The focus lies on governance features composed of various institutional and social capacities.

What follows is a brief overview on how urban climate governance is interlinked with urban sustainability concepts which have started in the late 80s and early 90s. It will then move on to the central question of this chapter: which are the underlying driving factors for ambitious local sustainability and climate policies in cities? How might they differ in comparing US and European cities? To answer these questions, we will firstly clarify the role of institutional and social capital and capacity building for sustainable policy outcomes in urban areas. Through a qualitative assessment of comparative US and European case studies, certain enabling or success factors emerged, which will then be depicted in more detail. Lastly, this chapter concludes with an overview of research limitations, conclusions and ways for future research.

SUSTAINABILITY, THE URBAN AND CLIMATE CHANGE

The sustainable city – as a term as such – is rooted in the general discourse of sustainable development. While the international diffusion of these concepts received major inputs from the World Commission on Environment and Development (WCED) report 'Our common future' of 1987, also known as the Brundtland report,[2] and the following United Nations Conference on Environment and Development (UNCED) in Rio de Janeiro in 1992 – better known as the Earth Summit – there have been also earlier notions on sustainable development. With regard to sustainable city development, authors such as Lewis Mumford coined visions of the re-development of the city from the1930s onwards as an 'organic community, designed on a human scale and oriented towards human needs, fuelled by a life-enhancing economy, surrounded by undeveloped lands, and streets filled with people instead of automobiles' (Wheeler, 2005, p. 21).[3] In retrospective, these ideas seem pretty similar to some of the vision statements in city plans today.

The Brundtland report is not only the central source for the understanding of sustainable development underlying both the Rio Conference and the follow-up process by the United Nations Council of Sustainable Development (UNCSD). It has also shaped the most common definition of sustainability as a development which 'meets the needs of the present without compromising the ability of future generations to meet their own needs' (WCED, 1987) and which is supposed to find a balance between the three pillars of environmental, social and economic sustainability. In fact, rarely are papers found *not* referring to this definition (including this one). This definition was and still is being contested, for instance because for being too anthropocentric, too general in terms or too lax about differentiating 'needs' of people (e.g. Baker, 2006; Wheeler, 2005). In the year of the 20th anniversary of the Earth Summit (which led organizers to call the conference in 2012 'Rio + 20'), the concept has shifted to focus on a new term: green economy or green growth. Some have argued that green economy represents the next 'oxymoron' (Brand, 2012), just as sustainable development was used to express squaring the circle (Sachs, 1999).While it remains to be seen how international negotiations on sustainable development will move forward facing rather challenging tasks, it is clear that cities have made it on the international agenda. Next to other themes, such as job growth and poverty, sustainable cities are one of the seven core themes of Rio + 20.

But what does a sustainable city actually imply and how does it relate to climate change? International organization and planners have come up with a variety of responses and definitions. All these conceptions are usually not sharply dividable from each other and they often share common characteristics. For instance, the idea of the 'eco-city' emphasizes urban infrastructure and design which corresponds to environmentally-friendly approaches, but it neither excludes aspects of social nor economic scope (OECD, 1996, Register, 2006). Early on, the European Commission (EC) has stressed this point as it states that 'ecology may be used as a metaphor or model for the social and economic as well as physical processes of cities, viewing the city itself as a complex, interconnected and dynamic system' (EC, 1996, p. 21). In the 21st century, it seems that climate change is dominating the ecologic bottom line of cities and states alike.

Based upon the general definition of sustainability, many researchers come up with similar definitions for the local scale. For example, Capello, Nijkamp, and Pepping (1999, p. 7) define local sustainability as 'a development which ensures that the local population can attain and maintain an acceptable and non-declining level of welfare, without jeopardizing the opportunities of people in adjacent areas'. On an institutional level, the European Union (EU) has taken on a leading role in shaping central ideas for the sustainable development of urban areas, but other multi-lateral organization have also influenced debates on urban sustainability. A few milestone publications and events deserve to be mentioned on this account.

In 1990, the EC publicized its Green Paper on the Urban Environment, which addresses necessary reduction efforts on energy consumption, CO_2-emissions and other waste streams. It also advocates the development of a compact city (EC, 1990). In the same year, a working group was established by the OECD which presented its status report 'Environmental Policies for Cities' (OECD, 1990). Two years later, as an output of the Earth Summit in Rio, a number of globally influencing action packages were launched. The most relevant reference for urban sustainability remains the Local Agenda 21 (LA21), which serves as a orientation for some local communities until today (Echebarria, Barrutia, & Aguado, 2009).

While LA21 activities have spread mostly around Europe, US municipalities took a difference stance on it. Although many activities in US cities are very similar to measures under the LA21 umbrella in other countries, these are often framed as 'smart growth' (Krueger & Gibbs, 2008). Yet, in terms of the role of climate change measures – and no matter how general strategies are being framed on the local level – 'cities have embraced sustainable development as a means to mitigate and adapt to climate change' (Slavin, 2011, p. 3). This seems to hold true equally in the United States and in Europe. Both regions are home to ever growing networks, which support and facilitate municipal action. The most prominent in the United States is the Mayors Climate Protection Agreement which was joined by roughly over 1,000 mayors up until 2012 (Warden, 2011). The European equivalent is the Convent of Mayors which started three years later in 2008. These emerging networks are often supported by other transnational organizations, such as International Council for Local Environmental Initiatives (ICLEI) or – in the European case – Energy Cities or the Climate Alliance. These associations have a long track record. For example, ICLEI was founded in 1990, even before the Earth summit, and often acts as a facilitator of various urban programs, such as the Cities for Climate Protection, which aim to support urban measures. Especially, in the early phase of urban climate governance, support through networks and association can be key (Bulkeley, 2010).

SUCCESSFUL GOVERNANCE FOR URBAN SUSTAINABILITY

The term 'success factor' for urban sustainability is used interchangeably as driving or enabling factors in this chapter. In principle, they can be defined as 'factors that are expected to have an effect on the policy outcomes themselves' (Evans et al., 2005, p. 31). Success or driving factors hence represent the process aspects, which deem relevant for supporting sustainability and climate change measures selected cities.

For Luhde-Thompson (2004, p. 483) the art of 'governing sustainable cities' is to 'create competent local governments that, in interaction with a highly responsible and responsive civil society, apply a form of governing that brings about the most sustainable solutions'. Thus, she points towards the importance of two elements for local governments: the competence of the local government or – the institutional capacities – and the involvement of an active civil society or – social capacities. This is similarly expressed by Lafferty (2002) who identifies political legitimacy and an active, ongoing commitment to integration of goals along with the involvement and mobilization of civil society as vital prerequisites for sustainable governance.

Complementing these basic characteristics, there are a number of different features which affect the setting of governance possibilities in cities (see also Echebarria et al., 2009).

Cities are dependent on higher superior decision-making levels and these regional, national or even international levels have great influence on the legal and institutional administration of the local level. Even though cities and communities are responsible for a number of governmental tasks at the local level which is closest to citizens as part of the subsidiarity principle (Kopatz, 1998, p. 74), many constitutive decisions fall into the regional and national sphere of policy making. As Bulkeley emphasize, it is important to 'step beyond the local as a frame of reference, and to engage with the processes which shape local capacity and political will for sustainable development at multiple sites and scales of governance' (Bulkeley & Betsill, 2005).

Having said this, the role of local governments in shaping certain sectors and their environmental output should not be underestimated.

Local authorities may use their influence, for example, to greatly reduce greenhouse gases due to their decision making in transportation issues, waste management and land-use planning. Differences between local achievements are also obvious in the results of climate change competitions or rankings (see, for instance, SustainLane.com [4] or the Green City Index by Siemens[5]). Through the use of rankings, it becomes very clear that some cities achieve better results than others in the same state or country. Even beyond environmental and economic features, such as a high share of industrial sites or dense urban settings, there seem to be other explanations responsible for better or worse outcomes.

As stated earlier, success of sustainable cities does not refer to an end status of outcomes, but rather to a continuing development of policies for sustainable programs within the particular city. Based on the approach by Evans et al. (2005) it is theorized that certain levels of institutional and social capital need to be present to achieve high policy outcomes for sustainability. Within this framework institutional capital refers to the 'knowledge, resources, leadership and learning that can make local governments effective and dynamic entities' (Evans et al., 2005, p. 2). This presents a practical approach which focuses on the capacities of municipal governments, which is arguably more specific than other definitions including also notions of general intellectual or social capital (e.g. Healey, 1998). Examples of high institutional capital include committed officers within the local administration, explicit political will, training offers for sustainable development, mainstreaming of sustainable policies in working practices, national and international network cooperation and support from regional-level networks. In comparison, the concept of social capital can be broadly defined as: 'those networks and assets that facilitate the education, coordination and cooperation of citizens for mutual benefit' (Evans et al., 2005, p. 14). Generally, the presence, character and effects of social capital are considered as a complex area, which are not easily identified.[6] Hence, for the following approach, the focus lies on tangible impact by examining the role of interest groups and civil society organizations. Relevant sectors which make up a part of social capital in relation to local authorities include (mainly environmental) non-governmental organizations (NGOs), the media, local businesses, industry organizations and universities.

The notion of 'capacity building' or 'institutional learning 'is closely connected to both institutional and social capital and can be seen as the instrumental link between them. Evans et al. define efforts of capacity building as 'all measures that strengthen the governmental structures to meet the demands of sustainable development, as well as measures that create these capacities in cooperation with civil society' (Evans et al., 2005, p. 26).[7] In this way, capacity building can strengthen the ability of local communities to deal with sustainable issues and at the same time reinforce the interconnection between institutional structures and existing social capital. Generally, the relationship between governments and civil society is complex on the local as well as on the national level and both affect each other. For example, the pure existence of local-interest organizations doesn't necessarily imply a purposeful interaction and eventually positive sustainable policy outcomes. Rather, the quality of partnership, exchange and trust level between stakeholders with different means of power is essential. It is assumed that the creation of long-term and effective partnerships within communities is a matter of complex societal tradition, democratic values, applicable and accepted rules, which are established over years (Adger, 2003; Hoff, 1998). According to Luhde-Thompson local government needs to think in long-term options if sustainable policy outcomes are to be achieved. This requires increased capacity building for both local government and civil society, so that a 'learning and listening local authority' is both: better informed and capable to better inform so that dialogue becomes a 'two-way process' enabling the civil actors to express their interests as well (Luhde-Thompson, 2004, p. 484). Based on these assumptions, it is precisely this interplay between the local government and the local community – which could be accordingly also defined as successful governance for sustainable urban governance.

There are different variations of integrating sustainability and climate issues into urban modes of governance. For instance, in some cases NGOs or individuals within local government try to push topics of sustainability, but haven't accomplished reaching a wider scale. This would represent a threshold situation generally rather in the beginning of a policy process in which decision about the future impact of sustainability issues are made. For putting certain topics on the policy agenda, problems are most often encountered where impacts exceed the local frame of actions. Wheeler mentions the difficulties when 'decision makers at any given level are also often not used to keeping broader perspectives in mind, and may have little personal stake in doing so' (Wheeler, 2005, p. 14). Reid and Satterthwaite transfer these problems to the field of climate change and energy reductions and state that 'too many policy makers at national and city levels see climate change as an environmental or global issue that is not their concern'. They also note the difficulties of adaptations towards sustainable changes in the face of opposition from 'powerful vested interests' (Reid & Satterthwaite, 2007, p. 2). The general situation on the local level regarding sustainable solutions is sometimes considered as a 'wicked problem'. This means that urban issues cannot be solved without creating others which is often because of powerful social dynamics with many conflicting interests (Wong, 2006, p. 163). In order to entangle these vested interests, some institutional and social capital features seem to be more important than others. In the next chapter, this relationship analysed more closely while comparing case studies from the United States and Europe.

COMPARING INSTITUTIONAL AND SOCIAL CAPITAL FOR SUSTAINABLE CITIES

As outlined above this comparison looks specifically at relevant characteristics of institutional and social capital in cities. It is hypothesized that high levels of institutional and social capital increase the chance of sustainable policy outcomes on the local level. This is also in line with the main hypothesis of Evans et al. who state that there are 'causal links between (good) governance and sustainable development policy success' (Evans et al., 2005, p. 102). Based on this, four typologies of governance in cities are proposed which emerge from a 2×2 matrix ranging from low to high levels of either institutional or social capacities. Within this categorization, the first type of 'dynamic governing', where both levels are high, is supposed to produce the most viable and effective sustainability policy outcomes. Through the co-operation of an active government and an equally supportive civil society the chances for successful sustainable cities are best.

A second category is termed 'active governments' and composed of high levels of institutional capacities, but low levels of social capacities. Cities falling into this category have generally highly structured and motivated local authorities providing for basic capacities for good policy outcomes for sustainability. However, capacity building for engaging local society and strengthening social capital is underrepresented and hence cities don't achieve equally high sustainability outcomes as compared to the first category.

Another category could be phrased 'passive governments'. These local government types basically represent cases where sustainability or climate-related policies are virtually non-existent. Even though local governments will still be exercising routine tasks, additional measures for sustainability are largely omitted. Besides, because social capital is also low, pressure from the civil society for changing activities towards sustainability would remain also low.

The last category termed 'voluntary governing' stands for low institutional and high social capital. In this situation the main driving forces for reaching sustainability would come from the civil society without or very little support from local governments. Because local authorities would again only stick to routine tasks, the chances for securing positive policy achievements would remain limited and capacity-building for sustainable cities would only take place within the action radius of civil organizations.

The following analysis is mainly based on two studies with a reasonable large set of case studies. One deals with 40 cities in Europe (Evans et al., 2005); the other with 44 (initially 24) cities in the United States (Portney, 2003, updates in 2007 and 2012). In the European case, cities with projects for sustainability from all over Europe have been rated in terms of their success of policy outcomes. These cities have either won a 'European Sustainable Cities and Towns Award' or have been identified as 'good practice' cases in the 'Local Authorities Self-Assessment of Local Agenda' (LASALA) project.[8]

In a similar study by Portney, large US cities with a range of sustainable policies were compared. The central piece of research represents Portney's book of 2003 (where he analysed 24 cities), but he has continuously updated his ranking (currently including 44 cities) and added different articles about the general research theme (e.g. Portney & Berry, 2010; Portney & Cuttler, 2010). Portney uses a list of 35 policies and program for sustainability which either fall into the environmental, social or economic realm. For the presence of a particular program, points are allocated.

Both studies take note of the hurdles of comparing difference cities and their governance features with each other. Evans et al. (2005, p. 14) state that 'it is never an easy task to measure abstract concepts such as policy success or governance' and Portney declares in view of analyzing different cities that 'efforts to compare cities are often difficult and fraught with methodological minefields' (Portney, 2003, p. 240). For the purpose of this chapter which studies patterns of governance, the comparison of these two studies is valuable, precisely because of their focus of governance features. When compared to other urban comparison studies, especially those of SustainLane or Siemens as mentioned above, governance characteristics fall often short and cannot be properly assessed.

METHODS

The subsequent analysis of institutional and social capacities draws on different sources. The main basis is secondary data of overall 84 cities presented in key publications of Evans et al. (2005) and Portney (2003). In each study general driving factors were described as a part of the background of sustainability measures in each city. These descriptions along with indepth case study reports from other organizations such as ICLEI and the results of a qualitative mail survey to relevant city staff from sustainability and environmental departments were used to identify specific institutional and social capacity features. As the original data source for cities varied initially from region to region, it was decided to only include cities where a comparable qualitative and quantitative amount of data was present. This reduced the detailed study of US and European cities to the ones listed in Table 1. Even though Portney's study features similar assumptions, the approach differed somewhat and hence comparative results should be handled with care. Based on different research methods and selection modes (e.g. size and geographic location of cities) in the US and the European cases it was methodologically not advisable to aggregate various elements of institutional and social capital into one index score for the US. In the European comparison, Evans et al. did indeed find a correlation for the above mentioned typologies and the policy outcomes for sustainable development. Most cities in the dynamic governance mode yielded equally high policy

Table 1. European and US Cities Selected for Analysis of Driving Factors.

European Cities (based on Evans et al., 2005)	US Cities (based on Portney, 2003)
Gotland, Sweden	Seattle, Washington
Munich, Germany	Denver, Colorado
Albertslund, Denmark	Albuquerque, New Mexico
Hanover, Germany	Minneapolis, Minnesota
Calviá, Spain	Boulder, Colorado
Ferrara, Italy	San Jose, California
Stavanger, Norway	Chicago, Illinois
Modena Province, Italy	Santa Monica, California
Gdansk, Poland	Portland, Oregon
Barcelona Province, Spain	Scottsdale, Arizona
Haarlem, The Netherlands	San Francisco, California
Dunkerque Urban Community, France	Kansas City, Missouri
Tampere, Finland	Sacramento, California
Ottignies-Louvain la Neuve, Belgium	Austin, Texas
Santa Perpetua de Mogoda, Spain	Chattanooga, Tennessee
Korolev, Russia	Grand Rapids, Michigan
Dunajska Luzna, Slovakia	Baltimore, Maryland
Kuressaare, Estonia	Boston, Massachusetts

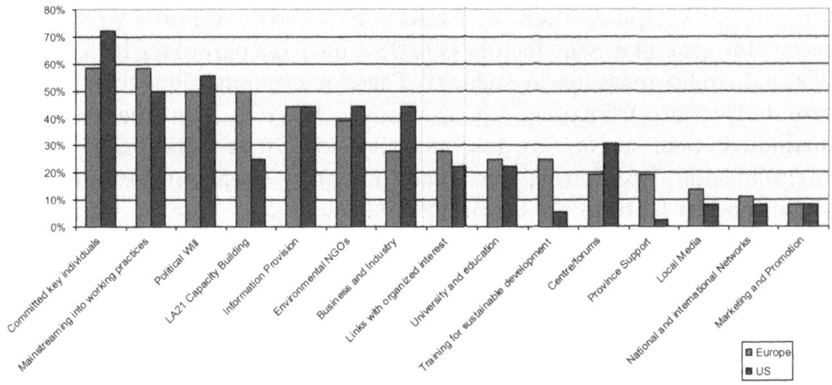

Fig. 1. Comparison of Driving Factors for Sustainable Policy Outcome in the United States and Europe: Percentage of sample cities in which driving factors were evaluated as an important condition to facilitate sustainable policy outcomes.

outcomes for sustainability whereas other in the 'passive government' mode only showed limited extent of progress (Evans et al., 2005, p. 107).

RESULTS

While the reproduction of Evans' framework was not fully possible for the cities in the United States for above stated reasons, it was feasible to study the relative importance of process factors falling into the institutional and social realm. The bars in Fig. 1 represent the percentage of cities in which a particular success factor was assessed as being of primary importance for sustainable policy outcomes. For instance, in more than 70% of the studied US cities, the notion of committed key individuals was directly mentioned to be key for the sustainable progress of the city. In a explorative fashion these results point to the salient characteristics of governance modes which need to be present for urban sustainable policies. Next, the leading success factors and their differences in the United States and in Europe are explained in more detail.

Committed Key Individuals

In all case studies described, the role of committed key individuals on the side of local governments was viewed as the salient, most effective, process factors in pursuit of local sustainable development. In some cases, key individuals were described as being charismatic, innovative and smart leaders with the commitment and will to set sustainable development on the agenda. A notable comment about the characteristics of their mayor was made in Grand Rapids, Michigan: 'Our mayor is a great leader and presenter and tells our story wherever he travels advocating for community, partnerships, balanced triple bottom line, and dialogue'. These key actors are not always necessarily the mayor or members of the city council. In many cases (e.g. Albuquerque, Seattle, Dunkerque or Durham), a strong team of city officers may also bring sustainability efforts forward. Equally positively notes were made about the influence of highly educated and motivated staff members (e.g. Tampere, Dunkerque, Boulder, Santa Monica) and the impact of having a diversity of ages and young officers within the local authority which tends to contribute to a 'learning culture' within the government (Evans et al., 2005, p. 67).

The role of key actors needs to be seen against the background of effective institutions supporting the innovation pushed forward by entrepreneurial leaders. Long-term success can only be achieved commitment of local authorities will be consistent even if key personel leave office and changes his or her job. The influence of policy drivers also depends on the prevailing culture within the institution itself. It is likely that innovative leaders are able to secure progress for sustainable action if there is a certain 'degree of receptiveness' in the local government or if there are drivers from outside, for example, from the national government or the European policies (Evans et al., 2005, p. 110).

In the US case studies, emphasis on the role of key officers tends to be slightly higher than in the European samples. Because of the small comparative case number, it is difficult to draw definite conclusions about this fact. Pertaining to other literature comments, one assumption could be that US local leaders are more 'needed' for sustainable policy successes because the national legislation is not as supportive on sustainability issues such as the EU (Sorensen, 2004, p. 8). However, this differentiation should not diminish the fact that committed leaders – at best with an efficient institutional background – are important in every regional context.

Mainstreaming Sustainable Development in Working Practices

Mainstreaming of policies towards sustainable development is viewed in European cities almost as important as the impact of committed key individuals. The ratings for the US cities are slightly lower, but in total also among the three most important factors. The issue of mainstreaming sustainable development is understood as 'integrating principles of sustainable development within the institutional culture and working practices of a local authority, and transferring this into actions' (Evans et al., 2005, p. 68). A wide range of examples can be collected here from both European and US cities starting from improved communication and collaboration between city departments on sustainability issues over green budgeting initiatives to implementing best available techniques for city owned premises and infrastructure. For the latter, an often noted example in US cities is the diffusion of 'Leadership in Energy and Environmental Design' (LEED)[9] certified building schemes which promote especially energy-efficient building. European cities started, for example, the integration of green procurement across departments (Ferrara, Stavanger) and in many cases LA21 plays a supportive role for the mainstreaming of sustainable policy action in different local departments.

Political Will

Political commitment to local sustainable development refers to two aspects of local authorities. On the one hand the political (green) engagement of key players in the local government was noted, which is closely connected to the already mentioned crucial role of innovative entrepreneurial individuals. On the other hand, the influence of political parties supporting sustainable development was also mentioned. The latter played a special role in European cities with 'Red-Green' coalitions which tend to favor local sustainable actions (Evans et al., 2005, p. 68). In contrast, Portney concludes that 'being a Democratic place does not seem to serve as a prerequisite for taking sustainability seriously' (Portney, 2003, p. 219). In fact, a number of leading cities such as Chattanooga and Jacksonville are primarily Republican dominated. Especially in the United States, it seems that independent of party politics, cities are more likely to initiate sustainable actions on their own, because national and regional support does not seem as high as in the European countries (Sorensen, 2004). This could also explain why the driving factor of political commitment is perceived as slightly more important in the US cities than in European cases.

Local Agenda 21 Capacity Building

The influence of the process factor capacity building of the LA21 shows the largest difference between the European and US cases. However, in this case this is due to a different reception of the LA21 movement. As explained above, LA21 in the United States is not as deeply rooted as in other regions of the world, especially in Europe where a leading sustainable city is nearly always implementing a LA21 program. The low dissemination of LA21 in the United States does not imply that there are no local movements for sustainable development. Rather US initiatives are framed differently and have not adopted the international action programs because of the US tendency to not follow international agreements (Geißel & Kern, 2000). In Europe different measures can be listed: for example, a LA21 participation plan (Santa Perpetua de Mogoda), a LA21 methodology to involve civil society (Barcelona) or general partnership through LA21 (Durham). Without the terminology of LA21 it is possible to find similar participatory actions in US cities, although because of the different research focus it is sometimes difficult to asses how much of these measures actually contribute to strengthening capacities of the civil society. Many US city officials point to initiatives such as roundtable meetings and public forums with different stakeholders (e.g. Albuquerque, Sacramento), decentralized 'community sustainability partnerships' (e.g. Grand Rapids) or sustainable indicator projects (e.g. Seattle, Baltimore). Since there are other different categories in this assessment like 'Centre/forums' or 'Information provision to civil society', these activities are more strongly weighted there.

In general, regarding capacity building for governing processes it is difficult to differentiate between strengthening capacities of civil society actors or local authorities. In the most ideal case this process is like an interchangeable 'double loop' where local governments profit from input of civil society actors and vice versa. This relationship can be best detected by extend of trust and stability between actors from government and civil society. Yet, in reality it is difficult to find equal amounts of capacity building actions and 'it seems that while some local governments have actively invested in building the capacity of their staff and politicians through training and educational programs, there is little evidence that this has been widely replicated in civil society' (Evans et al., 2005, p. 114).

Information Provision to Civil Society

Providing information about the environmental, social and economical status of the city is a crucial factor for starting sustainable development action, because 'without knowledge there is no (perceived) problem, no public awareness and consequently no policy process [...]' (Jänicke & Weidner, 1997, p. 7). In both European and US cities this factor is ranked similarly high among city officials and plays an important part for bridging information gaps between key actors. Information provision can also be considered as an important element for building-up capacities for sustainable development. In both city samples, the influence of electronic tools has increased over the years; the internet has become the central organ for dissemination of city status information (e.g. Hanover, Seattle, Santa Monica; Dunajska Luzna), announcements of city council meeting on sustainability issues, the development of sustainable indicators and some experimental information application. For example, in Alberslund the city council established green accounts for its residential areas, where local residents are enabled to check the improvements and actions for sustainability in their neighborhood. There is also a number of notable online facilities in the United States, for example, the cities San Jose, Cleveland, San Francisco and Baltimore feature comprehensive and easily accessible information about sustainability action on their homepages. Compared to European city websites, US cities tend to be slightly more user-friendly and transparent, e.g. by providing direct contact options and by publishing protocols. Telephone numbers for direct contact to city officials. However, finding the relevant departments or names of responsible officers can sometimes be a matter of luck on less well-structured city portals on both sides of the Atlantic Ocean. Here, city officials might start to improve urban information services with a view to best practice examples of other cities. In cities where the major or city council have lately taken on sustainable development as a key issue, links to further information is often displayed on the entry side of the city portal (e.g. Baltimore, Austin, Phoenix). Information provision should however not only be understood as a one-way street – in other examples, education projects with young people about sustainability issues were positively mentioned (e.g. Alberslund).

Non-governmental Organisations

Compared to available data material on institutional capacity, empirical data on the influence of social capacity is not as straightforward or extensive. As mentioned above, the notion of social capacity was framed by identifying a number of relevant actors whose influence in the city may account for high levels of social capacity. Here, environmental NGOs are most often mentioned for taking on a leading role in governance processes towards sustainable development.

In the US case studies, influences of NGOs or the business community were mentioned in the descriptions and sometimes also in the responses in the e-mail survey. In general, the involvement of civil actors in the US cities is perceived as more relevant than in the European cases. The role of NGOs and businesses are in the United States on the same level as the factor of 'information provision', whereas these indicators for social capacity rank lower for the European cases. In sum, the importance of including a broad range of actors in the discussion and implementation of local sustainable development is recognized in most of the cities in Europe and the United States alike, but it seems that emphasis on certain actors is more strongly perceived in the US samples. Portney identifies in 2005 that 'of the 41 cities studied as part of the broader assessment of sustainable cities, at least 34 have some form of participatory processes associated with their respective programs'. He also notes that naturally some of these participatory processes are more extensive than others (Portney, 2005, p. 584). The influence and role of NGOs in cities varies throughout the process of 'taking sustainability seriously' (Portney, 2003). In many cases, NGOs, especially environmental NGOs, started campaigns to raise the awareness of city officials and citizens towards certain pressing (environmental) issues in the city (e.g. Boulder; Seattle; Cleveland; Vantaa, Finland; Falkenberg; Sweden). In some cases, NGOs also take up certain themes from national or international networks (e.g. in Leicester, UK). Although, NGOs play also an important role for the further process of becoming more sustainable by strengthening the social capacities and links between civil actors and the local government, the leading part is usually taken on by local authorities who possess both power and legitimacy, if sustainable policies outcomes are high (Evans et al., 2005, p. 83).

Business and Industry

The second most important factor in terms of social capacity seems to be the involvement of the business community. Here, the difference between US and European case studies is among the highest: here the role of business is perceived as a more important factor in US cities than in Europe. A number of US case studies describe the leading and sometimes innovative role taken on by the business community. Members of the business community are not only often part of roundtable or town hall meetings but they are sometimes change innovators themselves like in the notable case of Chattanooga. Because of its high air pollution Chattanooga used to be among the most polluted cities in the United States. Through combined efforts of the city council, which was initially started and later actively supported by the local Chamber of Commerce, the city succeeded in drastically improving environmental quality in its urban area (Portney, 2003). Other noteworthy links between businesses and local government are notable e.g. in Boulder for their involvement of local SMEs, in Washington for a number of active public-private partnerships and in San Jose for their connection and promotion of a green business program. In Europe, Haarlem, Dubrovnik, Munich and Lahti are mentioned because of particular active links between business organizations and the local government on sustainability projects. These usually focus on energy efficiency measures and green areas in business parks or eco-industrial parks (Evans et al., 2005, p. 70).

The described process or success factors are those emerged as the most important factors from the data of the European and US case studies. The first three indicators – namely 'committed key individuals', 'mainstreaming' and 'political will' – falls into the realm of institutional capacities. The following two – 'LA21 capacity building' and 'information provision' are considered as capacity building measures and the involvement of NGOs and businesses in the sustainability process are examples of social capacity.

Other factors namely, 'links with organized interest', 'university and education', 'training for sustainable development', 'centre and forums', 'province-level support', 'local media', 'national and international networks' and 'city-marketing and promotion' are almost evenly spread over the three aspects of institutional and social capacities and capacity building. In single case studies, one or more of these factors can play an important role for the sustainability process of the city. Here, they will not be discussed in detail, because compared to the factors above they seem of less weight and relevance to the overall number of cities.

The results of Evans et al. (2005) emphasize the correlation between equally high institutional and social capacities and positive policy outcomes towards sustainability in the European cases. The same could be assumed for the studied US cities, where cities with remarkably high notions on institutional and social capital tend to rank among the top cases. The top US cities, such as Seattle, Portland, Albuquerque, Denver and Boulder all show exceptionally strong characteristics of the described capacities. Yet, relating to the whole sample of cities, it was not attempted to categorize all US cities according to European framework. This was avoided because of less availability of coherent information for the US part and the danger of 'bypassing' the experiences of cities with non-sufficient material and hence producing unstable conclusions.

While this study focused on individual driving factors, the sample of leading cities was selected, because of their high level of sustainability outcomes. This in turn affects the patterns of success factors. Some influences such as the impact of urban networks or the role of environmental NGOs would be arguably much stronger if other cities at an early stage of their sustainability efforts were studied. It is also beyond the scope of this exploratory study to gauge specific causal relationships between individual driving factors and outcomes. Future studies should also take account of interdependencies and trade-offs between urban climate mitigation, adaptation and sustainable development – relationships which were not included in this study yet.

Nevertheless, the fact that singular key driving factors have emerged in very similar ways in Europe and the United States alike can already be stated to be an interesting result as such. At the same time differences, in particular, with regard to the role of Local Agenda 21 measures and the role of businesses point to varying regional adoption practices in the United States and Europe.

CONCLUSIONS

In general, cities in the United States and Europe feature more similarities than differences relating to key driving factors. The most progressive cities in terms of sustainable policies are characterized by a strong interplay between institutional and social capacities enhanced by continuous capacity-building.

Those are often cities with a strong tradition of mutual trust and a positive correlation between civil society and local government. This is, for example, the case in the Scandinavian countries where cities are more successful than compared to those in the Eastern European countries where large-scale governments and low level of civic engagement seem to prevail.

In terms of the relative importance of individual success factors, European and US cities bear strong resemblance. There are a few driving factors, such as the impact of Local Agenda 21 capacity building or the role of businesses, which show interesting regional differences between the US and European cases. However, the overall pattern of prevalent driving factors is very similar. While taking into account possible variations in the order of important driving factors due to different research approaches and therefore different sets of data in the US and European case, it is noteworthy that the most prominent driving factors are linked to institutional capacities. Namely, the factors of 'committed key individuals', 'mainstreaming sustainable development into working practices' and 'political will' seem to dominate governance practices for sustainable development. Thus, the most important driving factor for local sustainable development seem to be most prominant proactive, capable and informative local governments, which are not only driven by leadership personalities staff but also organized in such a way that political mainstreaming of sustainability issues is effectively integrated in different governmental departments. Although in almost all case studies analysed, the role of the civil society was explicitly noted for improving acceptance levels and sometimes initiating sustainable initiatives, the importance of high institutional capacities seems to outweigh the means and influence of civil actors.

This can be partly explained by the power structure and legitimacy of local authorities which normally bundle the political tasks for sustainable policies. While low capacity of local governments is unlikely to be substituted by high levels of social capacities, it is also clear that a city relies upon its social capacities to play a role in the top league of sustainable and low-carbon cities. Even though institutional capacity seems to be key, it is only through the involvement, common learning and sharing of local governments with civil society that successful policies for local sustainability can be achieved in the long run.

NOTES

1. One recent exception to the general focus of small-scale research is the ongoing research project at the University of Durham 'Urban Transition and Climate Change' which undertakes a large-scale comparison of climate change initiates with a total number of 100 global cities and over 650 different initiatives.
2. Named after its chair-woman and former Norwegian Prime Minister Gro Harlem Brundtland.
3. In Mumford's own words, he reportedly advised his fellow countrymen in the United States to 'forget the damned automobile and build the city for lovers and friends' (1979, p. 54).
4. SustainLane.com was one of the first US websites which provided a ranking of sustainability outcomes in the 50 most populous cities in the United States, starting in 2005. According to Slavin (2011), in the age of the Internet, it is perhaps the most visible sign of the rise of sustainability to the top of public policies agendas in the urban milieu.
5. The reports by the Economist Intelligence Unit were commissioned by Siemens and cover the largest cities in almost all world regions: North America, Latin America, Africa, Europe and Germany and Asia (Siemens, 2012). The Green City Index seems to have received increasing attention from practitioners, but few academic evaluations on it exist yet. A few concerns regarding the methodology and the selection and assessment of indicators were raised by scientists in blogs (e.g. Benfield, 2011).

6. For example, Meadows raises the question whether 'capital' is the right conceptual framework at all, since neither human nor social capital may be adequately expressed in terms of materials, energy or money. Having the caveats of the social capital concept in mind, she concludes that it is yet possible and useful to create indicators for (both) human and social capital, which can 'accumulate over time, can be invested in, can depreciate and must be essential factor in sustainable development' (Meadows, 1999, p. 378).

7. Many researchers offer more theoretical background about the relationship of social, institutional capital and capacity building (e.g. Healey, 1998; Jänicke & Weidner, 1997; Putnam, 2000).

8. LASALA is a project initiated by ICLEI in order to assess cities' efforts and achievement in implementing LA21 capacity-building measures. It involved a self-assessment exercise of 230 governments all over Europe. The research project demonstrated 'the significant levels of commitment to the LA21 process [...] and notable achievements during a very short space of time' (Evans et al., 2005, p. 7).

9. LEED is an US label certifying high performance of green buildings. There are different quality levels, the best being 'Platinum Standard'. Although the LEED label has become widely accepted in the United States, it has also been criticized for concentrating too much on fossil fuels (half of applicable point scale is related to the (albeit) efficient use of fossil fuels in buildings and for focusing primarily on large US cities (Burr, 2008).

REFERENCES

Adger, N. W. (2003). Social capital, collective action, and adaptation to climate change. *Economic Geography*, *79*(4), 387–404.

Agyeman, J., & Evans, B. (2006). Justice, governance, and sustainability: Perspectives on environmental citizenship from North America and Europe. In A. Dobson & D. Bell (Eds.), *Environmental citizenship*. Cambridge, MA: MIT Press.

Alber, G., & Kern, K. (2008). Governing climate change in cities: Modes of urban climate governance in multi-level systems. In OECD (Ed.), *Competitive cities and climate change* (pp. 175–196). Paris: OECD.

Anguelovski, I., & Carmin, J. (2011). Something borrowed, everything new: Innovation and institutionalization in urban climate governance. *Current Opinion in Environmental Sustainability*, *3*(3), 169–175. doi: 10.1016/j.cosust.2010.12.017

Baker, S. (2006). *Sustainable development*. Abingdon: Routledge.

Barnes, M., Sullivan, H., Knops, A., & Newman, J. (2004). Power, participation and political renewal: Issues from a study of public participation in two English cities. *IDS Bulletin*, *35*(2), 58–66.

Barton, H. (Ed.) (2000). *Sustainable communities: The potential for eco-neighbourhoods*. London: Earthscan.

Beatley, T. (2000). *Green urbanism: Learning from European cities*. Washington, DC: Island Press.

Benfield, K. (2011). *'Green cities' in the US and Canada, kind of (conclusion: CO2, energy, waste, buildings, air, water, governance)*. Retrieved from http://switchboard.nrdc.org/blogs/kbenfield/green_cities_in_the_us_and_can_1.html

Brand, U. (2012). Green economy-the next oxymoron? *GAIA*, *21*(1), 29–32.
Brandon, P. S., & Lombardi, P. (2005). *Evaluating sustainable development: In the built environment*. Oxford: Blackwell Publisher.
Brotchie, J. (1999). *East west perspectives on 21st century urban development: Sustainable eastern and western cities in the new millennium*. Aldershot: Ashgate.
Bulkeley, H. (2010). Cities and the governing of climate change. *Annual Review of Environment and Resources*, *35*(1), 229–253. doi: 10.1146/annurev-environ-072809-101747.
Bulkeley, H., & Betsill, M. (2005). Rethinking sustainable cities: Multilevel governance and the 'urban' politics of climate change. *Environmental Politics*, *14*(1), 42–63.
Burr, A. (2008, November 12). *LEED's big market bias: Green building's first mainstream push landed it in the nation's biggest cities. But what about everywhere else*? Retrieved from http://www.costar.com/News/Article/LEEDs-Big-Market-Bias/100646
Camaren, P., & Swilling, M. (2011). *Sustainable, low-carbon cities in the 21st century – Making it happen!* Position Paper for the Summit for the Urban Environmental Accords (October 11-13 2011 in Gwangju, South Korea). Commissioned by UNEP.
Capello, R., Nijkamp, P., & Pepping, G. (1999). *Sustainable cities and energy policies. Advances in spatial science*. Berlin: Springer.
Carlye, M., Jenkins, P., & Smith, H. (Eds.). (2001). *Urban development and civil society: The role of communities in sustainable cities*. London: Earthscan.
Chance, T. (2009). Towards sustainable residential communities; the Beddington Zero Energy Development (BedZED) and beyond. *Environment and Urbanisation*, *21*(2), 527–544.
Corfee-Morlot, J., Lamia K., Donovan, M., Cochran, I., Robert, A., & Teasdale, P.-J. (2009). *Cities, climate change and multilevel governance*. OECD Environmental Working Papers No. 14, 2009, OECD, Paris.
Crosbie, T., & Baker, K. (2010). Energy-efficiency interventions in housing: Learning from the inhabitants. *Building Research and Information*, *38*(1), 70–79.
Dhakal, S. (2008). Climate change and cities. The making of a climate friendly future. In P. Droege (Ed.), *Urban energy transition. From fossil fuels to renewable power* (1st ed., pp. 173–192). Amsterdam: Elsevier.
Echebarria, C., Barrutia, J. M., & Aguado, I. (2009). The ISC framework: Modelling drivers for the degree of local agenda 21 implantation in Western Europe. *Environment and Planning A*, *41*(4), 980–995. doi: 10.1068/a40301.
European Commission. (1990). *Green paper on the urban environment*. Brussels. Retrieved from http://ec.europa.eu/environment/urban/pdf/com90218final_en.pdf
European Commission. (1996). *European sustainable cities*. Report, Expert Group on the Urban Environment, Directorate General XI Environment, Nuclear Safety and Civil Protection. Brussels.
Evans, B., Joas, M., Sundback, S. & Theobald, K. (2005). *Governing sustainable cities*. London: Earthscan.
Geißel, B., & Kern, K. (2000). Soziales Kapital, Lokale Agenda 21. Lokale umweltpolitische Initiativen in den USA. In H. Heinelt & E. Mühlich (Eds.), Lokale Agenda 21-Prozesse. Erklärungsansätze, Konzepte und Ergebnisse (pp. 257–276). Opladen: Leske + Budrich.
Girardet, H. (1999). *Creating sustainable cities*. Bristol: Schumacher Briefings.
Healey, P. (1998). Building institutional capacity through collaborative approaches to urban planning. *Environment and Planning A*, *30*(9), 1531–1546. doi: 10.1068/a301531.
Hoff, M. D. (1998). *Sustainable community development: Studies in economic, environmental, and cultural revitalization*. Boca Raton, FL: Lewis Publication.

Holden, M., Roseland, M., Ferguson, K., & Perl, A. (2008). Seeking urban sustainability on the world stage. *Habitat International, 32*(3), 305–317. doi: 10.1016/j.habitatint.2007.11.001

Inoguchi, T., Newman, E., & Paoletto, G. (1999). *Cities and the environment: New approaches for eco-societies* (Vol. 1023). Tokyo: United Nations University Press.

Jänicke, M., & Weidner, H. (1997). *National environmental policies: A comparative study of capacity- building*. London: Springer.

Kopatz, M. (1998). *Lokale Nachhaltigkeit: Vom internationalen Diskurs zur Umsetzung in Kommunen*. Oldenburg: Bibliotheks-und Informationssystem der Universität Oldenburg.

Krueger, R., & Gibbs, D. (2008). 'Third wave' sustainability? Smart growth and regional development in the USA. *Regional Studies, 42*(9), 1263–1274. doi: 10.1080/00343400801968403

Lafferty, W. (2002). *Adapting government practice to the goals of sustainable development*. Working Paper No. 1. Retrieved from http://www.oecd.org/gov/1939762.pdf

Luhde-Thompson, N. (2004). Governing sustainable cities. *Local Environment, 9*(5), 481–485. doi: 10.1080/1354983042000264919

Meadows, D. (1999). Indicator and Information Systems for Sustainable Development. In D. Satterthwaite (Ed.), *The Earthscan reader in sustainable cities* (pp. 364–393). London: Earthscan.

Moore, S. A. (2007). *Alternative routes to the sustainable city: Austin, Curitiba, and Frankfurt*. Lanham, MD: Lexington Books.

Mumford, L. (1979). *My works and days: A personal chronicle* (1st ed.). New York, NY: Harcourt Brace Jovanovich.

Newman, P. (2006). The environmental impact of cities. *Environment and Urbanisation* (18), 275–295.

Ooi, G. L. (2005). *Sustainability and cities: Concept and assessment*. Singapore: IPS Inst. of Policy Studies.

Organisation for Economic Co-operation and Development. (1990). *Environmental policies for cities in the 1990s*. Paris: OECD Publications and Information Centre.

Organisation for Economic Co-operation and Development. (1996). *Innovative policies for sustainable urban development: The ecological city*. OECD, Paris.

Portney, K. E. (2003). *Taking sustainable cities seriously: Economic development, the environment, and quality of life in American cities*. Cambridge, MA: MIT Press.

Portney, K. E. (2005). Civic engagement and sustainable cities in the United States. *Public Administration Review, 65*(5), 579–591. doi: 10.1111/j.1540-6210.2005.00485.x.

Portney, K. E., & Berry, J. M. (2010). Participation and the pursuit of sustainability in U.S. cities. *Urban Affairs Review, 46*(1), 119–139. doi: 10.1177/1078087410366122.

Portney, K. E., & Cuttler, Z. (2010). The local nonprofit sector and the pursuit of sustainability in American cities: A preliminary exploration. *Local Environment, 15*(4), 323–339. doi: 10.1080/13549831003677704.

Putnam, R. D. (2000). *Bowling alone: The collapse and revival of American community*. New York, NY: Simon & Schuster.

Register, R. (2006). *Ecocities: Rebuilding cities in balance with nature* (Rev. ed.). Gabriola, BC: Berkeley Hills Books; New Society Publishers.

Reid, H., & Satterthwaite, D. (2007). *Climate change and cities: Why urban agendas are central to adaptation and mitigation*. IIED Sustainable Development Opinion Papers, Vol. 12. Retrieved from http://www.iied.org/pubs/display.php?o=17025IIED&n=1&l=46&a=D%20Satterthwaite&x=Y

Rosenzweig, C., Solecki, W., Hammer, S. A., & Mehrotra, S. (2011). *Climate change and cities: First assessment report of the urban climate change research network*. Cambridge, MA.

Sachs, W. (1999). *Planet dialectics: Explorations in environment and development*. London: Zed Books.

Siemens. (2012). *Green City Index: Reports*. Retrieved from http://www.siemens.com/entry/cc/de/greencityindex.htm

Slavin, M. I. (Ed.) (2011). *Sustainability in America's cities: Creating the green metropolis*. Washington, DC: Island Press.

Sorensen, A. (2004). *Towards sustainable cities: East Asian, North American and European perspectives on managing urban regions*. Aldershot: Ashgate.

United Nations Human Settlements Programme (UN-Habitat). (2011). *Cities and climate change: Global report on human settlements 2011*. London: Earthscan.

Viguié, V., & Hallegatte, S. (2012). Trade-offs and synergies in urban climate policies. *Nature Climate Change*, *2*(5), 334–337. doi: 10.1038/NCLIMATE1434.

Warden, T. (2011). Viral governance and mixed motivations: How and why U.S. cities engaged in the climate change issues 2005–2007. In World Bank Group (Ed.), *Cities and climate change. Responding to an urgent agenda*. Washington, DC: World Bank.

Wheeler, S. (2005). *Planning for sustainability: Creating livable, equitable, and ecological communities*. London: Routledge.

Wong, C. (2006). *Indicators for urban and regional planning: The interplay of policy and methods*. London: Routledge.

World Bank. (2010). *Cities and climate change: An urgent agenda* (No. 10). Washington, DC. Retrieved from http://go.worldbank.org/FMZQ8HVQJ0

World Commission on Environment and Development. (1987). *Our common future*. Retrieved from http://www.un-documents.net/wced-ocf.htm

CHAPTER 2

GROWTH MACHINES AND CARBON EMISSIONS: A COUNTY-LEVEL ANALYSIS OF HOW U.S. PLACE-MAKING CONTRIBUTES TO GLOBAL CLIMATE CHANGE

Matthew Thomas Clement and James R. Elliott

ABSTRACT

Purpose – *To combine insights from urban and environmental sociology to examine local drivers of carbon emissions in the United States, with particular focus on demographic, economic, and consumptive dynamics.*

Design/methodology/approach – *Apply spatial regression analysis to a novel county-level data set to test hypotheses about how different conditions and activities relate independently and positively to total carbon emissions at the local level.*

Findings – *Results provide strong support for theoretically derived hypotheses, even after controlling for other factors, including spatial autocorrelation. The implication is that within a social system that treats land as a commodity, efforts to increase the exchange value of this*

Urban Areas and Global Climate Change
Research in Urban Sociology, Volume 12, 29–50

ISSN: 1047-0042/doi:10.1108/S1047-0042(2012)0000012005

commodity tend to drive up local carbon emissions, thereby contributing to global climate change.

Originality/value – *Complements previous sociological work on greenhouse gas emissions at the national level. Shows how local processes in general and urbanization in particular contribute to global climate change at and from the local areas where they occur.*

Keywords: Growth machines; carbon emissions; urbanization; United States; counties

Global climate change is nothing new, but its sources are. Before the rise of large industrial cities, significant shifts in Earth's climate were driven largely by natural processes beyond human manipulation and control (Allègre & Schneider, 1994). Since this time, however, sources have become much more anthropogenic (Intergovernmental Panel on Climate Change (IPCC), 2007; Karl & Trenberth, 2003; Weart, 2003). This historic shift and the challenges that it poses for future human well-being have spurred a new wave of social scientific inquiry into contemporary drivers of carbon dioxide and other greenhouse gases. To date, much of this research has focused on patterns and processes observable at the national and global levels, asking, for example, why some nations emit more harmful gases into the atmosphere than others (e.g., Dietz & Rosa, 1997; Fisher & Freudenburg, 2004; Grimes & Kentor, 2003; Jorgenson 2006, 2007; Ravallion, Heil, & Jalan, 2000; Satterthwaite, 2009; Shandra, London, Whooley, & Williamson, 2004; York, Rose, & Dietz, 2003) and how we can best develop and implement international accords aimed at slowing these trends (e.g., Beck, 2009; Betsill, 2005; Hasselmann et al., 2003; Hempel, 2006).

By contrast, locally oriented research has remained less prominent and more focused on reactions than causes. Norgaard (2006, 2011), for example, has brought much needed attention to the social psychology of global warming in her study of climate change, emotions, and everyday life in a small Norwegian community now experiencing warmer than usual winters. Relatedly, other researchers have investigated differences in local support for emission reduction policies within the oil industry (Pulver, 2007) and within the United States more generally (e.g., Dietz, Dan, & Shwom, 2007; Fisher, 2006; Krause, 2011; Lutzenhiser, 2001; McCright & Dunlap, 2000, 2003; O'Connor, Bord, Yarnal, & Wiefek, 2002; Zahran, Grover, Brody, & Vedlitz, 2008; see other chapters in this volume). To be sure, these local approaches to global climate change remain valuable and should continue;

however, they also leave us with relatively little understanding of how local processes in general and urbanization in particular contribute to global climate change at and from local areas where they actually take place.

The present study begins to fill this gap. It starts not with the political economy of the world system or nation state, as is common in most sociological research on drivers of carbon emissions, but with the political economy of local place-making, using Molotch's growth machine theory to understand how and why local coalitions organize to grow demographically and expand economically in ways that contribute to global climate change (Molotch, 1976; Logan & Molotch, 2007). After identifying key environmental processes within this framework, we examine each simultaneously using newly available carbon emissions data at the county level for the continental United States (Gurney et al., 2009; see also Parshall et al., 2010). Results parallel findings typically reported in national level research and highlight how the dynamic relationship between human society and environmental change is multiscalar and interdependent. On the one hand, different dimensions of local place-making produce greenhouse gases that contribute to global warming; on the other hand, global warming is expected to exert potentially negative pressures on some locales (Millennium Ecosystem Assessment, 2005), for example, in the form of urban heat islands (e.g., Harlan et al., 2008; Klinenberg, 2002; Lowry, 1967) and coastal flooding (Curtis & Schneider, 2011).

Below, we focus on the former dynamic, drawing attention to the "upward" influence of local political economies on carbon emissions that contribute to global climate change. Subsequent studies in this volume then illuminate the "downward" influence of global climate change on particular places and related efforts to engage and mitigate this influence. The underlying point of all these studies is that different scales and directions of society–climate interaction are connected and that, from an academic perspective, urban studies can provide unique insights into important but neglected elements of this interdependence.

LOCAL GROWTH MACHINES AND GLOBAL CLIMATE CHANGE

The theoretical basis for this study begins with what Molotch (1976) calls the basic stuff of place: the land. In this case, we are particularly interested in the commodification of land and how this commodification influences how cities, towns, and other places across the United States grow and contribute to global climate change. In considering these dynamics, it would be easy to presume that local growth simply reflects an aggregation of commercial interests and individual tastes for particular locales, but these factors constitute only part of the story. In order for businesses, individuals and families to actualize these geographic tastes, local areas must grow, and this growth requires broader political and economic forces to promote and legitimate ongoing development of local lands. Sociological efforts to comprehend these dynamics, past and present, span many theoretical traditions but among the most prominent is the idea of places as "growth machines" (Molotch, 1976; Logan & Molotch, 2007; for a review see Jonas & Wilson, 1999).

The first tenet of the growth machine thesis is that all settlements in market economies, including the United States, have a dual nature. On the one hand, they constitute "home," where people develop meaningful social relationships, deep attachments to place, and a fundamental sense of community. On the other hand, they also constitute commodities that are subdivided into parcels to be bought and sold, rented and leased for profit in the market. This duality of place creates conflict between groups primarily interested in preserving and improving local quality of life, or "use values," and groups primarily interested in maximizing profits, or "exchange values," associated with local lands that they happen to own or otherwise control. The second tenet of the growth machine thesis is that these two sides are unequal. Developers, realtors, bankers, utility companies, and other social actors that profit from continued growth tend to be more powerful than homeowners, neighborhood associations, and civic groups that advocate primarily for use values, and they use this power to "capture" local political officials and have them act in the interest of maximizing growth. This pro-growth alliance, in fact, is what the "growth machine" refers to: a coalition of business elites united with local political officials in pursuit of ongoing demographic growth and economic expansion in the local area. The third tenet of the growth machine thesis is that these pro-growth coalitions typically promote their agenda by asserting that local growth is good for all residents because it brings new jobs, taxes, and stature to the area. In this way, those who benefit most from continued development and land-use intensification present it as a public good to be pursued aggressively and with great civic pride by all. Just consider the advertisement placed by pro-growth interests in Louisiana: "Nature made it perfect. We made it profitable" (cited in Logan & Moloth, 2007, p. 59).

Viewed from this perspective, demographic and economic growth at the local level is more than a matter of geographic circumstance and individual choice. It is also a product of powerful local actors and institutions working together to generate and extract exchange values through ongoing land-use intensification. Local governments are instrumental to these efforts in the United States because they hold legal authority over zoning and land-use decisions and because they are well positioned to leverage capital investments that drive local growth. Municipal governments can, for example, build new roads and civic infrastructure into formerly undeveloped lands, disregard federal flood maps, facilitate drainage and landfill projects, create allowances for new commercial facilities, reduce taxes in locally defined enterprise zones, and generally shape where and to what extent infrastructural improvements will occur. In most areas, these pro-growth initiatives outstrip environmental protection and in the process erode wetlands, forests, and other natural buffers to environmental hazards such as global warming (Freudenburg, Gramling, Laska, & Erikson, 2009; Pais & Elliott, 2008). In this manner, local growth machines not only draw down natural resources of surrounding areas but also contribute to global climate change through the carbon wastes that these activities tend to produce.

Inserting this perspective into the study of global climate change moves us beyond the simple recognition that some groups and places are likely to be more vulnerable to such change than others to illuminate how this vulnerability is generated by ongoing and unequal struggles over local development in market-driven nations such as the United States. In turn, it also raises basic questions about which types, or dimensions, of local growth are most responsible for these sources of global environmental change. Below we review these different dimensions and their likely consequences for locally produced carbon emissions. The driving assumption is that while all of these dimensions contribute to local land-use intensification that raises carbon emissions, each has an independent effect that can be conceptually disentangled to facilitate empirical analysis.

Total and Urban Population

According to the growth machine thesis, local growth is best understood as a "syndrome" of related processes (Molotch, 1976, p. 310), and among these processes, none is presumed to be more central than population growth, both directly and indirectly. Directly, population growth expands local customer bases and local revenue streams; and indirectly, it increases local land rents, thereby raising exchange values associated with local parcels. These dynamics, moreover, are not environmentally costless. As the growth machine thesis asserts, adding more people to an area means not only adding more consumers and raising local land rents but also increasing air pollution and traffic congestion that can contribute to smog and other airborne emissions that drive global climate change in addition to local environmental degradation.

This line of argument is consistent with more general claims from human ecology (Catton, 1980) but with an important caveat. Although classic human ecology argues that population growth is a general driver of environmental degradation, the growth machine thesis makes no such transhistorical claim. Rather, it argues that ecological dimensions of population growth must be understood within the context of the specific sociocultural system in which they occur. Within the United States, this context is a system in which land is treated as a commodity and owners extract profits from lands under their control, regardless of any negative environmental consequences that might result from these arrangements. Moreover, this context is nested within a broader context of mass consumption that amplifies the environmental impacts of local population growth within the United States. To illustrate, (York et al., 2003, p. 295) point that, at current consumptions levels, slow but steady population growth within the United States may be equally or more detrimental to the biosphere than higher rates of population growth elsewhere, where consumption levels tend to be lower and local growth machines less aggressive.

Again, the general point is that national–historical context matters, especially when it comes to population growth and anthropogenic greenhouse gas production (Satterthwaite, 2009). In wealthy nations such as the United States, where profits are the dominant organizing principle for land development at the local level, the effect of population growth on carbon emissions is hypothesized to be positive and quite dramatic.

With these points in mind, the growth machine thesis goes further to argue that the "clearest indication of success at growth is a constantly rising *urban-area* population" (Molotch, 1976, p. 310). In other words, if local population growth is environmentally detrimental within the U.S. context, then urban population growth is expected to be additionally detrimental, above and beyond the negative externalities of local population growth generally (see also Clement, 2010). This distinct but related effect is expected to occur because growing urban populations indicate a significant shift in the types and degree of local activities available for consumption, which in turn require greater amounts fossil fuel to sustain them (Smil, 1994). Compared with rural areas, for example, urban places have more shopping centers, hotels, restaurants, gyms, espresso bars, zoos, and the like, which are often commodified to further enhance private exchange values. Indeed, compared with rural living, daily life in large urban areas tends to be accelerated and extended into all hours of the day and night, with much of the social reproduction required to sustain daily life occurring outside the domestic sphere in the realm of market exchange (c.f. Axinn & Ghimire, 2011; Melbin, 1978).

Total and Industrial Production

As with demographic dimensions of local growth, Logan and Molotch (2007) situate their analysis of economic production within a particular sociohistorical context. In the context of the capitalist mode of production, such production is understood to be driven by the profit motive, in which the exchange values of commodities are prioritized over their use values. In this way, expanding economic production helps local growth machines achieve their goals, and this is true despite the fact that capital is mobile and often not directly involved in local politics. In fact, such economic growth, according to Logan and Molotch (2007), triggers a chain reaction of local development projects that encourage, for example, the in-migration of labor, the construction of housing, and the provision of retailing. Consequently, although specific localities vary in their degree and type of economic specialization, all forms of economic production are presumed to lead "to the same pot of gold: more intense land use and thus higher rent collections, with associated professional fees and locally based profits" (p. 58). For these reasons, expanding economic production of any kind is understood to be a primary aim of local growth machines, which is also consistent with the treadmill of production theory developed by Schnaiberg (1980), an environmental sociologist whose work Logan and Molotch cite. The common logic is that as local growth machines develop, they accelerate local treadmills of production, increasing withdrawals of natural resources from other places, near and far. These resources are consumed in economic production for the sake of profit and then transformed into locally generated wastes, which increasingly take the form of carbon emissions released into the atmosphere.

As with demographic growth, however, Logan and Molotch (2007) also assert that some forms of economic production are more intensive than others when it comes to natural resource use and pollution. These "carbon-intensive" industries include construction, manufacturing, mining, utilities, and transportation, which despite new technological advances and government regulations, are expected to have particularly damaging environmental effects, above and beyond economic production generally. In this regard, the growth machine thesis differs from ecological modernization theory (e.g., Mol, 1995), which argues that the development of modern industry can become more environmentally efficient and sustainable over time. By contrast, Logan and Molotch point out that effluents from newer, hightech manufacturing still endanger local water supplies (c.f., Pellow, 2007) and that air pollution released from modern factories is often "more dangerous, if less visible, than the smoke of nineteenth-century factories" (p. 223). Our working hypothesis is that these same industrial activities have a similar effect on carbon emissions, rendering places with larger industrial workforces more locally carbon intensive (Zahran et al., 2008).

Local Affluence and Government Spending

In addition to demographic and economic sources of local growth, the growth machine thesis also draws attention to local consumer power and public spending. It does so by first pointing out that although local affluence and economic production are connected, the two are conceptually distinct and spatially organized. Indeed, within a capitalist system such as the United States, an urban hierarchy of places develops in which wealth from distant production flows into corporate headquarters located in bigger cities positioned further up the place hierarchy. Consequently, earnings from local economic production do not stay entirely in the localities where production happens, implying that economic production and local affluence make distinct contributions to the success of local growth machines. On one hand, economic growth creates jobs and brings more people to a locality; on the other hand, increasing local wealth escalates the consumption of commodities, which increases profits for growth machine actors in locally expanding retail sectors.

This dynamic constitutes the consumer side of the local growth machine, but Logan and Molotch (2007) recognize that its environmental consequences are not always straightforward. On the one hand, increasing local affluence translates into greater consumption of natural resources, implying a negative environmental impact, including the potential for greater carbon emissions. On the other hand, more affluent communities can also afford more resource-efficient products and infrastructure, and avoid industrial activities that tend to be dirty and carbon intensive. These latter processes amount to what Molotch (1976) calls "aristocratic conservation" (p. 328), that is, the privilege of the rich to act as environmentally conscious consumers. The ability of affluent communities to consume more resource-efficient products may seem to reflect aspects of ecological modernization (e.g., Mol, 1995), but the main concern of the growth machine thesis is how class inequalities among places unevenly distribute environmental outcomes, which means that wealthy localities are environmentally privileged while poorer ones are disadvantaged (see Lutzenhiser & Hackett, 1993). Considering that affluence can increase the consumption of natural resources, like fossil fuel, but also be used to purchase resource efficiency and to avoid carbon-intensive activities suggests that local affluence exhibits a curvilinear relationship with local carbon emissions, rising at lower levels of household income but then beginning to taper at higher levels as opportunities for aristocratic conservation increase.

In addition to these private sources of local consumption, there is also public consumption, or investment, through government spending and decision making. Compared with many other nations, local governments in the United States have extensive authority to make land use decisions. In a system where land is treated as a commodity, this authority has fostered intimate connections between local government leaders, politicians, and other pro-growth actors. Meanwhile, at the national level, the U.S. federal government also provides subsidies for public and private sectors to develop land in accordance with local growth machine goals, even if these subsidies ultimately do not attract capital investment. Therefore, according to Logan and Molotch (2007), while government spending accomplishes a variety of use values (e.g., education, police and fire protection, and highway construction), it ultimately serves the function of enhancing the exchange value of land and generating profit for capital. In this sense, and based on the growth machine thesis, local and federal government expenditures can be expected to have a positive effect on local carbon emissions because government spending helps to develop the physical infrastructure of local places, the use of which increases consumption and transformation of natural resources into carbon waste.

DATA

We test the above hypotheses using local data for all counties and county-equivalents in the continental United States ($N = 3{,}073$). A local-level analysis in the United States is worthwhile because the country as a whole generates approximately a quarter of the world's anthropogenic greenhouse gases, thereby contributing to what Freudenburg (2005) calls the "disproportionality of environmental privilege and harm." Yet, we still know relatively little about how carbon emissions within the United States vary, which is important because nearly half of all U.S. counties emit more carbon waste than many under-developed nations and show no sign of reversing this fact anytime soon (Satterthwaite, 2009). We describe these data in greater detail below; and Table 1 provides a summary of all variables and sources.

Dependent Variable: Total Carbon Emissions

The dependent variable for our study is the natural logarithm of *total carbon emissions* measured in metric tons at the county level for the year 2002. These data come from the Vulcan Project, formerly located at Purdue University and funded by NASA and the U.S. Department of Energy (Gurney et al., 2009; see also Parshall et al., 2010). The Vulcan Project combines data from a number of emissions monitoring and fuel consumption inventories for local electricity production, industrial manufacturing, transportation, and residential and commercial activities. Unlike prior subnational emissions inventories (e.g., Olivier et al., 1999), the Vulcan Project does not estimate total emissions using population as a predictor, which opens new opportunities for studying population–environment linkages at the local level. Another feature of the Vulcan dataset is that it excludes emissions from air and sea travel beyond those produced upon immediate departure and arrival. These accounting procedures mean that local emissions are measured in ways that are directly linked with locally observed processes and conditions, without the complication of assigning emissions discharged in transit thousands of feet in the air or en route along regional and international waterways.

Independent Variables: Growth Machine Dynamics

To examine the different dimensions of growth machine dynamics discussed above, we operationalize a series of variables. Each is drawn from *USA Counties* published by the U.S. Census Bureau and transformed into its natural logarithm to normalize its arithmetic distribution for statistical analysis. We measure *total population* and *urban population* as, respectively, the total number of people residing in a county and the number of people residing in urban areas within that county for the year 2000. *Gross domestic product* (*GDP*) represents the total earnings reported from all economic sectors in the county for the year 2002, and *industrial labor force* represents

Table 1. County-Level Variables: Descriptions, Sources, and Summary Statistics[a] ($N = 3{,}073$).

Variable	Description	Source	Mean (SD)
Total carbon emissions	Total carbon emitted from the combustion of fossil fuels in metric tons, 2002	The Vulcan Project (Gurney et al., 2009)	11.87 (1.53)
Population size	Total number of people, 2000	US Census Bureau *USA Counties*	10.25 (1.41)
Urban population	Number of people residing in urban areas, 2000	US Census Bureau *USA Counties*	7.58 (4.42)
GDP	Total earnings from all industries, 2002	US Census Bureau *USA Counties*	19.71 (1.77)
Industrial workforce	Total employees in construction, manufacturing, mining, transportation, and utilities, 2000	US Census Bureau *USA Counties*	8.36 (1.31)
Median household income	Median household income, 2000	US Census Bureau *USA Counties*	10.45 (0.24)
Median household income^2	Median household income squared, 2000	US Census Bureau *USA Counties*	109.34 (5.02)
Local government spending	Total direct general expenditures by local governments, 2002	US Census Bureau *USA Counties*	11.22 (1.56)
Federal government spending	Total federal expenditures, 2002	US Census Bureau *USA Counties*	12.01 (1.45)
Land area	Land area in square miles, 2000	US Census Bureau *USA Counties*	6.52 (0.75)
Very cold climate	Location in a very cold climate zone, 0 = No and 1 = Yes	US Department of Energy (2010)	0.03 (0.18)
Hot-humid climate	Location in a hot-humid climate zone, 0 = No and 1 = Yes	US Department of Energy (2010)	0.14 (0.35)
Spatial lag	Mean carbon emissions in adjacent counties using a queen, first-order spatial contiguity matrix	Geographic information software	11.92 (1.03)

[a]All variables are transformed into their natural logarithms, with the exception of dummy indicators of climate zone.

the total number of workers employed in construction, manufacturing, mining, transportation, and utilities – what other researchers have called the "carbon intensive" industries (Zahran et al., 2008).

To measure local affluence and potential consumer power, we use *median household income* for the county in the year 2000; and to assess its potentially nonlinear relationship with carbon emissions, we include its squared, or quadratic, form in our full model below. Finally, *local* and *federal government spending* represent, respectively, the total direct general expenditures by local governments and total federal government expenditures in a county for the year 2002.

Independent Variables: Controls

To account for spatial autocorrelation in our dependent variable, which can violate assumptions of independence in regression models, we compute and include a *spatial lag* for total carbon emissions using a queen, first-order spatial contiguity matrix. This procedure effectively computes the average total emissions in adjacent counties for each county of observation. This spatial lag helps to address the fact that counties do not always correspond to meaningful social or political boundaries and that nearby counties may affect one another through broader, supra-local dynamics emanating from a common regional industrial structure, electrical production profile, or transportation system. For all these reasons, we estimate all models using the spatially lagged dependent variable.

In addition to this spatial control, we also include three control variables derived from prior human ecology research on carbon emissions (e.g., York et al., 2003). One is *land area*, which is calculated as the natural logarithm of the total land area (in square miles) in a county for the year 2000. The other two variables are dummy indicators for a county's location in either a *very cold* (e.g., Cass County, ND) or a *hot-humid climatic zone* (e.g., Miami-Dade, FL), as designated by the United States Department of Energy (2010). Natural science research shows that energy consumption increases with both heating- and cooling-degree days (NOAA, 2010); therefore, location in an extreme climatic zone is expected to correlate positively with carbon emissions.

Analysis

We use OLS regression to test our hypotheses, regressing *total carbon emissions* on growth machine variables in four different models: Model 1 examines the effects of *total population* and *urban population* on the dependent variable, net of controls; Model 2 adds *GDP* and *industrial labor force*; Model 3 then adds *median household income, local government spending* and *federal government spending*; and finally, Model 4 adds the quadratic form of *median household income squared.* We report results in this nested fashion in order to provide readers with a full sense of how findings for total and urban population size shift with the addition of related covariates. In our discussion below, however, we focus on the final model, Model 4, because it offers the best empirical fit to the data.

In all models, we use logarithmic values for the dependent and independent variables of interest to render estimates similar to elasticity models in economics, wherein coefficients indicate the percentage change in the dependent variable in response to a 1-percent change in the respective independent variable. In supplemental analyses (not shown), we also control for state-level effects using a series of dummy indicators (with Texas as the reference). In no instance did these results change substantively from those reported in Table 2. Also, the Breusch–Pagan/Cook–Weisberg test for heteroskedasticity is significant for all models; therefore, we use robust standard errors to account for nonconstant variance when conducting significance tests for slope estimates.

RESULTS

Table 2 reports the results of our regression analyses, which include slope estimates, robust standard errors, and estimates of model fit (R^2). Overall, we find that the respective models explain approximately 70 percent of county-level variation in total emissions, and that slope estimates for all growth machine variables are statistically significant at the .05-level. Moreover, with the exception of federal government spending, all of these estimates are in the hypothesized direction. These results remain consistent even in supplemental analyses (not shown) where we include dummy indicators to control for fixed effects among states. Thus, the immediate and overarching finding from Table 2 is that the growth machine thesis offers an empirically robust framework for examining county-level variation in carbon emissions. We now discuss specific results for respective dimensions of the growth machine thesis.

First, to provide a better sense of the relative importance of each factor, Table 3 provides its rank order by standardized effect on carbon emissions, using results from Model 4. (Ranking of the two variables with negative

Table 2. OLS Regression Results Predicting Total Carbon Emissions (logged) at County Level ($N = 3,073$).

	Model 1		Model 2		Model 3		Model 4	
	b	*SE*	*b*	*SE*	*b*	*SE*	*b*	*SE*
Growth machine variables								
Population size	0.740	0.023***	0.222	0.098*	0.272	0.097**	0.339	0.098***
Urban population	0.036	0.006***	0.030	0.006***	0.027	0.006***	0.025	0.006***
GDP			0.258	0.073***	0.233	0.072**	0.221	0.071**
Industrial workforce			0.244	0.051***	0.185	0.052***	0.127	0.054*
Median household income					0.101	0.094	14.899	3.381***
Median household income^2							−0.703	0.160***
Local government spending					0.143	0.043***	0.146	0.042***
Federal government spending					−0.110	0.040**	-0.111	0.040**
Human ecology controls								
Land area	0.181	0.024***	0.186	0.024***	0.172	0.025***	0.164	0.025***
Very cold climate	0.191	0.072**	0.157	0.074*	0.126	0.074	0.102	0.074
Hot-humid climate	0.062	0.050	0.138	0.050**	0.151	0.050**	0.153	0.050**
Spatial control								
Spatial lag	0.158	0.023***	0.153	0.023***	0.131	0.023***	0.135	0.023***
Constant	0.942	0.268***	−0.833	0.668	−1.306	0.865	−79.06	17.752***
R^2	0.704		0.713		0.716		0.717	

*$p<0.05$; **$p<0.01$; ***$p<0.001$ (two-tailed tests).

Table 3. Variables Predicting Total Carbon Emissions at County Level by Descending Order of Predictive Importance.

Rank	Variable
1.	*Median household income* (+)
	[*Median household income*^{2} (−)]
2.	*Population size* (+)
3.	*Gross domestic product* (+)
4.	*Local government spending* (+)
5.	*Industrial workforce* (+)
6.	*Federal government spending* (−)
7.	*Land area* (+)
8.	*Urban population* (+)
9.	*Hot-humid climate* (+)
10.	Very cold climate (+)

Note: Ranking is based on standardized slope estimates from Model 4 of Table 2. *Italics* indicate statistical significance at $p < 0.05$ (two-tailed test). (+) denotes positive effect; (−) denotes negative effect.

slope estimates in Model 4 is based on the coefficient's absolute value.) At the top of this descending list is median income, in part because its quadratic form reveals a significant curvilinear, or ∩-shape, relationship with carbon emissions. This relationship indicates that emissions increase up to a median household income of approximately $40,000 and then begin to decline thereafter, with the upward-sloping portion of this curve accounting for approximately 75 percent of all cases, and the downward-sloping portion accounting for the remaining 25 percent. This finding is consistent with the growth machine hypothesis that increasing local affluence has a dual effect on environmental outcomes such as carbon emissions: on the one hand, greater county-level affluence results in more fossil fuel consumption; on the other hand, wealthier counties have the means to invest in more environmentally efficient lifestyles and to separate themselves from dirty, carbon-intensive activities in ways that begin to reduce carbon emissions, all else equal.

By contrast, the leading linear, or monotonic, driver of local carbon emission is population size. Here, results from Model 4 indicate that a one percent increase in total population corresponds with a 0.34 percent increase in local carbon emissions, all else equal. The effects of *urban* population in particular are much less pronounced but nonetheless present. These results indicate that for every one percent increase in the local urban population there is an additional 0.03 percent increase in total emissions, over and above those generated by total population. In other words, while total population drives emissions at the county level, having this population reside in urban areas amplifies this effect by an average of roughly 9 percent (0.03/0.34 = 0.88). So, both demographic size and its spatial organization matter for local carbon emissions.

To see how these top two factors – household income and population size – work together at different levels, we use Model 4 of Table 2 to estimate average county-level emissions at different levels of each variable, holding all other factors constant at their respective population means. Results appear in Fig. 1 in nonlogged form. Here, we see that while declines in total emissions are evident when moving from the 50th to 90th percentile of median household income, the overall impact of this decline pales in comparison with overall increases in total emissions when moving from the 50th to 90th percentile of total population size. In other words, population size's linear effect on emissions overwhelms the net curvilinear effect of

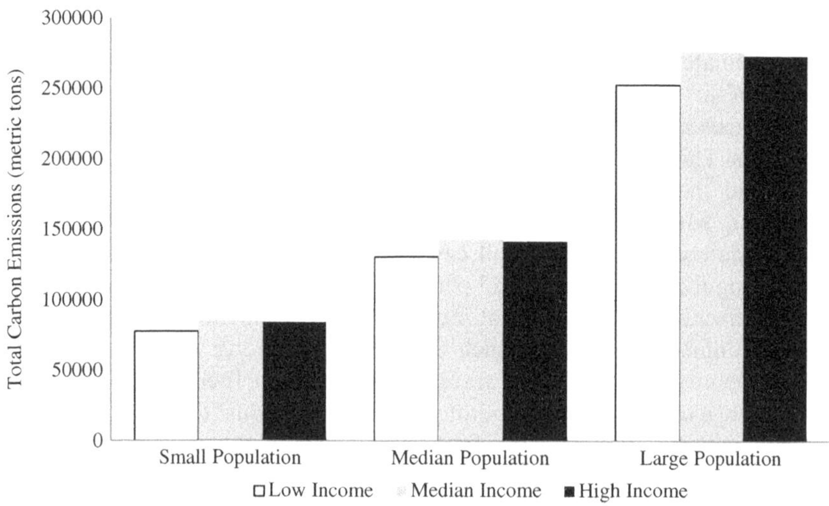

Fig. 1. Simulated Carbon Emissions by Population Size and Median Household Income*. *Source*: Model 4, Table 2; all other variables held constant at respective population means.

Population: Small, median, and large populations refer to the 10th, 50th, and 90th percentile of this variable, respectively (total population: 5,366; 25,261; 174,555).

Income: Low, median, and high incomes refer to the 10th, 50th, and 90th percentile of this variable, respectively (median household income: $26,108; $34,149; $46,938).

household income, once both components of the latter (squared and unsquared household income) are combined. (This assessment is further supported by the statistically insignificant coefficient for household income in Model 3, without its squared term present.)

Returning to Model 4 of Table 2, we also find strong, positive effects for total GDP and industrial employment. Specifically, results indicate that, a one percent increase in local GDP corresponds with a 0.22 percent increase in total emissions, all else equal. Similarly, a one percent increase in local industrial employment corresponds with a 0.13 percent increase in total carbon emissions. So, as with population size, there is evidence of both general and specific effects: In general, total economic production of any sort appears to increase local emissions; and specifically, this effect appears to be amplified the more it is linked with industrial labor.

Finally, results of Model 4 indicate a countervailing influence of government spending on local emissions. On the hand, a one percent increase in local government spending corresponds with a 0.15 percent increase in total emissions, as hypothesized. On the other hand, and by contrast, a one percent increase in federal spending corresponds with a 0.11 percent *decline* in total emissions. While the latter finding is unexpected in light of Logan and Molotch's original formulation, it corroborates more recent sociological research on local growth machines' effects on the environment. Rudel (2009), for instance, argues that under neoliberalism the federal government's role in regional development has diminished and that local governments have assumed primary responsibility for intensifying local land-use and exchange value. Moreover, federal spending at the local level is now largely oriented toward the (social) service sector and research and development projects; whereas, local government spending is oriented more toward infrastructural projects that tend to boost local emissions.

CONCLUSION

For most of human history, anthropogenic environmental degradation was locally generated and locally experienced (Ponting, 2007). However, with the rise of capitalism and major urban-industrial centers, the environmental impacts of local activities have become increasingly global in their effects (Foster, 1994); this is particularly the case with climate change (Clark & York, 2005; Gonzalez, 2009; Grimm et al., 2008). Indeed, in considering these developments, Logan and Molotch (2007) write, "Talk about change of scale!" (p. xvii). In the present study, we set out to learn more about how local dynamics of economic and demographic growth pursued by local pro-growth coalitions across the United States contribute to these developments. Results affirm that while some dimensions of local growth drive carbon emissions down, these effects are limited and small relative to the collection of forces that drive emissions up. In this way, we conclude that the syndrome of local growth within the United States currently contributes to global climate change.

The theoretical and practical implications of these results present challenges to modernist studies of human–environment relations (e.g., Mol, 1995) and to mainstream environmental policy (e.g., Newman & Kenworthy, 1999), both of which emphasize the roles that better design and technological innovation can play in ameliorating humanity's environmental impact. By contrast and based on results presented above, we conclude that the political economy of local place-making offers an indispensable conceptual framework for studying global environmental impacts and for developing practices to improve the relationship between human society and the biosphere. This framework does not ignore the roles that design and technological innovation can play in these processes, nor does it diminish the need to design local places in a more ecological manner. Rather, it refocuses discussion to include key sociological factors behind ecological change (e.g., population growth, economic expansion, and local government spending).

As a way of pointing toward future research in the vein, we consider a more nuanced example of the modernist approach in environmental social science recently offered by Sassen and Dotan (2011). These researchers acknowledge that local urbanization contributes to negative environmental change, and they recognize the sociological forces behind this change. Nevertheless, Sassen and Dotan ultimately come to highlight how the incorporation of technological development projects into local place-making can ameliorate modern society's environmental impact. These projects include, for example, algal wastewater processing, carbon sequestration, bacterial concrete, and bioreactor landfills (pp. 829–831). Developing these projects, they argue, should reduce the amount of natural resources consumed and pollution emitted by modern society and its urban centers. In this way, the negative environmental "articulations" of local place-making can be transformed into environmentally beneficial processes "by activating biospheric capacities in urbanized settings" (p. 823).

Yet, as these researchers acknowledge, technological innovation is only one step in creating a more sustainable relationship between modern society and the biosphere. Other important steps must emerge from ongoing discussions about the multiscalar and sociological properties of urbanization. This is where our study can contribute. In the analyses presented above, we examined how the sociological properties of local places relate to the environment. While our results are empirically similar to those reported from cross-national studies on greenhouse gas emissions (e.g., York et al., 2003), they offer a conceptually distinct contribution to the literature. This contribution emerges from the recognition that within advanced-industrial societies such as the United States, local areas are deliberately developed as sources of capital accumulation as well as sites, or locations, for capital accumulation through other productive means. This duality of place-making means that towns and cities will continue to attract powerful actors united in pursuit of expanding local demographic, economic, and consumptive activities, which we have demonstrated result in more carbon emissions at and from the local level. In this sense, a focus on the political economy of local place-making is as important, if not more, for addressing anthropogenic environmental change than a focus on technological innovation. Researchers such as Sassen and Dotan (2011) acknowledge this point, but they do not elaborate. Our research represents an initial effort to start developing this line of thought, empirically investigating how modern society is connected to global environmental change at and from the local level, where urbanization actively and actually takes place.

REFERENCES

Allègre, C. J., & Schneider, S. H. (1994). The evolution of the earth. *Scientific American, 271*(4), 44–51.

Axinn, W. G., & Ghimire, D. J. (2011). Social organization, population, and land use. *American Journal of Sociology, 117*(1), 209–258.

Beck, U. (2009). *World at risk*. Malden, MA: Polity.

Betsill, M. (2005). Global climate change policy: Making progress or spinning wheels? In R. S. Axelrod, D. L. Downie & N. J. Vig (Eds.), *The global environment: Institutions, law, and policy* (2nd Ed., pp. 103–124). Washington, DC: Congressional Quarterly Press.

Catton, W. R., Jr.. (1980). *Overshoot: The ecological basis of revolutionary change*. Urbana, IL: University of Illinois Press.

Clark, B., & York, R. (2005). Carbon metabolism: Global capitalism, climate change, and the biospheric rift. *Theory and Society, 34*(4), 391–428.

Clement, M. T. (2010). Urbanization and the natural environment: An environmental sociological review and synthesis. *Organization & Environment, 23*(3), 291–314.

Curtis, K. J., & Schneider, A. (2011). Understanding the demographic implications of climate change: Estimates of localized population predictions under future scenarios of sea-level rise. *Population & Environment, 33*(1), 28–54.

Dietz, T., Dan, A., & Shwom, R. (2007). Support for climate change policy: Social psychological and social structural influences. *Rural Sociology*, *72*(2), 185–214.

Dietz, T., & Rosa, E. A. (1997). Effects of population and affluence on CO_2 emissions. *Proceedings of the National Academy Sciences*, *94*(1), 175–179.

Fisher, D. R. (2006). Bringing the material back in: Understanding the United States position on climate change. *Sociological Forum*, *21*(3), 467–494.

Fisher, D. R., & Freudenburg, W. R. (2004). Postindustrialization and environmental quality: An empirical analysis of the environmental state. *Social Forces*, *83*(1), 157–188.

Foster, J. B. (1994). *The vulnerable planet: A short economic history of the environment*. New York, NY: Monthly Review Press.

Freudenburg, W. R. (2005). Privileged access, privileged accounts: Toward a socially structured theory of resources and discourses. *Social Forces*, *84*(1), 89–114.

Freudenburg, W. R., Gramling, R., Laska, S. B., & Erikson, K. (2009). *Catastrophe in the making: The engineering of Katrina and the disasters of tomorrow*. Washington, DC: Island Press.

Gonzalez, G. A. (2009). *Urban sprawl, global warming, and the empire of capital*. Albany, NY: State University of New York Press.

Grimes, P., & Kentor, J. (2003). Exporting the greenhouse: Foreign capital penetration and CO_2 emissions 1980–1996. *Journal of World-Systems Research*, *IX*(2), 261–275.

Grimm, N. B., Faeth, S. H., Golubiewski, N. E., Redman, J. W., Bai, X., & Briggs, J. M. (2008). Global change and the ecology of cities. *Science*, *319*, 756–760.

Gurney, K. R., Mendoza, D. L., Zhou, Y., Fischer, M. L., Miller, C. C., Geethakumar, S., & de la Rue du Can, S. (2009). The Vulcan project: High resolution fossil fuel combustion CO_2 emissions fluxes for the United States. *Environmental Science and Technology*, *43*(14), 5535–5541.

Harlan, S. L., Brazel, A. J., Jenerette, G. D., Jones, N. S., Larsen, L., & Prash, L. (2008). In the shade of affluence: The inequitable distribution of the urban heat Island. *Research in Social Problems and Public Policy*, *15*, 173–202.

Hasselmann, K., Latif, M., Hooss, G., Azar, C., Edenhofer, O., Jaeger, C. C., … Wokaun, A. (2003). The challenge of long-term climate change. *Science*, *302*, 1923–1925.

Hempel, L. C. (2006). Climate policy on the installment plan. In N. J. Vig & M. E. Kraft (Eds.), *Environmental policy: New directions for the twenty-first century* (6th ed., pp. 288–310). Washington, DC: Congressional Quarterly Press.

Intergovernmental Panel on Climate Change (IPCC). (2007). *Climate change 2007: The physical science basis*. Cambridge: Cambridge University Press.

Jonas, A. E. G., & Wilson, D. (Eds.). (1999). *The urban growth machine: Critical perspectives two decades later*. Albany, NY: State University of New York Press.

Jorgenson, A. K. (2006). Global warming and the neglected greenhouse gas: A cross-national study of the social causes of methane emissions. *Social Forces*, *84*(3), 1779–1798.

Jorgenson, A. K. (2007). Does foreign investment harm the air we breathe and the water we drink? A cross-national study of carbon dioxide emissions and organic water pollution in less-developed countries, 1975 to 2000. *Organization & Environment*, *20*, 137–156.

Karl, T. R., & Trenberth, K. E. (2003). Modern global climate change. *Science*, *302*(5651), 1719–1723.

Klinenberg, E. (2002). *Heat wave: A social autopsy of disaster*. Chicago, IL: University of Chicago Press.

Krause, R. M. (2011). Policy innovation, Intergovernmental relations, and the adoption of climate protection initiatives by U.S. cities. *Journal of Urban Affairs*, *33*(1), 45–60.
Logan, J. R., & Molotch, H. L. (2007). *Urban fortunes: The political economy of place*. Berkeley, CA: University of California Press.
Lowry, W. P. (1967). The climate of cities. *Scientific American*, *217*(2), 15–24.
Lutzenhiser, L. (2001). The contours of U.S. climate non-policy. *Society & Natural Resources*, *14*(6), 511–523.
Lutzenhiser, L., & Hackett, B. (1993). Social stratification and environmental degradation: Understanding household CO_2 production. *Social Problems*, *40*(1), 50–73.
McCright, A. M., & Dunlap, R. E. (2000). Challenging global warming as a social problem: An analysis of the conservative movement's counter-claims. *Social Problems*, *47*(4), 499–522.
McCright, A. M., & Dunlap, R. E. (2003). Defeating Kyoto: The conservative movement's impact on U.S. climate change policy. *Social Problems*, *50*, 348–373.
Melbin, M. (1978). Night as frontier. *American Sociological Review*, *43*(1), 3–22.
Millennium Ecosystem Assessment. (2005). *Ecosystems and human wellbeing*. Washington, DC: Island Press.
Mol, A. P. J. (1995). *The refinement of production: Ecological modernization theory and the chemical industry*. Utrecht, The Netherlands: Van Arkel.
Molotch, H. (1976). The city as a growth machine: Toward a political economy of place. *American Journal of Sociology*, *82*(2), 309–332.
National Oceanic and Atmospheric Administration (NOAA). (2010). *Residential energy demand temperature index*. Retrieved from http://www.ncdc.noaa.gov/societal-impacts/redti/. Accessed December 20, 2011.
Newman, P., & Kenworthy, J. R. (1999). *Sustainability and cities: Overcoming automobile dependence*. Washington, DC: Island Press.
Norgaard, K. M. (2006). 'We don't really want to know': Environmental justice and socially organized denial of global warming in Norway. *Organization & Environment*, *19*, 347–370.
Norgaard, K. M. (2011). *Living in denial: Climate change, emotions, and everyday life*. Cambridge, MA: MIT Press.
O'Connor, R. E., Bord, R. J., Yarnal, B., & Wiefek, N. (2002). Who wants to reduce greenhouse gas emissions? *Social Science Quarterly*, *83*(1), 1–17.
Olivier, J. G. J., Bouwman, A. F., Berdowski, J. J. M., Veldt, C., Bloos, J. P. J., Visschedijk, A. J., … Zandveld, P. Y. J. (1999). Sectoral emission inventories of greenhouse gases for 1990 on a per country basis as well as on $1^{\circ} \times 1^{\circ}$. *Environmental Science and Policy*, *2*, 241–264.
Pais, J., & Elliott, J. R. (2008). Places as recovery machines: Vulnerability and neighborhood change after major hurricanes. *Social Forces*, *86*(4), 1415–1453.
Parshall, L., Gurney, K., Hammer, S. A., Mendoza, D., Zhou, Y., & Geethakumar, S. (2010). Modeling energy consumption and CO_2 emissions at the urban scale: Methodological challenges and insights from the United States. *Energy Policy*, *38*(9), 4765–4782.
Pellow, D. N. (2007). *Resisting global toxics: Transnational movements for environmental justice*. Cambridge, MA: MIT Press.
Ponting, C. (2007). *A new green history of the world: The environment and the collapse of great civilizations*. New York, NY: Penguin Books.

Pulver, S. (2007). Making sense of corporate environmentalism: An environmental contestation approach to analyzing the causes and consequences of the climate change policy split in the oil industry. *Organization & Environment*, *20*, 44–83.

Ravallion, M., Heil, M., & Jalan, J. (2000). Carbon emissions and income inequality. *Oxford Economic Paper*, *52*(4), 651–669.

Rudel, T. K. (2009). How do people transform landscapes? A sociological perspective on suburban sprawl and tropical deforestation. *American Journal of Sociology*, *115*(1), 129–154.

Sassen, S., & Dotan, N. (2011). Delegating, not returning, to the biosphere: How to use the multi-scalar and ecological properties of cities. *Global Environmental Change*, *21*, 823–834.

Satterthwaite, D. (2009). The implications of population growth and urbanization for climate change. *Environment & Urbanization*, *21*(2), 545–567.

Schnaiberg, A. (1980). *The environment: From surplus to scarcity*. New York, NY: Oxford University Press.

Shandra, J. M., London, B., Whooley, O. P., & Williamson, J. B. (2004). International nongovernmental organizations and carbon dioxide emissions in the developing world: A quantitative, cross-national analysis. *Sociological Inquiry*, *74*(4), 520–545.

Smil, V. (1994). *Energy in world history*. Boulder, CO: Westview Press.

United States Department of Energy. (2010). *Building America: Climate regions*. Retrieved from http://www.eere.energy.gov/buildings/building_america/climate_zones.html. Accessed on May 26, 2011.

Weart, S. R. (2003). *The discovery of global warming*. Cambridge, MA: Harvard University Press.

York, R., Rosa, E. A., & Dietz, T. (2003). A rift in modernity? Assessing the anthropogenic sources of global climate change with the STIRPAT model. *International Journal of Sociology and Social Policy*, *23*(10), 31–51.

Zahran, S., Grover, H., Brody, S. D., & Vedlitz, A. (2008). Risk, stress, and capacity: Explaining metropolitan commitment to climate protection. *Urban Affairs Review*, *43*(4), 447–474.

CHAPTER 3

TACKLING CLIMATE CHANGE ADAPTATION AT THE LOCAL LEVEL THROUGH COMMUNITY PARTICIPATION

Geneviève Cloutier and Florent Joerin

ABSTRACT

Purpose – *Adapting local areas to climate change is a wicked challenge for local administrations. A participatory research is applied to explore how local experience shared by local experts can inform decision and adaptation planning by taking into account local area characteristics and their interrelationships.*

Methodology/approach – *We turned to local actors, who live or work in the city and who can be seen as urban experts. Their experiential knowledge has given us a better understanding of the characteristics of their communities. These experts are likely to possess a representation that reflects the local territorial sensitivities, which can help us determine how these characteristics might be impacted by climate change.*

Findings – *A participatory approach bears many benefits such as mobilizing local stakeholders to find collective solutions. It also allows us to focus on common practices in the urban context, which are likely to be altered by changes in mean temperatures, precipitations, etc. It offers the*

Urban Areas and Global Climate Change
Research in Urban Sociology, Volume 12, 51–73

ISSN: 1047-0042/doi:10.1108/S1047-0042(2012)0000012006

additional benefit of putting into perspective the relations between a variety of urban issues.

Research limitations – *A participatory approach means relying on subjective assessments of the possible effects of climate change, which could challenge the relevance of perceived risks and the scope and types of actions taken.*

Originality/value of paper – *The number of the available adaptation planning processes involving community stakeholders and assessments of these processes is very limited. A participatory process such as the cross-sectoral initiative organized in Québec City can have significant repercussions on local engagement in climate change adaptation. This provides evidence of the potential of deliberation or interaction of territorial actors to improve their understanding of the issues and their adaptive capacity. On a methodological level, the participatory process in itself and the steps to organize it offered a planning frame that can be reproduced.*

INTRODUCTION

Until very recently, research in the field of climate change has emphasized potential actions to stabilize and slow global warming (De Perthuis, Hallegatte, & Le Cocq, 2010). The overall goal of public policies was to determine ways to mitigate climate change instead of pursuing adaptation efforts. The IPCC defines adaptation as the adjustment in natural and human systems in response to actual or expected climatic stimuli or to their impacts, by using the opportunities offered by these climatic stimuli or by avoiding some to their harmful consequences (IPCC, 2001). The priority given to mitigation may result from its capacity to limit the impacts of climate change on the systems in cases where adaptation is both more contingent and dependent on available information and political will. It may also result from the preference of governments for single interventions that clearly define objectives and set targets (Klein, Schipper, & Dessai, 2003).

In fact, the main challenge with the adaptation approach to climate change is the uncertainty in estimating the likely impacts of climate change on local areas. When adaptation planning is undertaken at the local level, it is necessarily to consider how reorganization and relocation can help reduce the negative impacts while taking advantage of the beneficial effects of climate change on the territory. This reveals the complex nature of the issue. What does a 3°C temperature rise in summer represent for the management of health services? How does a 10% increase in snowfall affect transportation and inner city mobility?

Urban adaptation to climate change can be described as a wicked problem. The term wicked is used in so far as it refers to a problem for which little experience exists and available knowledge is limited, and that is neither structured nor one dimensional (Rittel & Webber, 1973). At the local level, climate change adaptation remains a wicked challenge especially because it is a moving target, for which no single and definitive solution can be applied: it is difficult to accurately predict how climate change will evolve through space and time.

Although many aspects of the potential impacts of climate change in metropolitan centres have yet to be fully addressed, one finding stands out clearly: climate change will continue even if we can achieve substantial reductions in greenhouse gas emissions (GES) and these changes will continue to affect natural and human systems (Hegerl & Zwiers, 2007; Meehl & Stocker, 2007). This lends support to the adaptation approach. The key issue therefore becomes how to develop ways to tackle the wicked problem of adapting to climate change in urban areas. Knowing that there are no true or false answers, good or bad solutions, how can we face this challenge? What are the frames of reference that can enable us to understand this problem?

In order to address increasingly complex socio-environmental issues, the supporters of the actor-network theory are calling for a change of approach, leaving conventional scientific frameworks aside and drawing on a variety of expert opinions (Latour, 1999; Callon, Lascoumes, & Barthe, 2001). By diversifying and widening scientific as well as practical and everyday life knowledge, a comprehensive and cross-sectional approach to knowledge building can be developed. Furthermore, when a range of actors participate in debate on these major social and environmental questions, decision makers widely believe that their participation can in fact legitimize decision-making mechanisms and provide a stronger basis for action (Boholm & Löfstedt, 2004).

Moreover, a cross-sectoral and interdisciplinary communication and collaboration effort would help foster collective learning and decision making regarding constantly evolving problems (Brown, Harris, & Russell, 2010). In addition, integrating various individual, organizational, collective, etc. perspectives and insight would help improve understanding of the potential interactions between the various causes of the problem on the one hand, and between the different ways to tackle it, on the other. Described as a moving target, researchers and decision makers should address the wicked problem of climate change by examining in detail the implications it may hold for the territory in order to determine how attempts at solving one aspect of the problem can bring consequences to bear on other aspects. A straightforward response to an increase in the number of extreme heat events, such as promoting the widespread use of air conditioning systems, does not succeed in resolving the problem in a satisfactory way: energy demand for air conditioning increases and generates more emissions. This sort of response merely shifts the problem.

Discussions related to the adaptation of climate change in urban areas must therefore take into account a wide range of dimensions (Tompkins et al., 2010). Since the dimensions involved in the wicked problem vary according to the context, an adaptation approach can benefit by taking into account local area characteristics. Placing greater emphasis on the contextual factors that condition the problem can promote an understanding of the values and rationales that have given rise to this problem (e.g. residential developments in flood-prone areas), and that may not provide a suitable response (e.g. the construction of dikes to prevent floods).

Given these two observations, that is, the multidimensional nature of the wicked problem of climate change facing local administrations and the advantages of considering the particular contextual factors in order to improve our understanding of the dimensions and their interactions, the ensuing question is how to assist territorial actors (the decision makers) in tackling this sort of problem and developing appropriate and sustainable management practices.

Developed as part of an action research initiative conducted under the joint responsibility of the Canada Research Chair in Territorial Decision Making and the Ouranos Consortium, the objective of this study is to collect information from local actors – community members, urban planners, social workers, public health specialists, etc. – in order to identify the particular aspects of the Québec City (Canada) metropolitan territory that are likely to be affected by climate change. These local actors (about 50 in all) live or work in the city and can be considered as urban experts. They are likely to possess a representation that reflects the local territorial sensitivities[1] and, in turn, they are able to provide an understanding of how and where local administrations would most benefit from adapting to anticipated climate change (Hallegatte, 2009).

It seems important to provide a brief overview of the Québec metropolitan area climatic and territorial context before addressing the participatory approach, which is the focus of our research initiative.

QUÉBEC CITY AND ITS 'ORDINARY' CLIMATE

The metropolitan area of Québec City (province of Québec, Canada) is home to slightly more than 750,000 people, the majority of which (about 500,000) lives in the urban area located within the Québec City boundaries (see Fig. 1). Situated along the banks of the Saint Lawrence River, the total land area of this eastern Canadian medium-sized region is 18,638 km^2. With a density of 37.2 inhabitants per km^2, it is one of the most sprawling regions in Québec. In November 2011, the employment rate was 65.2%. Since it is the seat of the provincial government, most jobs in Québec City are concentrated in public administration. The demographic forecasts indicate an increase of about 11% by 2030, but the region will also experience a significant ageing of its population.

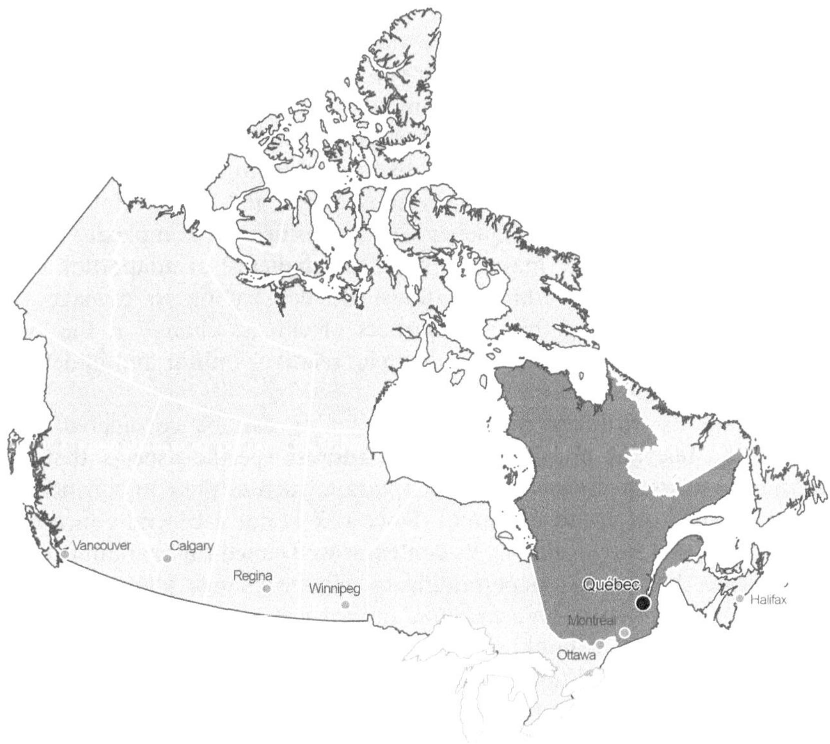

Fig. 1. Map of Canada.

The Québec City regional context shows no major vulnerability to climate change. Regarding climate, there is nothing extraordinary about Québec City. Even so, it is characterized by severe winters and important patterns of natural climate variability during the year. Indeed, Québec City experiences cold winters and relatively warm summers (19°C on average in July according to Environment Canada). The people of Québec City are well adapted to this variability.

It is estimated that by the year 2050, the Québec City metropolitan area will undergo a variety of climate changes, primarily an increase in mean precipitation, an increase in mean temperatures and more frequent extreme events. In fact, by 2050, temperatures in Southern Québec are expected to increase by 2.5–3.8°C especially in winter (Desjarlais et al., 2010). The increase in precipitation will be more significant in winter (from 8.6% to 18.1%) than in summer (3% to 12.1%) by the year 2050.

However, the climatic region of Southern Québec could experience a decrease in snow cover, which will negatively affect water supply. The region could also be subject to temperature variability: temperatures vary less in winter and more in summer. A major consequence would be an increase in the number of very hot days during the summer. In this regard, the frequency, intensity and duration of extreme climate-related events should increase. The Québec City region is also expected to experience a greater number of heat waves, but also more snowstorms and heavy rainfall.

The 'ordinary' climate of Québec City intensifies the complexity of the climate change issue and makes tackling the challenge of adaptation even more wicked for local administrations. Besides having to manage the uncertainty surrounding the consequences of climate change at the local level, administrations must try to anticipate relatively minor and ill-defined effects.

Yet, with its ordinary climate, Québec City can be considered as a particularly relevant object of study to address specific aspects that are seldom the focus of discussions on adaptation, such as the consequences of climate change on urban dynamics. Since risk is not a central concern in Québec City, but rather diffuse, its context is well suited for examining ways to integrate the discussion pertaining to climate change adaptation into urban planning practice on a medium and longer term basis.

Québec City region should so plan, as soon as possible, its adaptation to these new climate conditions. However, it has to be remained that Québec City, like all urban areas, is a constantly evolving social, political, economic and cultural organization. It has always adapted itself to cycles of production, migrations, new values and to a variety of stimuli. In which ways, therefore, can urban adaptive capacity be used to account for climate change? The answer to this question provides, in our view, fertile ground for thinking about the orientations public planning should take to adapt urban society by considering the changing nature of climate and our dynamic adaptation capability.

THE APPROACH – CROSS-TABULATING REPRESENTATIONS TO CONSTRUCT THE PROBLEM AT THE LOCAL LEVEL

How will building structures be affected by climate change? What could be the repercussions of the extension of summer and winter seasons on the urban fauna and flora? These questions are highly speculative and the available data is insufficient to yield an adequate response. Given the uncertain and dynamic nature of climate change, local adaptation plans should identify ways to adapt incrementally instead of trying to provide clear ultimate responses. Since adaptation is better understood as a bridging measure across space and time (De Perthuis et al., 2010), there is a keen interest in cross-tabulating the actors' perspectives and their distinctive expert opinions on a particular activity or sector of the territory.

Urban and regional planners, transportation, water or environmental technical consultants are key stakeholders to make information available on current practices in infrastructure management, project execution, etc. Community organizers and social service providers in general have an incomparable knowledge of particular social groups and of their needs. They are known for their basic knowledge of the repercussions on everyday life that the more vulnerable persons will face in diverse climatic contexts. For their part, public planning consultants, business people and interest groups have a clear understanding of the priorities for action, and are in a position to influence the outcome of decisions on a wide scale. Their ability to sense the urgency of issues that need to be addressed and represent them in terms of efficiency and needs satisfaction makes an important contribution towards adaptation planning. Whether they are specialists in water management, decision-making support, urban form or mobility, researchers are able to provide a critical analysis of particular territorial dynamics and their likely interactions with potential climate change effects.

All of these actors were considered and included in the 2010–2011 Québec City region participatory diagnosis initiative. Knowing that most had limited knowledge of anticipated climate change, the objective of this first of a three-phase action research initiative was twofold: to describe the particular local territorial sensitivities and appraise the participatory methodology used to diagnose climate change issues.

Accordingly, the initial step was to inform the future actions of decision makers about the territory. By undertaking a participatory diagnosis, our goal was to draw on the real situations lived by the territorial actors to describe the various components of the territory (infrastructures, populations, public places, etc.). By leading the participants into a discussion on urban space, activities and the population, the objective was to identify both the solid components and those that could be strengthened in order improve their resilience, in other words their capacity to cope with and adapt themselves positively to climate change impacts (Folke et al., 2002). This way of engaging in discussions on concrete aspects of the territory frees the deliberation from the inherent limitations of trying to identify rules that are applied to ensure a 'perfect' adaptation of the local territory. Lending support to the idea that even the most rigorous institutional framework could not ensure that all individual behaviours comply with collective goals (Sen, 2009), our initiative is therefore concerned with reality. It seeks to generate a comparison between the weak elements to be strengthened and to identify the key aspects of the urban environment for climate change adaptation planning. To this effect, the proposed participatory initiative provides local actors with the opportunity to construct the climate change problem in their context.

The following step of the initiative aimed at outlining a methodological framework for planning climate change adaptation at the urban scale. While the field of urban planning is increasingly recognized as a major tool for climate change adaptation, these stakeholders however seldom take advantage of this consideration (Blanco & Alberti, 2009). This is even more so when asking a variety of local actors to participate in a discussion on this question. The objective of the methodological part of the action research initiative was to show how investing in the implementation of such a participatory diagnosis instrument is relevant and effective. Through a translation process of individual representations of the local area made at discussion workshops, we describe the situation that forms the basis for the more objective and critical assessment of local risks, which is the following phase of the initiative.

Let us now examine more closely how the participatory initiative was carried out in Québec City. This will be combined with an appreciation of its value when tackling a wicked problem.

METHODOLOGY – EXPLORING THE TERRITORY AND ITS SENSITIVITIES THROUGH EXCHANGE AND DISCUSSION

The initiative organized by the research team in Québec City is based on a participatory diagnosis of climate change issues at the metropolitan level. A series of discussion workshops were held to examine local stakeholders' degree of knowledge and engagement about climate change adaptation. Moreover, these workshops served to organize a one-day forum, which represented the crucial phase of the participatory diagnosis and brought together about 40 local actors.

At this forum, the actors, whose knowledge about the climate question varied, were invited to freely examine the impacts of climate change in five different sectors of the Québec City metropolitan area. To select these sectors, researchers took a representative sample of the various urban patterns across the Québec City territory. Attributes of density, topography and socio-demographic composition were used to distinguish them. However, participants all share a similar proximity with the geographical sectors: they live and work there, or have a good understanding of their issues. By situating them in the context of the sector, a common frame of reference can therefore be provided to facilitate stakeholder discussions on the choices, values, priorities and orientations that should guide the adaptation planning process, at least for the time being.

More specifically, participants at the forum took part in simulations of the real life sectors: a neighbourhood located in the urban core, on the outskirts or in the suburbs. Caught in a kind of urban drift, the starting point was a specific place in the sector, which was identified beforehand as having a remarkable sensitivity. It could be, for example, a place with highly mineralized surfaces increasing the urban heat island effect; a low-lying area sensitive to floods; or a nursing home with a very sensitive population. From this point of entry into the sector, participants were then asked either to discuss climate-related events that generate these sensitivities, or to elaborate on other sensitive places or aspects of the sector. The focus is put on the sensitivities that result from other sensitivities: groundwater pollution could potentially lead, for example, to the contamination of a school's water supply. By travelling from place to place or from a climate-related event to another, participants are brought to examine the climatic strengths and weaknesses by geographical sector.

At the same time, a record is made of the systemic, essentially causal, relations with the purpose of holding a second workshop that aims at providing a generalization of the previously examined local situation. This record of the causal relations made by a member of the research team provides participants with a causal map of the main risk mechanisms under discussion (see Fig. 2). At this second workshop held during the forum, participants are asked to propose modifications to the causal map modelling the risk mechanisms spontaneously discussed in the first workshop. This will further strengthen their ownership of the statement analysis.

This one-day forum ends with the pooling of the sensitivities identified at the outset and with a look back on the modelling of risk mechanisms. The sharing of findings from the exchanges allows the participating actors to acquaint themselves with the generalizations that can be drawn from the discussion on Québec City.

The individuals attending the discussion workshops and the forum provide a valuable description of local realities by drawing on their own experiences. By voicing their representations of the strong and weak points of the territory of Québec City, they are translating a precise portrait of the factors to consider in the ongoing discussion on climate change adaptation. To ensure that this portrait makes sense and is relevant, pitfalls related to the participatory instrument used to diagnose local climate change issues must be avoided.

A Daunting Challenge

Bringing together a wide array of actors is a daunting challenge. The difficulties with establishing a dialogue across several disciplines are mainly due to the various frames of reference, value systems, interests and even choice of language (Aslin & Blackstock, 2010). The confrontation of

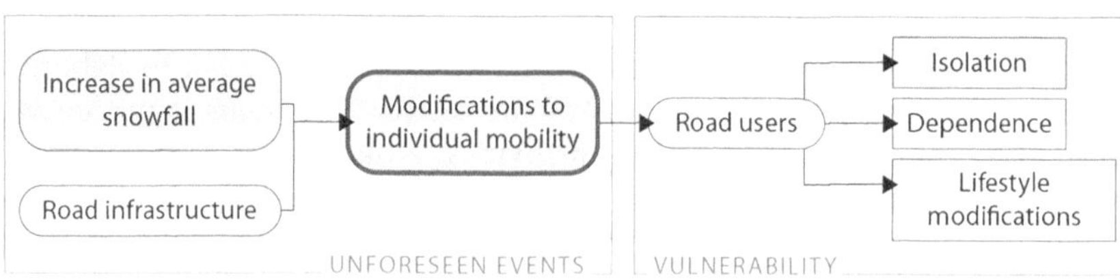

Fig. 2. Example of a Causal Pathway Based on the Statements Made by the Territorial Actors.

different values and interests can even undermine the engagement of some of the actors. The contrasting languages and frames of reference can be a barrier to knowledge sharing and mutual understanding among stakeholders (Füssel, 2007). The uneven capabilities of each participant to use an instrument properly can also call into question the instrument's potential effectiveness for collaboration (Dryzek, 2000; Mouffe, 1999).

In addressing the problem of anticipating the effects of climate change on the territory, particular challenges related to the shared contents must be addressed. A participatory approach for identifying the possible effects of climate change at the local level must also rely on subjective assessments of these effects. To what extent can this subjective assessment of the possible effects be considered reliable and valid? Is it relevant to distinguish between real and perceived risks identified by the territorial actors, knowing that a perceived risk has a bearing on decision making, regardless of its nature? Likewise, this presents the challenge of having to draw solely on the representations made by territorial actors of climate risk, while this risk is generally perceived, at least in North America, as remote and moderate (Leiserowitz, 2005). How can a dynamic of anticipation be created in a context where problems are intangible and pose little threat?

Furthermore, the different territorial actors' representations can give rise to potential confusion regarding responsibility for the intervention and the scale at which it should be carried out. Prior knowledge about climate held by individuals involved in the participatory process does not necessarily mean that they can take part in the discussion, other than in a private or personal capacity. Besides, positive outcomes of adaptation are usually more attractive and easier to perceive for private companies and households than for governments. The forms of action (protection of capital, response impacts, etc.) and the sectors covered by adaptation strategies (agriculture, transportation, etc.) primarily benefit individuals or private organizations. One example is houses built on stilts in flood-prone areas. This can sometimes lead to question the legitimacy of measures introduced by public administrations (De Perthuis et al., 2010).

A number of arguments are used in support of an adaptation planning approach that draw on the territorial actors' representations with an expert knowledge of the territory without being climate specialists. One argument is the real need for decision makers to access tools that have sufficient leverage to manage, even to some degree, the various climate risks. The two other arguments are the engagement effect and the distribution of relevant information, which are key principles of the participatory approach.

THE RESULTS – REPRESENTATIONS TO UNDERSTAND THE IMPLICATIONS OF CLIMATE CHANGE IN EVERYDAY LIFE

If the population of North America perceives the risk associated with climate change as relatively small, the contemporary social context is nevertheless in constant flux and marked by strong uncertainty, which shakes the foundations of social trust in the expert systems that seek to anticipate, even colonize, the future (Giddens, 1990). As Giddens remarks, decision makers and researchers of modern and hypermodern societies engaged in a process of change and innovation try to develop tools and methods for managing risk, to subdue it, by taking control over the future rather than be subjected to it.

Natural risk refers usually to the combination of potential natural disasters, unforeseen events, and local vulnerability (Adger, 2006; Thouret & D'Ercole 1996). The notion of risk is simultaneously multifaceted and context-based. It can be defined by anyone in a number of ways, which depend on each person's respective identity, culture, role, and position (Dessai et al., 2004). Further, linking the threat and the fragile social, physical, economic elements that are affected by it may infer a key aspect of risk management, namely its systemic dimension.

Indeed, although it is unconceivable to identify the possible effects of climate change while eliminating uncertainty, hazard mechanisms can nevertheless be considered and related directly to the events that they generate on the territory and to the consequences of these events at the local level. Linking the risk generating mechanisms with the components of the territory highlights the variety of anticipated events and the many potential responses to these events.

A Systemic Model Coupling Territorial Components with Climate Events

As part of our initiative, a conceptual framework combining the notions of events, consequences and territorial components leads us to present an overview of a systemic model of risk generating mechanisms. The participatory diagnosis, which is based on the experience and knowledge of local actors, provides details of this systemic model. The events reflect the observable modifications undergone by the system, whether territorial or climatic. The consequences refer to what has been affected on the territory by an event. As for the territorial components, they are simply basic elements of the territorial system such as road infrastructure, services, and a social group.

It is not necessary to identify the three notions in this way during the discussions. Besides, these territorial actors do not possess the skills necessary to consider in this way the risk mechanisms and are not expected to either. The research team will see that they are represented in the actors' statements. Above all, the important thing is the link made between the events, the territorial components and their consequences on the territory. Actors who are rooted in and directly involved in shaping the local context are in a position to address these links. While they are best suited to identify the links through the discussions, it remains elusive for a team of researchers to infer them without benefited from their participation.

Within this context, a significant aspect of the initiative lead in Québec City will have been to make use of current climate-related experiences rather than to try to direct the discussions only towards the potential climate change effects. Since the actors could not claim to be experts in climatology and since it is difficult for them to connect their own lives with climate change effects, bringing them to exchange on the topic of behaviours and climate effects in everyday life becomes more stimulating.

For example, building on their knowledge of a sector where the daily road traffic is particularly dense, the forum participants established links between the territorial components, whether infrastructure, organizational or social, which can be affected by an event such as heavy snow precipitation. On the one hand, the actors emphasized the likely results of a snowstorm on the physical environment and a city's strategic planning. Due to multiple heavy snowfalls in the same winter, more road salt is necessary and regular pavement maintenance practices need to be modified. The actors stated that spreading more road salt might increase the contamination of surface and groundwater and therefore affect both surface water and soil treatment, especially in areas where snow is piled up. Treatment is not conducted on a regular basis and should therefore increase in municipal service expenditures.

On the other hand, according to the forum participants, increased snow precipitation may also affect individual mobility. Disruption on the road network caused by heavy snowfalls could isolate people, especially the mobility impaired. Isolation itself has significant consequences on social interaction and personal integrity (Fougeyrollas, Cloutier, Bergeron, Côté, & St Michel, 1998), and the increased precipitation of snow can intensify isolation to the point where individual development could be more widely affected. Few articles in the scientific literature on climate change have addressed this type of psychosocial effect. The exchanges between stakeholders from various backgrounds and fields of activity provided such detailed information, which can have a significant impact on the conduct and development of urban society.

Statements made by the participants in responding this time to the issues surrounding the increase in mean temperature and number of heat waves also provided such detailed information. The analysis of the results from the forum show how territorial actors call attention to urban heat islands and to their possible effects on energy consumption on the one hand, and on citizen well-being on the other. For the territorial actors involved in the Québec City initiative, an increase in temperature will bring positive economic benefits as winter heating costs are expected to decrease. This will however be offset by increased summer energy use due to air conditioning and as long as concrete adaptation measures are not implemented.

The impact pathways, which arise from the statements made by the territorial actors during the discussion workshops, are based on the various forms of knowledge and concerns emerging from the participatory process. They do not take into consideration the full gamut of consequences related to climate and climate change at the local level. Scientific experts, in particular, could have identified other consequences from a more technical standpoint. Nevertheless, these climate–territory interactions, which were identified by a variety of social actors who are familiar with Québec City and its components, provides a picture that reflects both the sensitive elements to take into account and the subtleties of local arrangements.

In this respect, in order to formulate a local diagnosis, it seems particularly interesting to focus the exchanges on the territory as a whole. By grounding the discussion in a real territory and by encouraging the actors to place themselves in context, a number of responses can be elicited that would not have been possible if the focus of debate was limited to potential climate change. For an urban planner or a community organizer, it is far easier to consider spontaneously what constitutes the local area, its population, its physical layout than to consider the implications of an increase in the freeze-thaw cycles or altered precipitation patterns.

The abstract nature of these anticipated effects is a major challenge to overcome in order to engage the population and ensure that action on climate change adaptation is taken. Moreover, the complexity of the phenomena and uncertainties of climate change impacts are factors that keep citizens away from these questions. People are partially aware of the projected changes, but most feel that they have no bearing on their effects

(Lorenzoni & Hulme, 2009; Poortinga, Pidgeon, & Lorenzoni, 2006). In addition, at least in developed countries, the risk of being impacted in the short term by climate change is relatively small and, in general, underestimated. Hence, the added benefit of involving local actors in the initiative for planning climate change adaptation.

Different Urban Scales, Different Issues

Engagement of a diversity of actors in a participatory planning process provides them with an opportunity to contribute towards defining the problem of climate change. The face-to-face exchange allows actors to construct this difficult to see and perceive problem and, at the same time, to lay the foundations of their adaptive capacity (Hobson & Niemeyer, 2011). Seeking input from the actors in the discussion surrounding the possible interactions of climate and territory brings them to experiment the problem. This experimentation is an essential step in taking ownership of the issue.

The diagnosis paves the way for researchers and for the actors of the territory more broadly to engage in a process of understanding and identifying solutions to the climate issue, which is framed in terms of adaptation. The participatory diagnosis provides an overview of current issues to enable the research team to better understand 'the social experience' (Dubet, 2007) of climate change: by gaining insight on the ways of interpreting, techniques and the methods by which individuals address potential climate-related problems and try to resolve them.

In this respect, convergence in the statements made by actors and, in particular, the statements heard at the different discussion tables are promising avenues towards a better understanding of the social, organizational, material and natural sensitivities found in the Québec City context. Elements that were repeatedly identified irrespective of the geographical sector under study can be interpreted as indicators of the sensitivities of the territory towards climate change.

For example, during the forum, five groups out of five referred to a link between the increase in mean temperatures and the state of health of Québec City citizens, while only one group indicated a possible link between this temperature increase and the emergence of new parasites. Therefore, one of the major concerns that Québec City decision makers must take into account in adaptation planning could be the relation between the state of public health and global warming. The research team would benefit from including the issue of the emergence of parasites among the selected elements to be addressed and documented, even though this concern cannot be generalized on the entire territory.

Regarding the Québec City metropolitan territory, the other converging concerns, which were the focus of discussion for at least three of the five groups, are related specifically to increasing mean temperatures, precipitation and frequency of extreme meteorological events. A recurring theme in the statements made by forum participants is personal mobility. Their comments focus on the possible negative effects of warm weather and climate extremes on mobility capacity and the individual's choice of transport mode. It should be mentioned that sprawl in Québec City is endemic and public transport services are relatively scarce.

Another point of convergence reflected in the participants' statements is drinking water. Most of the groups linked the increase in mean temperatures with potential drinking water supply and quality issues. This link shows the importance of the critical work that the research team will have to conduct once the participatory process is completed. It is important to include the concern over the management of drinking water in a context of increasing temperatures in the diagnosis of the issues facing local adaptation since many territorial actors referred to it. Experts on the subject must therefore also consider this concern more 'objectively'. Participants involved in the diagnosis have expressed their interest to place drinking water-related issues at the top of the agenda of the local action plan for climate change adaptation. This is precisely what the research team must address even though the results from the critical or objective risk assessment indicate that such a link does not really exist.

Validation and Assessment of the Participatory Process

Once the phase of participatory diagnosis is completed, the research team members examined the points of convergences found in the statements based on the existing scientific literature. Consultations were also conducted with local experts. The experts in water supply and treatment emphasized that increases in mean temperature will not impact water quality in the Québec City region. What determines quality is the quantity of water available at source. In this case, quantity will essentially continue to increase except at certain times of the year and, according to the water experts, these crucial moments should be a matter of great concern.

Returning to the concerns raised by our experts on the territory, the fact that they have a shared representation of the issues to be monitored without being consulted beforehand at the discussion tables is a clear indication that a representation of the sensitive points of the territory is held in common. These points are essential elements of the problem of climate change for the Québec City region. This problem is steadily being constructed, and its elements identified during the exchange process reveal the first hints of its existence.

In addition to the links made in a particular territorial context, a cross-tabulation of a variety of experiences and points of view provides a space for discussion and debate. In doing so, the possible responses to climate change in and on the territory can be translated and interpreted. Although the uncertainty surrounding the true nature of climate change remains as before, it is no longer viewed as a challenge to action. Despite a number of unknowns, measures can be taken because some convergence was observed to help determine priority areas for public action. A subsequent problem will be to try to adjust these priorities to new effects and implications to be considered during the action implementation phase. Constructing this problem together with the local actors can also contribute towards their engagement in future adaptation efforts.

In spite of this, the extent to which the engagement of experts on the territory represents in terms of the anticipated effects of climate change remains unclear. Very little research has attempted to assess deliberation practice and its effect on the development of an adaptive capacity (Berkhout, 2010; Hobson & Niemeyer, 2011). In an attempt to understand the effect of participation in the forum on the Québec City region and climate change, participants were asked to complete a short survey.

Participation as an Adaptation Factor of Regular Practices

The survey was composed of 10 multiple-choice questions and was conducted online three months after the one-day participatory activity. Its purpose was to measure the participants' degree of understanding of the topic under discussion and the transfer of knowledge in their regular professional practice. In other words, to what extent did their participation in the discussion workshops enable the territorial actors to improve their understanding of climate change-related issues, and to what extent did their experience influence their work?

Table 1 presents the 10 questions addressed to the thirty-seven participants. The response rate to the online survey was 70% (26 over 37 participants).

Table 1. Québec City Forum Survey Results.

Questions
1. Before the March 4 forum, climate change was already a concern in my professional practice.
2. The March 4 forum on the city and climate change allowed me to become familiar with some of the climate change issues in the Québec City metropolitan area.
3. My participation at the forum led me to ask what my organization was doing to adapt to climate change.
4. My participation in the discussions was an opportunity for me to raise the question on how climate change could influence my activities, team and duties.
5. The discussions led me to better identify the elements of the territory of Québec City (populations, places, priority areas for action, etc.) that are more likely to be affected.
6. Interaction with the other participants allowed me to put into perspective the links between projected impacts of climate change on city life.
7. Climate change has become a more important factor in my professional activities since the March 4 forum.
8. Since March 4, I have integrated elements related to climate change adaptation in my professional activities (views, planning, intervention, etc.)
9. Resources (human and financial) in my organization could help in taking new measures to address climate change adaptation.
10. I would be interested in other meetings of this sort.
Answer choice
Strongly disagree
Disagree
More or less disagree
I do not know
More or less agree
Agree
Strongly agree

The results of the survey show that, for a majority of respondents (68%), the cross-sectoral participatory process improved their understanding of the possible effects of climate change on the various components of the territory of Québec City (Q2). Similarly, the majority of respondents (64%) considered that being able to engage with a diversity of participants brought them to see the links between the potential consequences of climate change on the city (Q6). The experience of the forum also prompted several participants to consider the actions taken by their organization in support of adaptation (40%, Q3). Moreover, participation resulted in incorporating slightly more climate change factors in regular professional activities (28%, Q7). A larger number of respondents (32%, Q8) said that after the forum they integrated elements related to adaptation in their regular professional activities. Although an equivalent proportion of respondents mentioned that after the forum they had not integrated elements related to adaptation in their activities (32%, Q8), we would argue that the proportion having done so is considerable.

It follows that a participatory process such as the cross-sectoral forum, even if it is a one-day event, can have significant repercussions on local engagement in climate change adaptation. This provides evidence of the potential of deliberation or interaction of territorial actors to improve their understanding of the issues and their adaptive capacity (Hobson & Niemeyer, 2011).

What seems to hinder the translation of the engagement effects of the participatory approach on the integration of concerns and adaptation practices is mainly the organizational framework. The resources (human and financial) are insufficient, according to a majority of respondents, to take new measures to address climate change adaptation (Q9).

CONCLUSION – A METHODOLOGICAL FRAMEWORK TO ADAPT LOCALLY TO THE MOVING TARGET OF CLIMATE CHANGE

As many North American cities, Québec City regional context shows no major vulnerability to climate change. Besides having to manage the uncertainty surrounding the consequences of climate change at the local level, city administrations must try to anticipate relatively minor and ill-defined effects.

Urban adaptation to climate change can be described as a wicked problem, a problem for which little experience exists and available knowledge is limited, and that is neither structured nor definitive. Given the uncertain and dynamic nature of climate change, local adaptation plans should identify ways to adapt incrementally instead of trying to provide clear ultimate responses.

The action research initiative presented in this paper sought on the one hand to gain a better grasp of the issues that Québec City must address and, on the other hand, to identify a framework to move forward in planning climate change adaptation. The initiative is comprised of three principal phases. The first is the diagnosis, which aims essentially at providing a better understanding of the problem but also at engaging some of the actors most likely to get involved. The second phase is one of prioritization, since it is not possible to have the entire city undergo adaptation all at once. The result is the identification of priority sectors that pose significant risks and create favourable conditions for adaptation. The third phase is one of detailed analysis at the neighbourhood scale of the climate change adaptation measures. The focus was here on the participatory diagnosis carried out during the first stage of the initiative.

The proposed participatory approach is based on the idea that a cross-sectoral and interdisciplinary communication and collaboration effort helps to foster collective learning and decision making regarding the constantly evolving problems of local adaptation to climate changes. It provides local actors with the opportunity to construct the problem in their context.

As we mentioned above, building on the participation of the territorial actors to aid in the diagnosis means that we must defer to these actors' points of view. This encourages us to think about ways to take advantage of these points of view in order to capitalize on their relevance while remaining critical. A forum's methodological framework was elaborated with this concern in mind.

On the one hand, we chose to confront different perspectives. Indeed, to improve the understanding of the potential multiple and chain effects of climate change, it seemed necessary that the interactive space engage experts' perspectives from different sectors in dialogue. Sectors include municipal management, citizen information and awareness, and provincial government programming.

On the other hand, the participatory structure was conceived so that the consulted actors could present their practices and particular ways of translating their concerns. By framing the discussion and debate within the existing geographical sectors, it became easier to elicit knowledge from each of the participants as well as their representation of the solid or more fragile components of the territory. By situating the participants in the real context of these sectors, a common frame of reference is provided to facilitate discussions. This approach also allowed the discussion to address the potential chain effects of climate change within the various social, physical, economic, organizational spheres, etc. Individual mobility and psychosocial effect have been, for example, pointed out as particular issues of Québec City adaptation to climate changes.

Once the phase of participatory diagnosis has been completed, some points of validation and assessment have been studied. First, points of convergences have been identified between the results of the participatory diagnosis and the statements based on the existing scientific literature. Consultations were also conducted with local experts. Second, in an attempt to understand the effect of participation in the forum on the Québec City region and climate change, participants were asked to complete a short survey.

The impact pathways, which arise from the statements made by the territorial actors during the discussion workshops, are based on the various forms of knowledge and concerns emerging from the participatory process. They do not take into consideration the full gamut of consequences. Scientific experts, in particular, could have identified other consequences from a more technical standpoint. Nevertheless, these climate–territory interactions, which were identified by a variety of social actors who are familiar with Québec City and its components, provide a picture that reflects both the sensitive elements to take into account and the subtleties of local arrangements.

Participants survey shows broadly that we have met our objectives. The abstract nature of the anticipated effects of climate change has been overcome and requirements to ensure that action on adaptation could be taken have been mainly gathered. This seems us particularly important in a context where people are generally aware of the projected changes, but most feel that they have no bearing on their effects. The diagnosis process brings them to experiment the problem of local climate change adaptation. This experimentation is an essential step in taking ownership of the issue.

NOTE

1. We use the term sensitivity to refer to the way a given structure, population or organization could be affected, positively or negatively, by unforeseen events. It designates a certain form of risk, without defining it precisely.

ACKNOWLEDGEMENTS

The authors would like to acknowledge the funding assistance of Social Sciences and Humanities Research Council of Canada, Ouranos, The Government of Quebec and Natural Resources Canada. We thank Martial Labarthe for his assistance.

REFERENCES

Adger, N. W. (2006). Vulnerability. *Global Environmental Change*, *16*, 268–281.

Aslin, H. J. A., & Blackstock, K. L. (2010). 'Now I'm not an expert in anything': Challenges in undertaking transdisciplinary inquiries across the social and biophysical sciences. In

V. A. Brown, J. A. Harris & J. Y. Russell (Eds.), *Tackling wicked problems through the transdisciplinary imagination* (pp. 117–129). Washington, DC: Earthscan.

Berkhout, F. (2010). Reconstructing boundaries and reason in the climate debate. *Global Environmental Change, 20*, 565–569.

Blanco, H., & Alberti, M. (2009). Building capacity to adapt to climate change through planning. *Progress in Planning, 71*, 158–169.

Boholm, Å., & Löfstedt, R. (Eds.), (2004). *Facility siting: Risk, power and identity in land use planning*. Risk, Society and Policies Series. London: Earthscan.

Brown, V. A., Harris, J. A., & Russell, J. Y. (2010). *Tackling wicked problems through the transdisciplinary imagination*. Washington, DC: Earthscan.

Callon M., Lascoumes, P. & Barthe, Y. (2001). *Agir dans un monde incertain. Essai sur la démocratie technique*. Paris: Seuil.

De Perthuis, C., Hallegatte, S. & Le Cocq, F. (2010). *Économie de l'adaptation au changement climatique*. Conseil économique pour le développement durable, Ministère de l'Écologie, de l'Énergie, du Développement durable et de la Mer.

DesJarlais, C., Allard, M., Bélanger, D., Blondlot, A., Bouffard, A., Bourque, A., Chaumont, D., Gosselin, P., ... Villeneuve, C. (2010). *Savoir s'adapter aux changements climatiques*. Ouranos, Montréal. Retrieved from http://www.ouranos.ca/media/nouvelle/document/54_sscc_21_06_lr.pdf

Dessai, S., Adger, W. N., Hulme, M., Turnpenny, J., Köhler, J., & Warren, R. (2004). Defining and experiencing dangerous climate change. *Climatic Change, 64*(1–2), 15.

Dryzek, J. (2000). Deliberative democracy and beyond. In W. Kymlicka, D. Miller & A. Ryan (Eds.), *Oxford political theory*. Oxford: Oxford University Press.

Dubet, F. (2007). *L'expérience sociologique*. Paris: Repères.

Folke, C., Carpenter, S., Elmqvist, T., Gunderson, L., Holling, C.S., Walker, B., Bengtsson, J., Berkes, F., ... Svedin, U. (2002). *Resilience and sustainable development: Building adaptive capacity in a world of transformations*. Environmental Advisory Council to the Swedish Government, Stockholm, Sweden.

Fougeyrollas, P., Cloutier, R., Bergeron, H., Côté, J. & St Michel, G. (1998). *Classification québécoise, Processus de production du handicap*. Québec, Réseau international sur le Processus de production du handicap (RIPPH)/SCCIDIH.

Füssel, H.-M. (2007). Vulnerability: A generally applicable conceptual framework for climate change research. *Global Environmental Change, 17*, 155–167.

Giddens, A. (1990). *Consequences of modernity*. Cambridge: Polity Press.

Hallegatte, S. (2009). Strategies to adapt to an uncertain climate change. *Global Environmental Change, 19*, 240–245.

Hegerl, G. C., & Zwiers, F. W. (2007). Understanding and attributing climate change. In IPCC (Ed.), *Climate change 2007: The physical science basis* (Ch. 9). Cambridge: Cambridge University Press.

Hobson, K., & Niemeyer, S. (2011). Public responses to climate change: The role of deliberation in building capacity for adaptive action. *Global Environmental Change, 21*, 957–971.

Intergovernmental Panel on Climate Change. (2001). *Third assessment report: Climate change 2001- TAR, synthesis report*. Geneva, Switzerland.

Klein, R. J. T., Schipper, E. L. & Dessai, S. (2003). *Integrating mitigation and adaptation into climate and development policy: Three research questions*. Working Paper No. 40. Tyndall Centre for Climate Change Research, Norwich.

Latour, B. (1999). *Politiques de la nature: Comment faire entrer les sciences en démocratie*. Paris: La Découverte.

Leiserowitz, A. A. (2005). American risk perceptions: Is climate change dangerous? *Risk Analysis*, *25*(6), 1433–1442.

Lorenzoni, I., & Hulme, M. (2009). Believing is seeing: Laypeople's views of future socio-economic and climate change in England and in Italy. *Public Understanding of Science*, *18*, 383–400.

Meehl, G. A., & Stocker, T. F. (2007). Global climate projections. In IPCC (Ed.), *Climate change 2007: The physical science basis* (Ch. 10). Cambridge: Cambridge University Press.

Mouffe, C. (1999). Deliberative democracy or agonistic pluralism? *Social Research*, *66*, 745–758.

Poortinga, W., Pidgeon, N. F., & Lorenzoni, I. (2006). *Public perception of nuclear power, climate change and energy options in Britain: Summary findings of a survey conducted during October and November 2005*. Technical Report, Working Paper No. 06-02. Centre for Environmental Risk, Norwich.

Rittel, H. W., & Webber, M. (1973). Dilemmas in a general theory of planning. *Policy Sciences*, *4*, 155–169.

Sen, A. (2009). *L'idée de justice*. Paris: Flammarion.

Thouret, J.-C., & D'Ercole, R. (1996). Vulnérabilité aux risques naturels en milieu urbain: Effets, facteurs et réponses sociales. *Cahiers de Science Humaine*, *32*(2), 407–422.

Tompkins, E. L., Adger, W. N., Boyd, E., Nicholson-Cole, S., Weatherhead, K., & Arnell, N. A. (2010). Observed adaptation to climate change: UK evidence of transition to a well-adapting society. *Global Environmental Change*, *20*, 627–635.

CHAPTER 4

CLIMATE PROTECTION AND CIVIL SOCIETY: DOES EFFECTIVE LOCAL CLIMATE POLICY NEED THE PARTICIPATION AND ENGAGEMENT OF CITIZENS? A COMPARISON BETWEEN THE CITIES OF POTSDAM AND MUENSTER

Mirjam Neebe and Fritz Reusswig

ABSTRACT

Purpose – *Many cities have taken action in order to reduce their carbon footprints. Moreover, the European city has historically been the home of democratic institutions, which have proven to be crucial for successful policy. The leading question of this chapter is whether or not this traditional link between democracy and active citizen participation also holds with respect to local climate policy.*

Urban Areas and Global Climate Change
Research in Urban Sociology, Volume 12, 75–104

ISSN: 1047-0042/doi:10.1108/S1047-0042(2012)0000012007

Design/methodology/approach – *In our chapter, we take a comparative look at two cities – Muenster in Rhineland Westphalia and Potsdam, the capital of Brandenburg next to Berlin. We have a look at the track records of both cities' carbon footprint and analyze the role of civil society in local climate policy. We develop a set of qualitative indicators, measuring local climate policy outcomes on the one hand and local climate policy performance on the other. We base our analysis on documents and on stakeholder interviews in both cities.*

Findings – *The findings show that Muenster has performed better in urban climate protection than Potsdam. Also the level of civil society engagement is higher in Muenster. Thus, the hypothesis that cities with a higher level of civil society engagement also perform better in urban climate policy can be confirmed. However, Muenster performs just slightly better than Potsdam. Both cities have failed to meet their climate goals. A closer look to the local climate policy performance leads us to the final conclusion that cities should be more active in supporting and including citizens in their local climate policies in all areas of life – including lifestyle politics and political consumerism.*

Keywords: Climate policy; civil society; mitigation; cities; democracy; lifestyle; political consumerism

INTRODUCTION

Cities and Climate Protection

As climate is a global commons, climate policy, preventing global warming to excess 2°C against pre-industrial levels, seems to be an exclusively global effort, and the *United Nations Framework Convention on Climate Change* (UNFCCC) the adequate political setting. However, recent international climate negotiations have demonstrated the limited efficiency of a consensus-oriented, large-scale, all encompassing, nation-state centered climate policies. Other actors, such as the business sector, civil society organizations, or local communities have started to take over responsibility by their own.

In this chapter, we are particularly interested in the interplay between local communities and civil society. Cities cover only about 2% of the

Earth's surface, but are home to more than 50% of the world's population, and a disproportionately high share of the global GDP. They are using 75% of the world's resources. In other words: cities are the planet's economic growth machines, besides of being its major knowledge centers and cultural laboratories. They are also major greenhouse gas (GHG) emitters and thus bear global responsibility. Greater London's GHG emissions, for example, exceed those of Greece, but only Greece is participating in the UNFCCC. At the same time, cities are highly vulnerable to climate change impacts, and thus cornerstones of any adaptation policy.

While at international climate policy many national states show symptoms of free-rider behavior–leading to sub-optimal outcomes, incremental progress, and a dominance of averaged national interests–at the local level we can observe that some cities move ahead and implement ambitious GHG reduction targets well beyond their national government's counterparts. Energy agencies are founded, emission reporting published. Meanwhile 77% of the cities in Germany have binding emission reduction goals (Sippel, 2010). This makes sense because the local level offers a great potential to deal with climate change. It's here where decisions taken on international or national level are implemented. For example, the share of cities of all public investments in Germany is 60% (Deutsche Bank Research, 2010).

According to Climate Alliance, there are at least four different roles or functions that local authorities can play in order to reduce GHG emissions:

- *Consumer and role model.* Cities and municipalities are important role models for concrete climate action, such as improving the energy efficiency of public buildings, introducing renewable energy to the local portfolio, purchasing green electricity, motivating their own employees, or greening public procurement.
- *Planner and regulator.* On a local level, communities do substantially influence the behavior of citizens and corporations by their manifold planning and regulation activities, by which the diffusion of climate-friendly buildings, technologies, and behaviors can be supported.
- *Provider and supplier.* Local authorities or their dependent enterprises supply water, energy, and other public services, and thus can reduce the local carbon footprint by their investment decisions and tariff structures.
- *Advisor and promoter.* Communities can exert indirect influence on citizens and businesses by providing information on climate-friendly behavior, possible funding, and reducing transaction costs by brokerage.

Civil Society and Climate Protection

Widening the scope of climate policy from national and international to local activities is a necessary step for a more holistic picture. It has to be accompanied by a focus on the individual citizen and civil society organizations. Even if local authorities would be endowed with perfect boundary conditions such as sufficient public funds, they would need the individual citizen to complement their different functions by co-operation and support. Today, many local communities suffer from public debt and understaffing, thus limiting their action capacity significantly. They thus even stronger need the support of citizens and civil society organizations. This has been proven in many other policy areas (Bode, Evers, & Klein, 2009; Herzberg, 2009; Shafir, 1998), and it holds true for local climate policy even more (Jänicke, 1990). For various reasons:

- *Effective climate policy needs the active participation of citizens.* The engagement of civil society actors contributes to the democratic quality of a given society (Klein, 2001) by opening up decision processes and the creation of responsibility. Weidner (2002, p. 1358) states that democracy "has significantly improved the opportunity structure for environmental proponents throughout the world by increasing the participatory, integrative, and cognitive-informational capacities of political systems. Democratization might therefore be considered a 'meta-capacity' for environmental capacity building." If climate change has the critical potential to challenge democracy (Leggewie & Welzer, 2010), broadening the social basis of climate policy can also help to stabilize democracy.
- *Effective climate policy needs democratic legitimization.* In democracies, political decisions have to be legitimized by the public. While elections and the parliamentary process are well-established procedures in mass democracies, the political engagement of individuals and groups forms the seedbed of democracy. And while citizen's participation in political decision making can lengthen procedures, it usually creates a sense of ownership and can improve acceptance (Geißel, 2007). As climate policy at the urban level has the potential to affect the everyday life of people – if not their lifestyle – participation in decision-making processes is both a question of political justice and efficiency (Newig, 2011; Newig & Fritsch, 2009; Zilleßen, 1993).
- *Effective climate policy needs the citizen-consumer.* In industrialized countries, private households are responsible for 35–50% of all GHG emissions, 40–70% if also indirect emissions are considered

(Vringer, 2005; Weber & Perrels, 2000). City administrations and public utilities can provide regulation and make offers, but without the active involvement and support of private households, these boundary conditions will not translate into a substantial reduction in the urban carbon footprint. The citizen has to be supplemented and supported by the consumer. Only a like-minded "citizen-consumer" can bring about a low-carbon community, reflected in various concepts in the literature, such as *ecological citizenship* (active participation and engagement in environmentally relevant politics), *political consumerism* (applying political criteria to consumption decisions), or *lifestyle politics* (influencing lifestyles as the sociocultural basis of consumption processes) indicate (Spaargaren & Oosterveer, 2010).

It is important to remember that the European city has historically been the home of democratic institutions and habits – not only in Ancient Greece but also in the medieval city, as Max Weber (1987, 1212pp.) has reminded us. Even after the modern concept of democracy has evolved at the nation state level in the United States and France, the local community level has remained a special one due to its closer connections between those governing and those governed, and due to a multifaceted active participation of citizens in local affairs.

The leading question of this chapter is whether or not this traditional link between democracy and active citizen participation does also hold with respect to local climate policy. Is local climate policy different from national or international climate policies, and is this difference influenced by a higher degree or different character of citizen involvement? Do local government climate policy initiatives find more support from citizens than national ones? Do different historic trajectories and actual institutional settings in different cities lead to different outcomes of local climate policy? Can local and may be even national climate policy learn from good practice examples in other communities?

We would like to answer at least some of these questions by a city comparison in Germany: the cases of Muenster in West and Potsdam in East Germany. We are fully aware that this very limited case study basis is totally insufficient for drawing far-reaching conclusions for local climate policy in general, even if we limit our scope to Germany. Nevertheless our modest comparison effort can be helpful for further research for at least two reasons: (1) the variance of cases is sufficient enough to develop a methodology of case study comparisons that can be used as a starting point for much more encompassing studies. (2) While cities across the planet do vary in many respects, they do share some commonalities, especially if we compare them to the national or international level of climate policy. Even small sample size studies could thus be informative with respect to cross-level comparisons of climate policy.

The rest of the chapter is organized as follows: after a brief description of our comparative methodology (second section) we briefly describe our two case study cities Potsdam (third section) and Muenster (fourth section). In the final discussion (fifth section) we try to answer our original questions, together with an outlook at the national and international levels of climate policy.

METHODOLOGY

Studying local climate policies and citizen participation is not an easy task. While there are some figures and data that might be indicative for policy outcomes (such as total GHG emissions of a city), the core object of study are social and political processes that are hardly covered by "objective data." We usually do have written documents (e.g., city council debates and resolutions), but they usually do not reveal underlying interests, problems, and the social dynamic that have led to these outcomes. In addition to the interpretation of written documents, we thus had to rely on stakeholder interviews. These interviews were based on a semistructured interview manual and recorded. The interviewees were recruited from both the city administration and the civil society organization domains.

While uninterpreted data are both hard to find and lack explanatory power, we developed a set of qualitative indicators, measuring local climate policy *outcomes* on the one hand (A – in a more formal interpretation: the dependent variables), and local climate policy *performance* on the other (B – the independent variables). Civil society participation forms a core element of the performance variables.

Cities could reach a total value of 16 points with respect to policy outcome, and 20 points with respect to policy performance.

Outcome of Local Climate Policy

(A-1) The indicator *achievement of climate targets* measures whether or not a city is on course to reach its own CO_2 reduction target. Most cities measure their own climate policy success in avoided tons of CO_{2eq}

emissions, compared to a particular baseline. They do continuously monitor urban emissions, report, and publish them. Following the approach of Sippel (2010), who analyzed the performance of climate action of 40 German municipalities against their targets, we control whether the city's current emission levels are in line with what would be a linear reduction pathway from base year emissions to target year emissions. From the emission target, target year emissions are calculated. By assuming a linear emission pathway, average annual emission development for the time period covered by emission reporting data is calculated and compared with the needed reduction pathway.

(A-2) The indicator *integration of climate protection in local polity* is measured against the existence of a *binding emission target*, the existence of *emission reporting*, the institutionalization of climate protection via *specific organizational units and procedures*, the existence of a *municipal climate protection strategy*, and the existence of a *concrete action plan* to implement the climate strategy. Referring to previous studies analyzing success factors of local climate policy (Gruber, 2000; Kern, Niederhafner, Rechlin, & Wagner, 2005), the indicator is based on the idea that climate policy should be integrated in all relevant political sectors. At the same time, we know from the same studies that success chances are higher if local climate policy is supported by the head of administration, for example, by establishing an energy or climate protection agency at the upper end of the hierarchy.

(A-3) The indicator *implementation level of the municipal climate protection strategy regarding the potential reduction of* CO_2 *emissions* is measured against the reduction potential of the already implemented measures. The potential can be low (1 point), middle (2 points), and high (3 points), and is based on the evaluation of the measures in the climate protection strategy of the case cities. There are many measures that a hard to quantify in terms of avoided emissions, for example, awareness campaigns or climate awards.[1]

(A-4) As has already been stated, cities can become active in climate protection in many action domains: energy provisioning, public transport, traffic planning, urban planning, waste management, procurement, public relation, and comprehensive measures. This broad array of arenas is crucial, given the diffuse nature of GHG sources and drivers in a city (Kern et al., 2005). Therefore, the indicator *implementation level of the municipal climate protection strategy regarding the diffusion in different fields of action* is measured against the implementation of measures in all fields of action.

Local Climate Policy Performance

(B-1) Effective climate policy needs the active participation of the citizens, who engage for the common good – even if this engagement might also pay out for their own interests. The indicator level of *civil society activity* shows how active civil society actors in the case cities are engaged in climate protection. Based on Eberhardt (2000), we distinguish between more conventional (press releases, lectures, discussions, flyers, expositions, consulting) and more citizen involving (manifestation, agreement, interview, study, petition) forms of communication, as well as discursive models of participation (round tables, energy-tables, public meetings, future workshops, mediation). Simply distributing information on climate change and possible action is not sufficient in order to motivate people to active engagement (Leggewie & Welzer, 2010), or to involve them effectively (Newig & Fritsch, 2009). More bidirectional, open, and discursive models of participation are needed, especially if new and creative solutions are to be found (Eberhardt, 2000). Because of their importance, we evaluate the implementation of discursive models of participation with two points, and more conventional forms with one point each.

(B-2) In global modernity, the classical distinction between the "citizen" and the "consumer" has been blurred. Instead, we observe new forms of merging the two into the "citizen-consumer" (Spaargaren & Oosterveer 2010), expressing itself in ecological citizenship and political consumerism. As local climate policy covers different action arenas, we would like to follow Barthel (2006) and take different forms or modes of life-politics into account (such as living, mobility, or food consumption). Taken together, these indicators measure the *level of diffusion of civil society activities*.

(B-3) Effective climate policy needs the self-activation and empowering of the citizens, who participate in decision-making processes. The indicator *level of municipal activity in integrating citizens* points out how active the two cities integrate their citizens in climate protection activities, and how active they motivate them to engage in climate protection. Like for indicator B-1, we distinguish between more conventional and more inclusive forms of communication as well as discursive models of participation. Because of their importance, we evaluate the implementation of discursive models of participation with two points again.

(B-4) Because of the blurred distinction between "citizens" and "consumers," municipalities should not only integrate their citizens in decision-making processes but also motivate consumer-citizens to actively engage in modes of political consumerism and life-politics. The indicator *diffusion of municipal activities in integrating citizens in climate protection* measures activities in the modes of ecological citizenship, political consumerism, and life-politics.[2] Further distinction for the mode life-politics are living (electricity use, heating, appliances), mobility, private consumption, and food. While B-2 indicators refer to the level of civil society activities exclusively, B-4 indicators focus on the interaction between the city and its civil society.

CITY CASE STUDY NO. 1: POTSDAM

City Characteristics

Potsdam, the capital of the German state of Brandenburg, is home to 155.000 inhabitants (2010). Located at the confluence of the Rivers Nuthe and Havel, Potsdam is a direct neighbor of Berlin, the German capital. Historically, the Prussian kings have developed Potsdam to their major residential city besides Berlin, and they engaged generations of famous landscape architects to create a unique urban–rural dreamscape, inspired by British and Italian role models. Today, the endowment with water bodies, parks, and green space, and an attractive cultural landscape (UNESCO world heritage since 1990), has made Potsdam a highly demanded place to work and live, especially for young families and retirees. While many other East German cities have to deal with deindustrialization and a shrinking population, Potsdam grows by 1,000–2,000 additional inhabitants annually (Fig. 1).

Fig. 1. Potsdam and River Havel as Seen from the South (Wikimedia Commons, 2002).

For the city's economy, tourism does play an important role. Historically a place of the court, the administration, and the military, industry did never play a central role in Potsdam – one of the reasons why the deindustrialization after German reunification in 1990 did not hit the city as hard as many others in the former German Democratic Republic. Today, small and medium enterprises in the service sector are much more important. Potsdam hosts three universities and many research institutions, is home to a globally competitive film production in the famous *Babelsberg* studios, and offers many jobs in the public administration sector (Landeshauptstadt Potsdam, 2007). Compared to both the East German average and to the situation in adjacent Berlin, Potsdam's economic situation is rather positive, which among other things is reflected in a relatively moderate unemployment rate of 8.3% (Landeshauptstadt Potsdam, 2011a). While there is a remarkable share of rich and even famous people that have chosen Potsdam as their home due to its high quality of life close to the much more hectic Berlin, the majority of Potsdam inhabitants have to make ends meet, and there is a substantial share of the population that lives in affordable housing schemes in the outer quarters of the city. Debates about growing costs of living, especially with respect to renting homes and flats, have come up more recently.

Climate Protection in Potsdam

Immediately after reunification, environmental problems, a nonissue under the GDR regime, have become public and made their way to the political space of the city. Among other things, local climate protection gained prominence during the early 1990s, leading the city administration – led by a Social Democratic Lord Mayor – to formulate climate policy goals, and to join (in 1995) the *Climate Alliance*, a European union of municipalities that shared the concern for the global climate. The member cities of the Climate Alliance have agreed upon a continuous GHG reduction strategy of 10% every five years, with the goal of 2.5 tons per capita in 2050.

In 2000, the city published its first GHG inventory report, being a duty for Climate Alliance members. In the same year, a special position in the city administration has been created, coordinating the local climate policy activities. In 2000, the city administration published its first GHG inventory as part of its Climate Alliance membership duties. First attempts in reducing the energy use of municipal building such as schools have been made, and from 1999 to 2001 a first energy round table tried to broker the efficiency knowledge and seize opportunities in the business sector.

One of the most important decisions had been taken a little earlier: the replacement of an old coal (lignite)-fired power plant by a modern, combined heat and power plant based on natural gas, serving both the extended district heating system and the electricity generation of the city owned energy utility company EWP (*Energie und Wasser Potsdam*). As a result, the urban carbon footprint dropped from 1.59 million tons of CO_{2eq} (11.2 t per capita and year) in 1990 to 0.843 million tons of CO_{2eq} (5.9 t) in 2005 (Landeshauptstadt Potsdam, 2009).[3] As climate friendly as it might have been, the substitution of the lignite power plant was politically a very difficult decision, as the state of Brandenburg has for decades been economically dependent on coal (lignite) mining in its Southern regions, and Potsdam is the capital of this "coal state." Opting against coal as a fuel for the city seemed to many – especially to the coal and energy industry workers from Southern Brandenburg – as a slap in the face of the interests of the state.

In 2008, the city administration created a new branch, the "Coordination Unit for Climate Protection," then located at the department for youth, environment, and health. At the same time, a "Climate Council" was founded, overlooking the city's climate policy and advising the Lord Mayor, composed of city administration, business, science, and NGO experts. The council is supported by four special task force groups on mobility, energy, urban planning, and building/housing (Reusswig, Altenburg, Neebe, Schmidt, & Peters, 2010).

In 2007 – the global climate debate had reached a historical peak during the publication of IPCC's Fourth Assessment Report that very year – local climate policy gained momentum in Potsdam. The city council decided to further reduce the city's emissions by 20% until 2020 – based on the 2005 emission values, not on 1990 values. Since the largest reductions were accomplished after the German reunification, this was an ambitious goal (Landeshauptstadt Potsdam, 2010). This translates into a goal of 0.693 million tons in 2020, or 4.33 tons per capita. In 2010, a consortium of city planners, energy consultants, traffic planners, and engineers led by PIK developed an "Integrated Climate Protection Concept," including about 100 concrete measures to accomplish the climate protection goal of Potsdam (Landeshauptstadt Potsdam, 2010), together with a CO_2 and cost calculation. Based on these findings, the city council in spring 2011 adopted the concept as a policy guideline (Beschlussvorlage Integriertes Klimaschutzkonzept, 2011). Until early 2012, a couple of the proposed measures have been realized, such as the reorganization of the Coordination

Unit, a solar rooftop website, a climate protection award, and first ideas for a new energy strategy of EWP.

Success of Local Climate Policy in Potsdam

Measuring the success of policies is not an easy task, as many factors do influence positive outcomes, including mere accident. While the causal link between policies and outcomes shows some uncertainties, the outcome dimension in itself can be targeted more thoroughly.

Based on our set of indicators the success of local climate policy in Potsdam is assessed with seven out of 16 criteria met.

(A-1) Potsdam has not yet been on the way to achieve its climate protection goals until 2020. The average annual reduction needed is 11.237 tons. But emissions increased between 2005 und 2008 by 1.3% (842.800 t–853.000 t). Thus, we judge the indicator *achievement of climate targets* not to be met.

(A-2) Potsdam has a climate protection goal, publishes GHG inventories regularly, and with the Climate Protection Coordination Office, the Climate Council, and the professional groups climate protection is institutionalized in Potsdam's administration. In 2010, the Integrated Climate Protection Concept was published, but until now not specified by a concrete action plan.[4] Thus, four of five criteria from the indicator *integration of climate protection in local polity* are met.

(A-3) The measures implemented since the publishing of the Integrated Climate Protection Concept are all soft measures, which cannot be quantified in exact CO_2 reduction figures (Landeshauptstadt Potsdam, 2010). Thus, the indicator *implementation level of the municipal climate protection strategy regarding the potential reduction of CO_2 emissions* is not met.

(A-4) While the more or less "soft" measures (public relations, administration) do not directly translate into GHG reductions, they still have a quite high potential to permeate different action arenas. The indicator *implementation level of the municipal climate protection strategy regarding the diffusion in different fields of action* is met with three of seven criteria.

Civil Society Engagement and Participation

While environmental activism has been a "no-go area" during the GDR period, environmental concerns of the citizens were made public immediately during and after the 1989 revolution – not only in Potsdam but also in all parts of Eastern Germany. Environmental NGOs did form, and environmental issues made their way to the agendas of most political parties. With climate change, a rather complicated and scientifically mediated issue made its way to everyday concerns – and to the issues that environmental NGOs did raise. However, while NGO activists did engage and push for the topic, the majority of citizens did also have to care for a plethora of socioeconomic risks and changes in the course of German reunification. Even in 2010/11, when we did two survey waves in Potsdam and its adjacent county, one can observe that despite some general concerns the majority of respondents still perceives climate change as a rather distant problem (both in time and space), and many do not see an immediate reason to act personally, but rather feel that the state or the corporate sector should deal with it.[5] While this might sound disappointing to the ears of those who expect major social changes to be brought about by social majorities that actively engage in politics, it still sends out an encouraging message to those who assume that even major social changes often start with smaller groups of people, able to create a "critical mass" for transformation processes (Olivier, Marwell, & Teixeira, 1985). We thus observe small, but very active and even influential groups of climate activists in the city, with good networks and some mass media attention.

Environmental NGOs in the city (mostly as local nodes of regional or national organizations) do actively push the issue. While the local branches of nation-wide organizations (such as BUND, WWF, or Greenpeace) do also follow the issue, we observe specialized local organizations that do play a more important role in the city's climate policy, namely the "Energy Forum Potsdam" (EFP) and the "Potsdam Solar Association."

The "Energy Forum Potsdam" (http://www.energie-forum-potsdam.de/) was founded in 2007 as a civil society organization dedicated to promote climate policy and clean energy for the city, actively and critically engaging with the city administration and the corporate sector. Members are ordinary citizens, but also experts from the public administration, the public utility sector, or environmental NGOs. There is some overlap with the members of the city's "Climate Council." Activities focus on public events, public statements, studies, and workshops. The EFP perceives itself as an assembly of independent expertise at the interface of the city and its inhabitants, and even appointed administration members can take a different stance toward certain issues. The "Solar Association" is mainly collecting private money for larger PV panels on private and public buildings, and is propagating renewable energy in the city.

More recently, a student group initiated an association that tries to green the University of Potsdam, especially in terms of improving its climate performance ("UniSolar"). This initiative has been taken up by the University administration. Potsdam's University of Applied Science (Fachhochschule) has also started to tackle urban climate policy issues, that is, by staging an interactive public exhibition on climate change and climate policy in Potsdam in fall 2010.

We see two major routes of civil society engagement: either the citizens become active by themselves, putting the political and business sphere under some pressure, or they get activated by opportunities created by the public administration, that is, by participation in public decisions. As we would like to focus on the city's climate policy performance in this chapter, we take both routes into account here. This does exclude activities initiated by national institutions or nation-wide foundations.

A major focus of municipal activities is energy consulting. The administration as well as municipal enterprises (such as public utility or public housing corporations) inform their clients via various channels (brochures, newsletters, Internet, direct contacts) about ways to reduce energy consumption. As one measure of the Integrated Climate Protection Concept, the city administration in 2012 brought on its way the "Potsdam Climate Award," offering 3 × 1.000€ for good ideas and best practice examples in climate protection. Some Potsdam households also participate in a Europe-wide "Energy Neighborhood" contest (Beratungs- und Service-Gesellschaft Umwelt mbH; Landeshauptstadt Potsdam, 2011b). The Local Agenda 21 process, part of the 1992 Rio Declaration (UN Department of Economic and Social Affairs. Division for Sustainable Development), did also play a role in Potsdam, but came to an end in the early 2000s (Landeshauptstadt Potsdam, 2011c; Reusswig et al., 2010). Participation in environmental issues did again play a role during the preparation phase of the city assembly's 2011 resolution to implement the Climate Protection Concept, but it did not attract many people. In 2011/12, the city and local housing companies set up a participatory process in order to redevelop a large prefabricated building quarter into a low-carbon "Garden City Drewitz," with more participants and tangible results.[6]

Level of Civil Climate Protection

Based on our set of indicators measuring the level of civil climate protection in Potsdam, we assess it with nine out of 20 criteria:

(B-1) Civil society groups mainly use conventional forms of communication. Only the "Energy Forum" uses broader forms of communication like studies and workshops. But no discourse oriented models of participation have been widely used so far. The indicator *level of civil society activity* is thus assessed with two out of four criteria.

(B-2) Civil society associations in Potsdam are mainly active in the mode of ecological citizenship. The "Energy Forum" tries to influence local policies. Within the "Bürgersolarverein" and "UniSolar," citizens are engaged in lobbying renewable energies. None of these groups directly addresses their human fellow in the modes of political consumerism or life-politics. Thus, the indicator *level of diffusion of civil society activities* only meets one out of six criteria.

(B-3) In order to integrate the public in climate protection activities and in order to motivate them to become active in their daily life, the municipality of Potsdam uses conventional and broader forms of communication as well as a discourse oriented model of participation (public meeting to discuss the climate concept). The indicator *level of municipal activity in integrating citizens* met four out of four criteria.

(B-4) The municipality of Potsdam is mainly active in the modes of life-politics – living (energy consultancy service) and ecological citizenship (climate award, public meetings). The indicator *diffusion of municipal activities in integrating citizens in climate protection* meets two out of six criteria.

CASE STUDY NO. 2: MUENSTER

City Characteristics

Muenster is situated in the federal state of Northrhine-Westfalia, and is the regional center of the "Muensterland." On a surface of 302.92 km^2, Muenster is divided into six districts. The city center is known for its historical buildings. As with Potsdam, the municipality also includes rural areas, result of an extension in the mid-1970s. Nearly half of the city's area is used for agriculture. Muenster is home to 285.180 inhabitants (in 2010), and continues to grow by about 400 people per year. With several universities and 45.834 students, Muenster is one of the biggest university towns in Germany. The city's economy is characterized by small and medium industry, financial services, bio- and nanotechnology, and administration. The unemployment rate in 2010 was 5.7%. Eighty percent of the 132,000 employees work in the service sector (Stadt Münster, 2011c) (Fig. 2).

Despite hosting many students, Muenster is a traditional and conservative city, with a catholic majority, and well-functioning networks. It thus offers

Fig. 2. Muenster as Seen from the South (*Source*: Stadt Münster, 2011b).

an interesting contrasting case to the traditionally left-lending Potsdam, where Social Democrats and the Left Party are strong.

Climate Protection in Muenster

Muenster has a long-standing tradition in environmental policy and climate protection, and is well known for its activities. It was back in 1992 that local politicians had to react on growing pressure from local associations engaged in energy transition issues. A "Climate and Energy Council" was formed as a consequence. The Council did the city's first climate inventory, and set the goal of reducing GHG emissions by 25% until 2005. Base year was 1990. A first package of climate protection measures was developed to meet this goal.

In 1995, the "Climate and Energy Coordination Office" was established with two staff members and one secretary subordinated to the office of environment (today the city department for open space and environmental protection) (Duscha, Dünnbeil, Gugel, & Kutzner, 2009). The Climate and Energy Council originally asked for at least five staff members, but this did not materialize. Thus not all measures could be implemented (Wuppertal Institut für Klima, 2003).

At the beginning, a concrete action plan (38 recommendations) how to achieve Muenster's climate goal was developed. In 1996, the action plan was accepted by a resolution of the city parliament, and many activities were initiated, such as an overall concept and funding program for retrofitting, a building planning with low energy house standards for municipal buildings, energy controlling for local enterprises, and heating contracting. Projects at schools and kindergartens to save energy and reduce waste were launched. Information and consulting regarding energy saving, the use of renewable energies, and combined heat and power were given to the inhabitants and local enterprises of Muenster. Since then, public relation and consulting are central activities in Munester's portfolio of local climate policy. The municipality collaborates with the Consumer Advice Center, the local energy supplier, and networks of other municipalities. A milestone for climate protection in Muenster has been the construction of a natural gas based combined heat and power plant in 2005. In the waste-disposal plant Coerde energy is generated from landfill and sewage gas. In the traffic sector, promoting the bicycle was a focal point, building on a long-standing tradition of cycling in the city.[7] The bus traffic system has also been improved. Despite some problems – ticket prices increased, leading to reduced public transport passenger numbers – the city also tried to optimize traffic flows, and the management of parking space (Duscha et al., 2009).

In 2008, the city council decided in a resolution a second emission reduction goal for Muenster: minus 40% until 2020 – with 1990 as base year. In 1990, GHG emissions were about 2.5 million tons CO_{2eq} (9 t per capita and year). This translates into an annual reduction of 32.500 tons per year until 2020 (1.3%). Until 2006 emissions decreased about 220,000 t to 2.3 million t (8.2 t per capita). This translates into an average reduction of 9%, and an annual reduction 0.5% per year. Thus, the first climate goal of an emission reduction by 25% until 2005 was not achieved.

To commission external expertise, the *Institute for Energy and Environmental Science* (ifeu) and the engineering company *Gertec* were asked to develop a "Climate Protection Concept." In 2010, the concept was presented to the public. Again, the "Climate and Energy Coordination Office" developed a concrete action plan that was accepted by the city parliament in December 2010. The action plan contains three sections. The first section describes measures that can be implemented directly. The measures of section two need a further resolution by the city parliament. The measures of third section are planned for the coming years.

In May 2011, the Muenster "Climate Alliance" was launched in order to establish a climate protection platform of local business companies. The "Climate and Energy Coordination Office" got two more staff members. In September 2011, a "Climate Council," including members of local enterprises, associations, and science was founded to supervise climate protection activities in Muenster and to advise the city administration. The funding program for retrofitting, which exists since 14 years, will continue with an additional budget of 350,000 Euro per year. Within the "Climate Change Business Activity Support Program," the municipality is in charge to retrofit two buildings per year. Activities in public relation are intensified.

Success of Local Climate Policy in Muenster

Based on our set of indicators, the success of local climate policy in Muenster is assessed with nine of 16 criteria met.

(A-1) The climate protection goal of Muenster aims to reduce GHG emissions by 40% until 2020 (base year 1990). Under the assumption of a linear reduction pathway, emissions should be reduced by 1.3% per year. Between 1990 and 2006, the year of latest emission reporting, emissions were reduced by 9%. This corresponds to a reduction of 0.5% per year. Thus, Muenster is not on the way to achieve its climate protection goal. Thus, we assess the indicator *achievement of climate targets* not to be met.

(A-2) Muenster has a climate protection goal and publishes GHG inventories regularly. With the Climate and Energy Coordination Office and the Climate Council, climate protection is institutionalized in Muenster's political system. In 2010, the Climate Protection Concept was presented to the public and specified by a concrete action plan. Thus, all five criteria from the indicator *integration of climate protection in local polity* are met.

(A-3) Two of the measures implemented since the publishing of the Climate Protection Concept can be quantified in exact numbers. The funding program for retrofitting has a high emission reduction potential. The "Climate Change Business Activity Support Program" has a low emission reduction potential (Duscha et al., 2009, p. 60). Thus, the indicator *implementation level of the municipal climate protection strategy regarding the potential reduction of CO_2 emissions* is met with one point.

(A-4) The measures already implemented focus on the action fields energy, comprehensive measures, and public relations. The indicator *implementation level of the municipal climate protection strategy regarding the diffusion in different fields of action* is met with three of seven criteria.

Civil Society Engagement and Participation in Muenster

Muenster is active in environmental and climate protection for more than 20 years. This is also a result of the engagement of several civil society initiatives. Already back in the 1960s, Muensters citizens prevented the construction of a commercial area in order to protect a bird sanctuary. In the 1990s, the privatization of the municipal energy supplier was prevented. In the same period, local environmental associations pushed the local government to establish the "Climate and Energy Council." This active intervention and participation in local politics was partly unwillingly tolerated, partly advocated complaisantly. Without this civil society engagement, today Muenster would not be known for its climate protection policies, as was stated by a civil society group member (Döbel, 2011). Nevertheless, similar to Potsdam, also in Muenster climate policy engagement is restricted to a small group of active citizens. For the general public the issue of climate change is quite distant to everyday life (Döbel, 2011).

In addition to classical environmental NGOs, which push the issue of climate change beside other environmental issues, we also find specialized associations, targeting climate protection exclusively. These groups form a well-connected network and pursue common activities. Members are private persons interested in the issue, not local politicians or members of municipal enterprises, as in Potsdam. One of our interviewees stated that they act in critical distance to the city administration: "We do not allow them to use us for lobbying, but pursue our own interests" (Döbel, 2011).

The "Working Group Local Agenda 21" is the most important civil society association in Muenster. Activities started in 1995 when it was still heavily supported by the local government (a coalition of the social-democratic and green party). The municipality made available four staff members to coordinate the process. Several working groups met regularly, and the general public was invited to participate. In 1990 – when the Christian-Democratic party took power – circumstances changed. The staff members were called back, the agenda process changed into an "Integrated City Development and City Marketing Process," and citizens' participation was restricted. As a consequence, the civil society groups leaved the process and continued their own agenda process without the municipality. Today the Agenda Group unifies four umbrella organizations ("Forum One World," "Woman Action League," "Peace Forum," and "Environmental Forum") with 110 civil society associations and initiatives. One of the main activities of the Agenda Group is the annual "Citizen Forum," which was also to be stopped when the municipality stopped the agenda process, but is now organized by the purely civic agenda initiative. The first "Citizen Forum" in the new constellation in 2003 had the issue of climate change. Further issues during the following years were faire trade, energy consumption, ecological farming, city budget, sustainable economy etc. In 2011, it was again climate change and in 2012 "Economies Beyond Growth" (Döbel, 2011; Stadt Münster, 2011f).

Environmental NGOs and groups active in the field of energy and climate are unified under the umbrella organization "Environmental Forum," which comprises 15 associations with 7,000 members, existing since 1991. Regarding energy and climate issues, the "Environmental Forum" offers renewable energy consultation hours with experts, promotes the issues via networking between science, planning, politics, citizens and investors, and sustains public relations (Umweltforum Münster e.V.).

Further, there is "Fairplanet eG," a cooperative organization collecting members and money in order to realize renewable energy projects in the global north as well as in the south and east (faiplanet eG). 'Nütec e.V.' (natural survival technology Münsterland) promotes environmental-friendly technologies, renewable raw materials and ecological agriculture via Internet, information material, expositions and events. "Nütec" developed the so called "Solarmobil," a mobile trailer with tools to illustrate the functioning of renewable energies. Another mission of "Nütec" is "Solarnet," a school project that is dedicated to push climate and energy issues in school curricula and to build a learning and qualification network concerning renewable energies (nütec e.V.).

The "Solarmobil" in Front of a School in 2010 (*Source*: nütec e.V.).

In July 2011, the initiative "Muenster eats veggie" was founded by Greenpeace, Muenster's "Climate and Energy Coordination Office," the

"Association of Vegetarians," and others. With information material, public relation and photo-events the initiative aims to introduce a meat free day per week. Citizens, restaurants, canteens, schools, kindergartens etc. were asked to participate (Initiative "Veggietag Münster"). In August 2007, several local initiatives installed a mobile sculpture, called "Climate Killer Engine Idle" in the city center of Muenster. The aim was to point out that the municipal climate policy is not sufficient to substantially reduce GHG emissions, especially in the traffic sector, and to convince people to turn of the engine when waiting at a traffic light (Verkehrsclub Deutschland (VCD) Kreisverband Münster e.V.).

In addition, the municipality of Muenster promotes civil society engagement in climate protection. The campaign closest to the people is the "Climate Protection Citizen Pact." Until now, 4,500 people signed the commitment to implement some easy measures[8] in their daily life (Wildt, 2011). Since these measures are low-threshold, the initiative is seen critically by some civil society associations who state that climate protection is more than buying some energy saving light bulbs (Döbel, 2011). In contrast, a member of the city administration declared in an interview that the measures have to be low-threshold in order to find people who participate in the initiative (Wildt, 2011). Further, the campaign "Climate Runs for Shelter" is the antecedent of the "Climate Protection Citizen Pact" and today combined with it. Citizens from Muenster commit to climate protection with a picture of them and a short statement how to combat climate change. Ca. 200 people participated until today. The pictures and statements are published in local newspapers, as posters in buses or wallpaper on advertising pillars (Stadt Münster, 2011a).

Since 1998, the project "Climate Partnerships in Muenster" gives financial support to local initiatives dedicated to renewable energies, energy efficiency, and climate change adaptation in the global south. Until now projects in Africa, India, South America, and East Europe were supported. In 2010 the municipality of Muenster invested 4,000 euro in a photovoltaic facility in Uganda, and 2,000 euro in a photovoltaic facility for the hospital in Muona, Malawi (Stadt Münster, 2011e). Further, energy and wastage saving projects in schools exist since 1998. Today, 100 of the 119 municipal schools and kindergarten participate (Stadt Münster, 2011d). And of course also energy consulting is a task of the municipality of Muenster. On the city's climate webpage interested people can check their electricity and heating demand, there are links to energy consultants, climate protection tips are published and the most important loan programs listed and explained (Stadt Münster, 2011a).

Level of Civil Climate Protection in Muenster

Based on our set of indicators the level of civil climate protection in Muenster is assessed with fourteen out of 20 criteria met.

(B-1) Muenster's civil society associations mainly use conventional and broader forms of communication. With the "Citizen Form" also a discursive model of participation is used. The indicator *level of civil society activity* is met with four of four criteria.

(B-2) Civil society associations in Muenster are mainly active in the modes of ecological citizenship and life-politics – living. They try to influence local politics and distribute information about renewable energy and energy saving to the fellow citizens. The campaign "Muenster Eats Veggie" brings up the issue of life-politics – food. The mobile sculpture "Climate Killer Engine Idle" brings up the issue life-politics – mobility, and tried to influence local politics. Thus, the indicator *level of diffusion of civil society activities* met four of six criteria.

(B-3) In order to integrate the public in climate protection activities and in order to motivate them to become active in their daily life, the municipality of Muenster uses conventional and broader forms of communication. Discursive models of participation are not used. The indicator *level of municipal activity in integrating citizens* met two of four criteria.

(B-4) The municipality of Muenster is mainly active in the modes of ecological citizenship ("Climate Partnerships in Muenster," "Climate Protection Citizen Pact") and life-politics – living (energy consulting, school projects). Further, the "Climate Protection Citizen Pact" includes also measures regarding mobility, and the campaign "Muenster Eats Veggie" brings up the issue of food. Thus, the indicator *diffusion of municipal activities in integrating citizens in climate protection* is met with four of six criteria.

DISCUSSION

If we compare the two cities, Potsdam and Muenster, we generally find that Muenster has performed slightly better in urban climate protection: 9 points as compared to 7 points in Potsdam. At the same time, the level of civil society engagement in climate policy is higher in Muenster (14 points) than in Potsdam (9 points). This leads us to conclude that our initial hypothesis can be confirmed: cities with a higher level of civil society engagement do also perform better in urban climate policy.

Several reasons might have contributed to this outcome: Muenster, located in the former Western part of Germany, has a long-standing tradition of civil society engagement, especially in the environmental domain, while in Potsdam, situated in the former East, it was not before 1990 that citizens were able to freely and actively engage for the environment. In addition, the political system of the former GDR did not support local authorities in positively responding to civil society engagement, quite the opposite: any self-organization of the civil society sector has a priori been perceived as critical, if not dangerous to the system (Markham, 2008). In Munster, the local political system had since the 1970s been accustomed to the environmental and other social movements.

This different local (and national) political tradition influences the styles and the outcomes of local climate policy until today, although new, common boundary conditions for both cities have emerged since then. While the city of Muenster is also very active in implementing and communicating its climate goals, the focal point of urban climate policy in Potsdam up to now clearly resides with the city as a political and administrative unit. Even important civil society actors in Potsdam, such as the Energy Forum, are oriented toward the city politics. One could even speculate that this might be an expression of the traditional Prussian positive stance toward the state.

However, even in the "Prussian" city of Potsdam[9] today, we find no straightforward top-down implementation of climate protection. As with the case of Potsdam, climate protection and adaptation to climate change are still voluntary areas of local politics. The majority of mandatory urban tasks comprise areas such as social politics, building licensing, or urban planning. These tasks have to be handled with tight, sometimes loss-making public budgets. The more cities are dependent on the active participation and support of their inhabitants, and should thus welcome civil society activities (Table 1).

While in direct comparison to Potsdam Muenster performs slightly better, there is no reason to qualify Muenster as a "top runner" in urban climate policy. As we have seen, Muenster has failed to meet its 2005 goal of a 25% reduction in the urban carbon footprint. In Potsdam, the major downturn in emissions occurred due to the substitution of a lignite fired by a gas fired heat and power plant in the mid-1990s. Since then, as in Muenster, the emissions did by and large stagnate. A comparative study on the state of implementation of urban climate policy in Germany (Sippel, 2010) has shown that only a quarter of all cities that do have climate targets did also manage to meet them. Both Muenster and Potsdam are thus no exceptions.

Table 1. Comparison Between Potsdam and Muenster.

	Potsdam 7 Points	Muenster 9 Points
(A-1) Achievement of climate targets (0–1 points)		
(A-2) Integration of climate protection in local polity (0–5 points)	+ + + +	+ + + + +
Binding emission target	+	+
Emission reporting	+	+
Institutionalization of climate protection via specific organizational units and procedures	+	+
Municipal climate protection strategy	+	+
Concrete action plan		+
(A-3) Implementation level of the municipal climate protection strategy regarding the potential reduction of CO_2 emissions (0–3 points)		+
(A-4) Implementation level of the municipal climate protection strategy regarding the diffusion in different fields of action (0–7 points)	+ + +	+ + +
Comprehensive measures	+	+
Energy	+	+
Traffic		
Planning		
Waste management		
Procurement		
Public relation	+	+

	Potsdam 9 Points	Münster 14 Points
(B-1) Level of civil society activity (0–4 points)	+ +	+ + + +
Conventional forms of communication	+	+
Broader forms of communication	+	+
Discursive modes of participation		+ +
(B-2) Level of diffusion of civil society activities (0–6 points)	+	+ + + +
Ecological citizenship	+	+
Political consumerism		
Life-politics – Living (electricity use, heating, appliances)		+
Life-politics – Mobility		+
Life-politics – Private consumption		
Life-politics – Food		+
(B-3) Level of municipal activity in integrating citizens (0–4 points)	+ + + +	+ +
Conventional forms of communication	+	+
Broader forms of communication	+	+
Discursive modes of participation	+ +	
(B-4) Diffusion of municipal activities in integrating citizens in climate protection (0–6 points)	+ +	+ + + +
Ecological citizenship	+	+
Political consumerism		
Life-politics – Living (electricity use, heating, appliances)	+	+
Life-politics – Mobility		+
Life-politics – Private consumption		
Life-politics – Food		+

Our conclusion from this rather sobering result is that cities should be more active in supporting and including their citizens in their own local climate policies – both at the level of "hard" (such as the extension and compaction of district heating systems) and of "soft" measures (such as public campaigns). While it remains true that decisions of the city council, the city administration, and the city's public utilities do play a major role in reducing the urban carbon footprint, cities need the public support for these decisions, and they also need active eco-citizenship in all areas of life, even if only performed by minority groups. We assume it to be crucial that cities create a public space for these initiatives, even if they not necessarily need to proclaim climate-friendly lifestyles of the "green vanguards" as the new rule. It would suffice to give them public visibility, and to symbolically honor them (e.g., by public awards). And of course it remains crucial to critically examine the way in which (public) decision makers calculate the costs and benefits of certain measures, for example, by questioning whether the assumed future cost development for energy is realistic or not.

NOTES

1. While we are fully aware of their high relevance, we nevertheless decided to not take them into account in the A indicators, assuming that their ultimate success will have to be measured in avoided emissions, while their immediate relevance lies more with the performance of urban climate policy. In order to avoid double counting, we thus excluded these "soft" or communication oriented measures from the outcome variable set.
2. Supporting and fostering the citizen-consumer is not a common role for the majority of local communities. But there are examples. The Italian city of Citadella, for example, did issue a ban on fast food restaurants, justifying it with their aesthetical and environmental drawbacks (Zöller, 2011).
3. The urban carbon footprint is calculated according to the CO_2 budget principles of Climate Alliance, thus excluding air travel, food, and consumption-related emissions. This explains why German city carbon footprints are lower than those calculated by the Environmental Protection Agency (UBA), resulting in an annual value of about 11 tons per capita per year. Potsdam's drop between 1990 and 2005 has several reasons, including the modernization of the building stock. However, the new CHP plant did a lot to reduce not only the city's carbon footprint but also the air pollutant load.
4. We do not count the city assembly vote from spring 2011 as a concrete action plan, but acknowledge it as a general expression of will.
5. This at least is a preliminary conclusion we can draw from an ongoing research project funded by the EU, in which 10 urban and rural regions from 5 European countries are compared with respect to attitudes to climate change, local climate

policy, and energy saving (http://www.gildedeu.org/). As final results are still pending, we will have to substantiate this hypothesis.

6. The redevelopment process was at the brink of total failure in the first place, as many citizens did not accept the restrictions for car parking. After the redesign of the participation process (including the establishment of an elected citizen's council), Drewitz turned into a real success story.

7. The share of environmental-friendly transport of the modal split in Muenster is 63.7% (2007) (Stadt Münster, 2011), a very high value for large German cities.

8. Measures included in the commitment are buy energy-efficient light bulbs, avoid stand-by, one short distance trip per week with bike or by foot, convince one more person to participate in the initiative. Within the following measures two had to be chosen: replace household machine with a new more efficient one, by a water saving appliance, dry laundry in air, reduce room temperature, shock ventilating instead of tilt ventilating, turn off PC and monitor when not used, switch to clean energy supplier, turn of light when leaving the room, dispose a car, check the personal CO_2 footprint, make an energy check for privately owned home.

9. The former German state Prussia was dissolved by the Allies in 1947. Today's state of Brandenburg, however, is – together with Berlin – a heartland of the old Prussia, and in some traditions still alive.

REFERENCES

Barthel, C. (2006). Der European Way of Life: Spielraum der Konsumenten bei ihren CO_2 Emissionen. *Energie & Management, 14*(3).

Beratungs- und Service-Gesellschaft Umwelt mbH. *Energienachbarschaften*. Retrieved from http://www.energyneighbourhoods.eu/de/home

Beschlussvorlage Integriertes Klimaschutzkonzept, Stadtverordnetenversammlung Potsdam 2011.

Bode, I., Evers, A., & Klein, A. (Eds.) (2009). *Bürgergesellschaft und Demokratie: Vol. 28. Bürgergesellschaft als Projekt: Eine Bestandsaufnahme zu Entwicklung und Förderung zivilgesellschaftlicher Potentiale in Deutschland* (1. Aufl.). Wiesbaden: VS Verlag für Sozialwissenschaften/GWV Fachverlage GmbH Wiesbaden. Retrieved from http://dx.doi.org/10.1007/978-3-531-91356-8

Deutsche Bank Research. (2010). *Kommunalfinanzen-zukunftssicher aufgestellt?* (Deutschland-Aktuelle Themen No. 482). Frankfurt am Main.

Döbel, R. (2011). *Bürger mischen sich ein-Bürgerhaushalt und Bürgerbeteiligung in Münster.*

Duscha, M., Dünnbeil, F., Gugel, B., & Kutzner, F. (2009). *Klimaschutzkonzept 2020 für die Stadt Münster: Endbericht*. Heidelberg; Essen.

Eberhardt, A. (2000). Partizipationsmodelle zur breiten Einbindung gesellschaftlicher Akteure in Prozesse der Nachhaltigkeit. In U. Böde & E. Gruber (Eds.), *Technik, Wirtschaft und Politik: Vol. 44. Klimaschutz als sozialer Prozess. Erfolgsfaktoren für die Umsetzung auf kommunaler Ebene* (pp. 119–139). Heidelberg: Physica-Verl.

faiplanet eG. *fairplanet: Die internationale Genossenschaft für Klima, Energie und Entwicklung.* Retrieved from http://www.fairpla.net/frame-123-Klimaformel.html

Geißel, B. (2007). Nachhaltige, effektive und legitime Politik durch Netzwerke? Fallbeispiel Lokale Agenda 21. In K. Jacob (Ed.), *Politische Vierteljahresschrift Sonderheft: Vol. 39. Politik und Umwelt* (1st ed., pp. 479–498). Wiesbaden: VS Verl. für Sozialwiss.

Gruber, E. (2000). Kommunale Aktivitäten und Erfolgsfaktoren: Ergebnisse einer Befragung von Städten und Gemeinden. In U. Böde & E. Gruber (Eds.), *Technik, Wirtschaft und Politik: Vol. 44. Klimaschutz als sozialer Prozess. Erfolgsfaktoren für die Umsetzung auf kommunaler Ebene* (pp. 87–95). Heidelberg: Physica-Verl.

Herzberg, C. (2009). *Von der Bürger- zur Solidarkommune: Lokale Demokratie in Zeiten der Globalisierung*. Univ., Diss.–Potsdam. Hamburg: VSA-Verl. Retrieved from http://www.vsa-verlag.de/books.php?kat=ap&isbn=978-3-89965-372-4

Initiative "Veggietag Münster". *Münster isst Veggie*. Retrieved from http://www.veggietag-muenster.de/veggietag-muenster/

Kern, K., Niederhafner, S., Rechlin, S., & Wagner, J. (2005). *Kommunaler Klimaschutz in Deutschland-Handlungsoptionen, Entwicklung und Perspektiven: Discussion Paper SPS IV 2005-101* (No. SPS IV 2005-101). Berlin.

Klein, A. (2001). *Der Diskurs der Zivilgesellschaft. Politische Hintergründe und demokratietheoretische Folgerungen*. Reihe Bürgerschaftliches Engagement und Zivilgesellschaft, Band 4. Opladen.

Landeshauptstadt Potsdam. (2007). *Integriertes Stadtentwicklungskonzept (INSEK)*.

Landeshauptstadt Potsdam. (2009). *Klimaschutzbericht Potsdam 2008*.

Landeshauptstadt Potsdam. (2010). *Gutachten zum integrierten Klimaschutzkonzept 2010*.

Landeshauptstadt Potsdam. (2011a). *Arbeitsmarkt: Arbeitslose und Arbeitslosenquote (Jahresdurchschnitt) seit 1997*. Retrieved from http://www.potsdam.de/cms/beitrag/10035657/400366/

Landeshauptstadt Potsdam. (2011b). *Energienachbarschaften: Die Stadtwette zum Klimaschutz: 9% Energie sparen*. Retrieved from http://www.potsdam.de/cms/ziel/1419818/DE/

Landeshauptstadt Potsdam. (2011c). *Lokale Agenda 21*. Retrieved from http://www.potsdam.de/cms/beitrag/10003483/27314

Leggewie, C., & Welzer, H. (2010). *Das Ende der Welt, wie wir sie kannten: Klima, Zukunft und die Chancen der Demokratie* (4. Aufl.). Frankfurt am Main: Fischer.

Markham, W. T. (2008). *Environmental organizations in modern Germany: Hardy survivors in the twentieth century and beyond*. New York: Oxford.

Newig, J. (2011). Partizipation und neue Formen der Governance. In M. Groß (Ed.), *Handbuch Umweltsoziologie* (pp. 485–502). Wiesbaden: VS Verlag für Sozialwissenschaften/Springer Fachmedien Wiesbaden GmbH Wiesbaden.

Newig, J., & Fritsch, O. (2009). Der Beitrag zivilgesellschaftlicher Partizipation zur Effektivitätssteigerung von Governance. Eine Analyse umweltpolitischer Beteiligungsverfahren im transatlantischen Vergleich. In (1st ed., pp. 214–239 I. Bode, A. Evers & A. Klein (Eds.), *Bürgergesellschaft und Demokratie: Vol. 28. Bürgergesellschaft als Projekt. Eine Bestandsaufnahme zu Entwicklung und Förderung zivilgesellschaftlicher Potentiale in Deutschland* (pp. 214–239). Wiesbaden: VS Verlag für Sozialwissenschaften/GWV Fachverlage GmbH Wiesbaden.

nütec e.V. *nütec Natürliche Überlebenstechnik Münsterland e.V.* Retrieved from http://www.nuetec.de/component/option,com_frontpage/Itemid,1/

Olivier, P., Marwell, G., & Teixeira, R. (1985). A theory of the critical mass. Interdependence, group heterogenity, and the production of collective action. *American Journal of Sociology*, *91*(3), 522–556.

Reusswig, F., Altenburg, C., Neebe, M., Schmidt, P., & Peters, V. (2010). Energy demand, governance and infrastructure in Potsdam and Potsdam-Mittelmark. The German case. In N. Gotts & I. Kovách (Eds.), *Studies in Political Science: Vol. 3. Climate Change and Local Governance. Alternative approaches to influencing houshold energy consumption. A comparative study of five European countries*. (pp. 110–184). Budapest.

Shafir, G. (Ed.) (1998). *The citizenship debates: A reader*. Minneapolis: University of Minnesota Press. Retrieved from http://www.gbv.de/dms/sub-hamburg/233495533.pdf

Sippel, M. (2010). *Cities in Germany and their climate commitments: More hype than substance?* Stuttgart. Retrieved from http://mpra.ub.uni-muenchen.de/23011/

Spaargaren, G., & Oosterveer, P. (2010). Citizen-consumers as agents of change in globalizing modernity: The case of sustainable consumption. *Sustainability*, *2*(7), 1887–1908.

Stadt Münster, A. f. S. S. V. (2011). *Gesamtverkehr*. Retrieved from http://www.muenster.de/stadt/stadtplanung/gesamtverkehr.html

Stadt Münster. (2011a). *Klimaschutz in Münster*. Retrieved from http://www.muenster.de/stadt/klima/allianz-fuer-klimaschutz.html

Stadt Münster. (2011b). *Münster am See: Der Aasee-Bildergalerie*. Retrieved from http://www.muenster.de/stadt/umwelt/muenster-am-see.html

Stadt Münster. (2011c). *Zahlen, Daten, Fakten für Münster*. Retrieved from http://www.muenster.de/stadt/stadtplanung/zahlen.html

Stadt Münster, A. f. G. u. U. (2011d). *Klimaschutz macht Schule*. Retrieved from http://www.muenster.de/stadt/umwelt/energie_abfallsparen.html

Stadt Münster, A. f. G. u. U. (2011e). *Klimaschutzpartnerschaften in Münster*. Retrieved from http://www.muenster.de/stadt/umwelt/klimaschutzpartnerschaften.html

Stadt Münster, A. f. G. u. U. (2011f). *Lokale Agenda 21*. Retrieved from http://www.stadt-muenster.de/umweltamt/umweltwegweiser/index.php?keywordID=344&task=searchByID

Umweltforum Münster e.V. *Umweltforum Münster*. Retrieved from http://www.umweltforum-muenster.de/index.php

UN Department of Economic and Social Affairs. Division for Sustainable Development. *Agenda 21*. Retrieved from http://www.un.org/esa/dsd/agenda21/

Verkehrsclub Deutschland (VCD) Kreisverband Münster e.V. *Aktionen: Klimakiller Lehrlauf-automobile Skulptur der Umweltverbände*. Retrieved from http://www.muenster.org/vcd/aktionen.htm

Vringer, C. R. (2005). *Analysis of the energy requirement for household consumption: Proefschrift Universiteit Utrecht*. Enschede: Febodruk BV.

Weber, C., & Perrels, A. (2000). Modelling lifestyle effects on energy demand and related emissions. *Energy Policy*, *28*(8), 549–566.

Weber, M. (1987). *Economy and society. Outline of interpretative sociology*. London: University of California Press.

Weidner, H. (2002). Capacity building for ecological modernization. Lessons from cross-national research. *American Behavioral Scientist*, *45*(9), 1340–1368.

Wikimedia Commons. (2002). *Potsdam view from above*. Retrieved from http://commons.wikimedia.org/wiki/Potsdam

Wildt, B. (2011). Interview by M. Neebe. Münster.

Wuppertal Institut für Klima, U. E. G. (2003). *Energie- und Klimainventur der Stadt Münster: Bilanzen des Energieeinsatzes und der Treibhausgas-Emissionen für das Jahr 2000*. Münster. Retrieved from http://www.muenster.de/stadt/umwelt/pdf/Klimainventur2000_muenster.pdf

Zilleßen, H. (1993). Die Modernisierung der Demokratie im Zeichen der Umweltproblematik. In H. Zilleßen, P. C. Dienel & W. Strubelt (Eds.), *Die Modernisierung der Demokratie. Internationale Ansätze* (pp. 17–39). Opladen: Westdt. Verl.

Zöller, M. (2011, August 9). *Italienische Kleinstadt verbietet Kebab & Co.* Retrieved from http://www.kleinezeitung.at/freizeit/lokalerezepte/2805277/italienische-kleinstadt-verbietet-kebap-co.story

CHAPTER 5

MILAN'S ANSWER TO THE CLIMATE CHANGE PROBLEM

Ilaria Beretta

Purpose – *This chapter shows the politics, plans, strategies, initiatives Municipality of Milan is putting in place, in the European Union context, in order to reduce its green gas emissions and to mitigate the negative effects of climate change.*

Design/methodology/approach – *The study was conducted on the basis of primary and secondary sources, in particular the analysis of municipal official documents.*

Findings – *The city of Milan structured its strategy against climate change in two branches: on one hand Milan takes part of national and international networks, on the other hand it acts at local level through the adoption of an ad hoc plan (called 'Piano Clima').*

Originality/value – *Among other things, this study shows the relevance of cooperation with different and similar urban areas. In fact, in order – both – to plan and to implement the most part of its initiatives against climate change, the contacts, exchanges, relations with other cities have been and will be of vital relevance for the city of Milan. From the experiences of peers, you can learn, for example, what measures are most*

Urban Areas and Global Climate Change
Research in Urban Sociology, Volume 12, 105–133

ISSN: 1047-0042/doi:10.1108/S1047-0042(2012)0000012008

effective, what measures are the easiest to implement and most accepted by citizens, how to implement measures.

Keywords: Networks; international cooperation; environmental politics; Environmental planning

THE REFERENCE CONTEXT AND THE CITY OF MILAN

In a scenario distinguished by the gradual concentration in urban areas of over half the world's populations, the role of the city takes on increasing importance both as a prime mover of growth of economies and as a place in which to tackle and win the struggle against global warming.

As can be seen in particular from the Thematic Strategy on the Urban Environment (2005), the European Union also acknowledges the central role of cities in the struggle against climate change and, more in general, in the pursuit of sustainable development. The document does in fact state that most cities are faced with the same environmental problems, such as bad quality air, heavy traffic and traffic jams, high noise levels, bad quality of built-up areas, abandoned land areas, greenhouse gas emissions, urban sprawl, wastes and wastewater production. Among the causes of such problems are changes in lifestyle (growing dependency on private vehicles, increase in the number of single-member families, increase in per capita rate of use of resources), and rising population numbers, which must be taken into account in the identification of any solutions. The Thematic Strategy then goes on to identify a number of priority areas of intervention, including adaptation to climatic change and the reduction in greenhouse gas emissions, and in this respect adds that the sectors where efforts must be concentrated are those of buildings and transport.

Nevertheless, in view of the different urban environmental emergencies identified in the document, the European Union has decided not to intervene 'from above' in regulating the urban context. Given the heterogeneous nature of urban areas and applicable national, regional and local obligations, it was decided that the best thing was not to legislate in order to achieve the document's goals. The best way of supporting the member states and local authorities it thought was to promote best practices, and favour their diffusion, and promote effective networking and exchange of experiences between cities.

In this context, the city of Milan is particularly active both as regards participation in national and international networks, and as regards the experimenting of plans and projects conceived and implemented on its own account.

The city of Milan extends over a surface area of nearly 18,200 hectares, most of which (14,200 hectares, about 80%) consists of built-up areas and infrastructures; the resident population amounts to 1,295,705 people (the second municipality in Italy after Rome), 588,197 families and a population density of 6,700 inhabitants per square kilometre (Siemens, Istituto Piepoli, 2010). The average Milan family income is € 41,665, much above the average income of Italian families (€ 22,470).

The Lombardy capital occupies a relatively small and densely populated surface area, but the actual city, the urban continuum of buildings, squares, dwelling houses, factories and accessory services, extends far beyond the municipal boundaries, above all to the north and east for many kilometres. The first belt of municipalities surrounding the Municipality of Milan, also called First Belt, does in fact create a single city made up of tens of independent municipalities and has a population of around 3.9 million. If, instead, reference is made to the metropolitan area as a whole, which besides Milan also includes other neighbouring provinces, this has about 5 million inhabitants and a total surface area of 2,944.53 km^2. While the areas to the north and east of Milan are densely populated and urbanised, to the south instead a huge agricultural area of considerable size (47,000 hectares) extends, which embraces the city like a huge green belt (Siemens, Istituto Piepoli, 2010).

From a geographic viewpoint, the city of Milan occupies part of the plain in the western area of the Lombardy region. The rivers that flow through it are the Olona, the Lambro, the Seveso and a number of canals. The climate is semi-continental. The Lombardy capital, like many other cities of the Po plain, owes its fortunes, but also its fragility, to its geographic position. In fact, the 'bowl' which forms between the Alps and the Apennines prevents good air ventilation and favours the build-up of air pollution and the formation of winter fogs, which in turn bind with the pollutants and make the air even less breathable. Moreover, Milan, despite being one of the Italian cities with the best public transport system, because of its sprawling hinterland, which is not always conveniently linked with the regional and national transport system, and because of its economic importance, is also a real catalyser of traffic, both private and commercial (Siemens, Istituto Piepoli, 2010).

As will be seen better later on, as specifically regards the problem of climate change, the Milan City Council recognises the importance of the role which local bodies can play in the struggle against global warming through the reduction in climate-affecting emissions and believes it ought to help by integrating a mitigation policy into its development strategy (cf. para. 3).

THE NATIONAL AND INTERNATIONAL NETWORKS

In the absence of any binding objectives for towns and cities, numerous international networks exist that promote the voluntary undertaking of commitments in favour of climate protection at different area levels and, in general, in favour of sustainability issues. Cooperation between local authorities plays a crucial role as regards the comparison and exchange of experiences between administrators and experts in relation to the results achieved and the difficulties found in implementing policies and measures, whenever conferences and presentations are periodically staged. Organisations normally rely on experts, who have developed instruments of diagnosis and analysis and catalogues of already-tested measures and place these resources and their supporting and advisory skills at the disposal of participants. The logic at the bottom of these networks is to achieve significant improvements in global environmental conditions, especially the reduction in greenhouse gas emissions, through the combined impact of local actions (Michele Betsill, 2001). The national networks, in particular, group together authorities referring to a common legislative and political context. We could mention in this respect the 'Red Española de Ciudades por el Clima' (http://www.redciudadesclima.es), the Swedish network of municipalities 'Klimatkommunerna' (http://www.klimatkommunerna.es) and the 'U.S. Conference of Mayors Climate Protection Agreement' (http://www.usmayors.org) which groups together numerous mayors of U.S. towns and cities. At international level, we can mention the International Council for Local Environmental Initiatives (ICLEI) (http://www.iclei.org), Local Governments for Sustainability and the 'Rete di Metropoli C40' (http://www.live.c40cities.org), while at European level we have the 'Climate Alliance' and the 'Covenant of Mayors' (http://www.eumayors.eu). The latter is in fact the major network dedicated to achieving the combined objective on energy topics and the reduction of '20 '20 '20 emissions at local level.

Thanks to the experience acquired by the network participating bodies, various associations have begun monitoring plan implementation and have provided the first considerations as regards the type of instruments used and the activities performed and critical aspects encountered along the road towards the reduction in emissions. The Climate Alliance, for example, carries out a periodical survey on the progress made by the cities taking part, on the basis of the data collected by means of a questionnaire divided into 'Ten steps', that is, the 10 stages recommended by the association to define a local mitigation strategy.

As has already been said, the Municipality of Milan also intends contributing to the struggle against climate change by drawing up a mitigation policy of its own. This awareness has been expressed by the administration within a General Development Plan 2006–2011, whereby the City Council has undertaken to develop a plan to reduce emissions at local level called 'Climate Plan' (cf. para. 3). This undertaking has also been confirmed by the signing of various international agreements (hereinafter briefly recalled) which define a series of objectives consistent with the approach outlined by the European Union in the '20, 20, 20' package.

In particular, the Municipality of Milan is part of the European Eurocities network (http://www.eurocities.com) and of the ICLEI international network, as part of which it has joined the 'Cities for Climate Protection Campaign' (CCP) and, during the COP13 in Bali, the 'World Mayors and Local Governments Climate Protection Agreement' (http://www.globalclimateagreement.org). This agreement underscores the importance of involving local authorities in mitigation policies and emphasises the need to implement these in a perspective that goes beyond the Kyoto undertakings, and to ensure cities adopt measures to guarantee the reduction in greenhouse gases at global level by 60–80% by 2050 compared to 1990 levels. The Milan City Council is also taking part in the 'SlimCity' initiative of the World Economic Forum (http://www.weforum.org), which promotes cooperation between local authorities and private players – in the energy, transport and building sectors – for the sharing of best practices and the implementation of green purchase campaigns in the public administrations, able to benefit from the 'critical purchase mass' represented by several urban communities.

In the energy field, the City Council joined the European Commission's Sustainable Energy Campaign (Sustainable Energy Europe) (http://www.sustenergy.com) in January 2007. This aims at involving a broad range of stakeholders-citizens, institutions and companies – in the promotion of sustainable energy use and production. The City's decision to join the SEE campaign is closed tied to its candidature to host the universal exposition of 2015, in order to place the lines of development of the city planned for the event in a perspective of infrastructural, environmental and energy sustainability. In September 2007, the City Council signed a plan agreement with the Ministry of the Environment and Land and Sea Protection, a national focal point for the SEE campaign, which sums up the commitment undertaken by the City Council to develop the Expo candidature along the lines of energy and environmental sustainability and identify the priority measures to be implemented consistently with the sustainable development initiatives already begun. The agreement also defines a series of projects to be implemented in this field, especially the evaluation of the possible implementation of 'urban sustainability areas', to be promoted with the backing of the European Commission, and the promotion of experiences of dissemination of good practices as regards renewable energy, energy saving and energy efficiency, including the opening of an 'Energy and Environment' counter at the disposal of the public and small and medium enterprises.

In early 2008, Milan expressed the desire to take part in an SEE campaign initiative specifically dedicated to local authorities, the Covenant of Mayors. This voluntary agreement answers the indications given by the European Union Action Plan on energy efficiency, with as priority action the drawing up of an agreement between the representatives of local administrative levels, and acknowledges the role which cities can have in achieving national emission reduction goals. Significant decentralised action potential is identified thanks to the close cooperation between local institutions and members of the public and by the ability possessed by the former in coordinating different actions and players. The Covenant requires the authorities to undertake to achieve the European goal of reducing emissions by 20% by 2020 through a Sustainable Energy Action Plan, which in the case of Milan is an integral part of the Climate Plan. The Action Plan must start from a territorial baseline and involve the territorial players both as regards the definition and the implementation of the contemplated measures. The formal signing of the Covenant took place in Brussels in February 2009 within the European Sustainable Energy Week.

THE CLIMATE PLAN[1]

The 'Boundaries' of the Plan

The Milan City Council's Climate Plan represents the synthesis and strategic framework of the commitments undertaken by the council administration as part of the aforementioned agreements and initiatives. This instrument intends providing a single reference framework for energy and environmental policies so as to plan measures consistent with future scenarios and optimise synergies existing between interventions in different sectors.

The Climate Plan focuses on CO_2 emissions, which represent about 92% of the greenhouse gas emissions of the Milan Municipal area; the emissions of CH_4 and N_2O, which are clearly lower than those of CO_2, are not taken into account.

The Plan includes all emissions produced by urban activities, which can be split into

- direct emissions, produced in the Municipality of Milan, which in turn include emissions caused by the internal production of electricity and heat, as well as those caused by energy consumption (fossil sources) for end uses (home and tertiary heating, passenger and goods transport);
- indirect ('shadow') emissions, produced outside the municipal area for the production of goods and services (e.g. electricity, steel, cement) imported and consumed by the Milan population.

As can be seen from the following chart, the direct emissions taken as a reference base for the Climate Plan amounted, in 2005, to 4,763 $ktCO_2$, while the shadow emissions estimated for the same year totalled 2157 $ktCO_2$ which, added to the former, result in a total of nearly 7000 $ktCO_2$ for the Municipality of Milan (Table 1).

The Reduction Goals

With the Climate Plan, the Municipality of Milan has set itself the goal of cutting its direct and indirect CO_2 emissions,[2] split up as follows:

a) Reduction in direct CO_2 emissions, that is, those relating to the municipal area, by 20% by 2020 compared to 2005 emission levels. This goal must be achieved with 100% domestic reduction measures.

Table 1. Summary Chart of CO_2 Emissions of the Municipality of Milan (Calculated in the Climate Plan).

From civil sector	3,035 $ktCO_2$
From industrial tertiary sector	134 $ktCO_2$
From transport sector	1,400 $ktCO_2$
From production plants not subject to European Directives *on emission trading* (non EU-ETS)	194 $ktCO_2$
Shadow emissions	2,157 $ktCO_2$
TOTAL	6,920 $ktCO_2$

Source: Our processing of Climate Plan data (2009).

b) 20% reduction in total emissions (including 'shadow' emissions), 75% of which to be achieved with reduction measures in the regional area, referable to Climate Plan actions, and 25% of which with projects such as Clean Development Mechanism (CDM) or Joint Implementation (JI).

With regard to the quantity estimate of emission reduction expected in order to achieve the aforementioned goals, a *Business as Usual* (BaU) scenario has been defined, on the basis of the evolution of the demand for energy services, which in turn has been imagined to depend mainly on the pattern of a number of fundamental variables, such as population growth, the demand for electricity and the demand for mobility.

After defining *Business as Usual* as the reference scenario, it was then possible to proceed to quantify the reduction in emissions until 2020. On the basis of the data collected, the 20% reduction target at 2020 of the direct emissions of CO_2 means a drop from 4,760 kt/year of CO_2 calculated in 2005 to 3,810 kt/year of CO_2 in 2020. The actual reduction goal nevertheless had to take into account the fact that, on the basis of *Business as Usual* scenario estimations, emissions are on the rise and consequently commitment will have to be above or equal to 1,340 kt/year of CO_2, to be achieved solely by means of domestic measures. As regards indirect emissions on the other hand, the reduction required on a yearly basis to achieve the goal in 2020 amounts to about 2,380 kt of CO_2, as already said, to be achieved 75% by domestic measures and 25% through CDM and JI.

The Identification of Measures

For the purpose of a more specific definition of the mitigation plan, an analysis was first of all performed of the planning and programmatic framework of the Milan City Council, so as to make the Climate Plan consistent with regard to the actions and objectives of the council administration in the different sectors. In this respect, the Climate Plan has been conceived as a tool for supplementing, correlating and strategically directing the different sector plans of the Milan City Council, aimed at quantifying the council greenhouse gas emission reduction objectives and successfully defining measures in the light of expectable developments concerning scenario variables.[3]

More specifically, for the Milan Climate Plan, the decision was taken to perform the sector analysis in two stages. The first stage contemplates the recognition of the plans and major provisions with respect to the areas for which high emission abatement potential has been estimated, in particular:

- Private transport
- Public transport
- Residential
- Tertiary
- Energy production
- Wastes
- Agriculture
- Public administration

Recognition was made through a short description of the plans, programmes and provisions of the sector, in relation to their goals, main contents and time validity; for each instrument the links were highlighted with the Climate Plan. The second stage of the sector analysis consisted in an analysis of the consistency of the Climate Plan measures: from each sector instrument, the measures were extrapolated consistent with the fields of actions defined in the Climate Plan together with any information, at goals and forecast level, useful for defining and specifying the Plan measures.

At general level, the estimate of emission reduction was made using the following method:

$$R = (\Delta I2020 - 2005) * C * FE$$

where
R = Reduction in the emissions expected from the measures between 2005 and 2020
ΔI2020–2005 = Variation in the value of the indicator between 2005 and 2020
C = Coefficient required for estimate
FE = Factor of emission

In other words, to calculate the contribution which each single measure is able to give to overall emission reduction, the implementation of each measure has been represented by the change in value of an indicator (e.g. passengers transported by underground railway lines) between a 'state of fact' value of 2005 and a 'forecast' value of 2020. The indicator value at 2020 of each measure depends on the strategic decisions of the administration, and represents one of the goals of climatic policies (for more details, see the attached sheets).

Altogether, a series of actions were identified for a total of 28 measures.

As regards the *transport macro-sector*, the expected measures are:

T1 – Development of underground railway lines
T2 – Development of public surface transport
T3 – Cycling mobility
T4 – Car sharing
T5 – Car pooling
T6 – Call systems
T7 – Private vehicle efficiency
T8 – Other reductions of car routes (e.g. ecopass, regulated parking, telecommuting, pedestrian precincts)
T9 – Use of motorcycles instead of cars
T10 – Reduction in commercial vehicle routes
T11 – Commercial vehicle efficiency.

For the *residential macro-sector*, the following measure have been identified:

R1 – Increase in energy efficiency of existing homes
R2 – New higher energy efficiency homes
R3 – Fuel change (diesel fuel → natural gas) (also includes tertiary sector)
R4 – Energy efficiency saving – end use efficiency

For the *tertiary macro-sector*, the following measures have been identified:

Z1 – Increase in energy efficiency of existing buildings
Z2 – New higher energy efficiency buildings
Z3 – Energy efficiency saving – end use efficiency

For the *Milan Municipality Public Buildings macro-sector*, the following measures have been identified:

E1 – Increase in energy efficiency of existing buildings
E2 – Energy efficiency saving – end use efficiency-buildings
E3 – Energy efficiency saving – end use efficiency – public lighting

For the *Energy Production macro-sector*, the following measures have been identified:

P1 – Upgrading of energy production efficiency
P2 – Solar photovoltaic
P3 – Centralised community heating
P4 – Centralised community heating – energy efficiency cogeneration

For the *Waste macro-sector*, the following measure has been identified:

W1 – Wastes disposed of by incineration – heat component

Finally for the *Agriculture and planting out micro-sector*, the following measures have been identified:

A1 – Surfaces cultivated with conservative techniques
A2 – Tree planting

By way of example, sheets are attached drawn up with reference to two measures (cycling mobility and energy efficiency of existing homes), but consider that the same job has been performed for all 28 envisaged measures.

Estimate of Achievable Results

In consideration of the reduction estimates of the single measures, and the internal constancy analysis, it seems that the goals set by the Climate Plan can be achieved as long as all the described measures are implemented.

In fact, the total sum of the reductions in all the macro-sectors, considering direct emissions only, amounts to around 1,344 ktonCO_2/year which fully achieves the 20% goal to be reached with domestic measures. The sum of the reductions in total emissions (direct and indirect) amounts to 1,786 ktonCO_2/year, equal to 75% of the envisaged goal; the remaining 25% can be covered by the realisation of CDM and JI projects.

It is interesting to see the contribution of the different macro-sectors to the implementation of the emission reductions, both as regards direct emissions and indirect emissions (following illustrations).

As regards the reduction in direct emissions, the transport sector is in lead position with 54% of the total, followed by the home macro-sector with 28% and the tertiary macro-sector (7%) and waste macro-sector (5%).

As regards the reduction in total emissions (direct and indirect), the transport sector is in any case in lead position with 41% of total reductions followed by the energy production macro-sector with 26%, by the home sector with 22%, and finally by the tertiary sector (6%) and waste sector (4%) (Table 2).

Table 2. Summary Picture of the Reductions Expected from the Climate Plan.

	Reduction ($ktCO_2$/year)
(a) Direct emissions only	
Transport	840
Home	452
Tertiary (shops, offices, businesses)	118
Public Assets Milan Municipality	26
Energy production	65
Wastes	80
Agriculture and planting	7
Total reductions	1.588
Minor reduction from double calculations	244
Actual reductions	1.344
Climate Plan reduction goal	1.340
Distance from goal	4
(b) Direct + indirect emissions only	
Transport	840
Home	452
Tertiary (shops, offices, businesses)	118
Public Assets Milan Municipality	26
Energy production	532
Wastes	80
Agriculture and planting	7
Total reductions	2.055
Minor reduction from double calculations	269
Actual reductions	1.786
Climate Plan reduction goal	2.380
Distance from goal	594

Source: Comune di Milano (2009).

Emission Reduction Costs Estimate

As regards the costs relating to the implementation of the CO_2 emission reduction measures, estimating these is rather complicated and has a far-from-negligible margin of uncertainty. The instrument used concerns a reduction curve estimate approach consistent with the analyses performed by McKinsey (2009) for the global and national level, aimed at quantifying the costs of a series of greenhouse gas 'reduction levers'.

Such approach assesses the extra costs associated with the implementation of a certain number of actions which can be traced back to the measures proposed in the Climate Plan, that is, the fixed and variable investment costs needed to reduce emissions. The costs can relate to the purchase of new technologies or to the setting up of infrastructure or requalification projects. The costs of the single reduction actions are therefore expressed as marginal costs compared to the *Business as Usual* scenario, do not include any incentives provided at national, regional or local level, and are expressed net of benefits obtainable by the action in terms of energy saving.

The average cost of emission reduction per tonne (€/tCO_2) is represented by the differential cost of the action, in relation to the emission reduction potential identified for that measure in the Climate Plan. A negative average reduction cost thus identifies an action which allows achieving annual savings greater than the extra costs of implementation[4] (by way of example, see annexes).

The actions able to reduce emissions, identified in relation to the Climate Plan measures, are listed on the following chart in growing-cost order. For a whole series of reasons, with reference to various actions of the transport sector, it has not in fact been possible to provide an estimate of the average reduction cost (Table 3).[5]

With regard to the measures for which the exercise was performed of identifying a simplifying action and attribution of costs and benefits, though necessarily taking into account that estimates further more in-depth investigation, from a preliminary assessment, it appears that the net cost of implementing the mitigation strategy proposed by the Plan is negative (−150 million €/year); altogether, the actions are therefore able to produce benefits for the community, quantified through energy savings, which are greater than their costs of implementation.

The reduction actions with negative cost correspond to 1,255 ktCO_2, equivalent to about 60% of the reductions expected by the Plan.

The per capita cost which appears from the analysis, considering the positive cost actions and the current population of the Municipality of Milan, is 10 €/year, much less compared to other evaluations made of mitigation strategy implementation costs, such as the McKinsey analysis relating to national level (45 €/year per capita). If we also include in the calculation the negative cost actions, the net cost for the population as a whole is – 115 €/year per capita (–25 €/year national level).

Expo 2015

The area to be occupied by the Expo 2015 exhibition halls is located in the north-west part of Milan, in the municipalities of Rho and Pero,

Table 3. Mitigation Actions Ordered by Growing-Reduction Costs.

Measure No.	Description	Reduction Potential $ktCO_2$/year	Average Reduction Cost Euro/tCO_2
T3	Cycling mobility	77	−508
R3	Change of fuel (diesel fuel to natural gas)	49	−437
R4	Energy saving – end use efficiency homes	113	−320
E3	Energy saving – end use efficiency public lighting	13	−202
T4	Car sharing	8	−191
T9	Change from car to motorcycle	21	−184
R2	New more energy-efficient homes	116	−80
T7	Private car efficiency	476	−65
T11	Commercial vehicle efficiency	99	−65
Z3	Energy saving – end use efficiency buildings	65	−60
E2	Energy saving – end use efficiency buildings	6	−60
R1	Increase of energy efficiency in existing homes	173	−33
Z1	Increase in energy efficiency in existing buildings	25	−33
E1	Buildings with energy certification	6	−33
A1	Land areas cultivated using conservative methods	5	−30
Z2	New more energy-efficient buildings	27	31
P3	Community centralised heating	8	40
W1	Wastes disposed of by incineration (heat)	80	40
P4	Community centralised heating – cogeneration energy efficiency	31	70
P2	Solar photovoltaic	26	191
A2	Tree planting	2	931
T1	Development of underground railways	48	Not quantifiable
T1	Public surface Transport	29	Not quantifiable
T5	Car pooling	11	Not quantifiable
T6	Call systems	Included in T2	Not quantifiable
T8	Other reductions in car journeys	Included in T1, T2, T3	Not quantifiable
T10	Mileage of commercial vehicles	72	Not quantifiable
P1	Better energy production efficiency	467	Not quantifiable

Source: Comune di Milano (2009).

substantially where the current Milan Exhibition Centre is situated, and will cover a surface area of 1.1 million square metres. By means of this event, many interlocutors expect Milan to be once again able to demonstrate its leadership, not only economic and technological but also cultural and environmental, over the rest of Italy, with sights set, as others underscore, on Europe, which must be the real point of reference for Milan in terms of its capacity to be a city with a strong international vocation, where multinational companies invest and bring wealth, unlike the situation of other Italian towns and cities (Comune di Milano, 2007).

In the Expo 2015 application dossier, climate change is presented as one of the major topics of the event, which must represent a chance to promote innovative programmes and integrated policies which permit significantly cutting the city's CO_2 emissions and improving the management of the energy and environmental system. The efforts which the city authorities will have to make start with a strategic plan for the coordination of greenhouse gas emission mitigation and reduction measures, involving the application of new technologies, the building of infrastructures and the wider use of sustainable practices in the urban context. The dossier does in fact present a packet of measures involving the major emission sectors, that is, civil (54.4% of total) as regards energy efficiency and the use of alternative sources for home uses, and transport (28.7%) in terms of public service, new mobility systems, infrastructures for facilitating the transit of road vehicles (use of private vehicles) and rail transport (underground railways). The city has also undertaken to implement a structured project support programme to be included in the flexible mechanisms of the Kyoto protocol, with the ultimate aim of generating credits to be used for offsetting the emissions produced by the event.

The Climate Plan does not have any intermediate reduction goals, inasmuch as only a decade has passed since it was first started. The universal exposition of 2015 will however represent a high impact event in terms of emissions produced by new infrastructures, the movement of people and start-up of the organising machine. The Plan therefore considers the structural measures contemplated for the EXPO at urban level (new underground railways), while a complete calculation of the emissions produced by the event is not at the moment feasible due to the vagueness of the projects, the precariousness of the forecasts, the lack of objective data (a first estimate in any case points to a 1.5% rise in the amount of CO_2 emissions produced by the city between 2008 and 2015 (Milan City Council, 2007).

CONCLUSIONS

By means of this contribution, we have tried to provide a more detailed picture – within the limits of available space – of the strategy implemented by the Milan City Council in the struggle against climate change. As we have seen, in this respect, the European Union has chosen to give local council authorities total freedom of action, inasmuch as it considers ineffective the issuing of one or more directives – for cities – which would then have to be adapted to a number of very different contexts. Milan, therefore, though 'not forced' to implement this strategy, has nevertheless decided to do all possible to combat global warming, in particular through two different kinds of activity: participation in international trade networks and the sharing of *best practices*, and the drawing up of an *ad hoc* strategic plan, called 'Climate Plan'.

Despite the numerous endeavours being made by different towns and cities in Italy and abroad, climate change nevertheless continues to represent an emergency, as is shown by the recent and increasingly more frequent environment disasters. For this reason, we must work harder in the direction of commitments and goals undertaken at different governance levels.

From this point of view, the city of Milan has been and wishes to continue to be in the front line; not only, in the past, has it chosen to take part in various networks and has represented one of the major supporters of the Covenant of Mayors but it also continues to contribute to the struggle against climate change, by setting itself increasingly more ambitious and binding goals, in a perspective of continuous improvement. In this respect, for example, in the World mayors summit on climate (Wmsc) staged in Mexico City just a few days before the United Nations COP16 (http://unfccc.int) in Cancun (November–December 2010), the participating mayors – including the mayor of Milan – signed an agreement in which they undertake to create a common and uniform system of calculation of CO_2 emissions. At the same time, the Milan City Council is taking part in an innovative project launched a few months ago and called LAIKA (http://www.life-laika.eu), which groups together 10 Italian cities and contemplates the definition and subsequent experimentation of an emission trading system at municipal level.

The harmful effects of climate change thus continue to be strongly felt, but the world's major towns and cities have already taken a path towards recognising the problem and tackling and combating the emergency situation it has already produced and, we can expect, it will continue to produce in other more numerous and peripheral urban contexts. We therefore believe that the positive effects of the decisions taken will soon make themselves felt and humankind will once again manage – as it has so often done in the past – to overcome this environment and social crisis.

NOTES

1. It should be remembered that in the 'Energy/Climate Package' which recently received the final approval of the European Parliament and the Council of Europe in December 2008, a 13% reduction is required from Italy in the 2005–2020 period, to which must be added the reductions in the European trading system, calculated not at single-country level but altogether at European level (–21% from 2005 to 2015).
2. The Climate Plan reduction commitment, which reflects the commitment undertaken as part of the Covenant of Mayors, is far greater than that required from Italy by the European Commission for 2020, which consists in a 20% reduction goal referred to 1990 emissions and not those of 2005.
3. The plans, programmes and provisions taken into consideration are the following: Milan City Council General Development Plan; Urban Policies Framework Document; Area Government Plan; Green Plan; Strategy for sustainable mobility for the protection of health and the Milan environment; Urban Mobility Plan (PUM); General Urban Traffic Plan; Urban Car-Park Programme; Regulation of parking; Triennial public transport Services Programme; Cycling Mobility Plan; Ordinances limiting goods transport vehicle traffic; Ecopass; Incentives for the promotion of low environmental impact vehicles; Directions for the implementation of the City of Milan Car Sharing service development project; Agreement with taxi drivers; Agreement of a pilot project for the promotion of electric vehicles within the municipal area; Building regulation; Reduction in the urbanisation costs for measures aimed at energy saving; Voluntary memorandum of intent concerning the technological requalification of centralised community heating plants; House Plan; City Council-AEM frame agreement for energy diversification in the municipal area by means of the promotion of centralised community heating; Jobs contemplated as part of the heat Contract; Light Plan; Traffic lights agreement.
4. The adopted approach would have implied a prompt definition of the actions and interventions contemplated for each reduction measure, which however it has not been possible to provide for all the measures proposed by the Plan. The evaluation was therefore changed according to the different measurements, in the following way:

1) lacking detailed indications as regards the content of the measure, possible actions have not been defined; for some measures therefore, no cost estimate is provided;
2) in the event of availability in literature of average reduction cost data on actions that can be potentially traced back to the Climate Plan measures, the cost estimate has been made by making reference to these data and to the reduction potential identified in the Climate Plan;

3) in the event of its having been possible to simulate a specific action in the municipal context, the approximate cost of the intervention has been estimated (yearly investment cost and running costs) and the yearly benefit in terms of energy resources not consumed.

5. This is mainly due to problems, as regards the single measure, of identifying specific actions in a complex sphere such as urban mobility, distinguished by the simultaneous presence and interaction between different means of transport. Problems also appear as regards attributing investment costs in local transport infrastructures to the reduction in emissions of greenhouse gases, inasmuch as mitigation represents one among the many aims pursued through such investments.

REFERENCES

AMA. (2007). *Indagine sulla mobilità delle persone nell'area milanese 2005/2006.*

AMA. (2009). *Spostamenti nel Comune di Milano nello scenario attuale, 2015 con a offerta invariata e 2015 con realizzazione delle infrastrutture programmate.* Comunicazione Agenzia Mobilità Ambiente del comune di Milano, 2 aprile 2009.

Betsill, M. (2001). Mitigating climate change in US Cities: Opportunities and obstacles. *Local Environment, 6*, 393–406.

COM (2005) 0718 def. Comunicazione della Commissione al Consiglio e al Parlamento europeo relativa ad una Strategia tematica sull'ambiente urbano.

Comune di Milano (2007). *Dossier di candidatura – Milano Expo 2015.*

Comune di Milano (2009). *Piano Clima del Comune di Milano*, rapporto finale.

EEA. (2008). *Climate for a transport change. TERM 2007: Indicators tracking transport and environment.*

ISTAT. (2001). *Censimento generale della popolazione.* Istat, Roma.

McKinsey (2009). *La riduzione delle emissioni di gas serra in Italia: opportunità e costi.* Retrieved from http://climatedesk.content.bymckinsey.com

Siemens, Istituto Piepoli (2010). *Città sostenibili: Milano.*

APPENDIX A: SHEET ON CYCLING MOBILITY

T3 – Cycling Mobility

Increase in the use of bicycles to move around the Milan urban area due to the extension of cycling lanes, better interlinking between these and the use of a new bike-sharing service.

An estimate of daily cycling journeys in Milan is available from the data of the survey on Mobility carried out by AMA in 2005. AMA also estimated the growth of bicycle journeys in 2015, on the basis of already-scheduled jobs. The measure considers other jobs, to be implemented in the 2015–2020 period, aimed at further increasing cycle journeys in 2020 to achieve the goal set by the 'Brussels Charter'.

It should also be remembered that a greater use of bicycles could be stimulated by the development of the ecopass, regulated parking and extension of pedestrian areas, as shown in the T8 measure.

Reference in City Council Planning
The development of cycling mobility is contemplated by various city council planning documents:

- Mobility Survey, AMA 2005
- PUM 2001–2010
- PUM State of implementation 2006
- General Development Plan 2006–2011
- Strategy for sustainable mobility for the protection of health and the environment 2006–2011
- Signing of the Brussels Charter

City Council Sectors Involved
Department of Mobility, transport and environment – Planning Mobility Transport and Environment Sector

Basic Value 2005
A figure has been estimated for 2005 of 132,000 bicycle journeys in the municipal area, based on the data shown on chart A9, page 133, of the AMA Mobility survey (AMA, 2007).

The journeys were taken into account made within municipal boundaries and those moving outside and then returning within municipal boundaries.

The journey figure per 100 inhabitants, equal to 10, is much below that recorded in many other European cities (following chart).

Total yearly journeys in the Municipality of Milan (43 million) have been estimated by imagining the measured figure as average over a period of around 320 days.

The AMA survey estimated, in 2005, a 2.6% modal share of bike use in the Municipality of Milan.

Table A.1. Daily Bicycle Journeys in 2005: Comparison Between Milan, Strasbourg and Copenhagen.

	2005 Journeys		Journeys/100 Inhabitants
Milan	1,300,000	132,000	10
Strasbourg	450,000	120,000	27
Copenhagen	500,000	396,491	79

2020 Target

AMA has estimated an increase in the use of bicycles in 2015, both as regards unchanged offer and that which foresees the building of scheduled public and private infrastructures, of 16% and 17% respectively, compared to 2005, until 48 and 50 million journeys/year will be achieved respectively. On these scenarios the bicycle's modal share remains substantially constant at 2.6%.

For purposes of the Climate Plan, the implementation is expected of a real policy of promotion of cycling mobility aimed at significantly increasing the use of this means of transport.

The estimate of bicycle journeys in 2020 was carried out – more precise data lacking – by considering the achievement of the goal set by the Brussels Charter (15% model share as regards journeys inside the Municipality) recently signed by the Municipality of Milan. Considering an internal journey share of 50% of the total, the goal corresponds to the achievement of a 7.5% modal share of total journeys.

This goal, which consists in achieving a number of daily bicycle journeys totalling 460,000, results in a yearly journey forecast of 148 million, over 3 times the 2005 journeys.

This is a very ambitious goal which, to be achieved, requires decisive and extraordinary action both in terms of cycling mobility infrastructures and for the promotion of cycling mobility as a whole.

In 2020 therefore, there will be 100 million additional yearly cycling journeys compared to the 2015 scenario, with offer unchanged, reference for the emission in the BaU scenario.

The figure expected at 2020, 462,500 journeys/day, corresponds to a figure of about 32 journeys/100 inhabitants, taking into account an increase in population expected according to the BaU scenario, a figure equal to less than half today's figure for Copenhagen.

The average distance covered by a Milan cyclist has been estimated by the AMA at around 3.5 km in 2005, and is similar to that estimated for other urban contexts; in Copenhagen, for example, the yearly survey estimates a daily average distance of around 5.7 km, which includes a prevalence of two daily journeys for work purposes corresponding to about 3 km/journey.

The following chart shows how the figure expected at 2020 for the number of kilometres/year covered in Milan (357 km/inhab./year) is below that already existing in 2000 in some cities and countries of northern Europe.

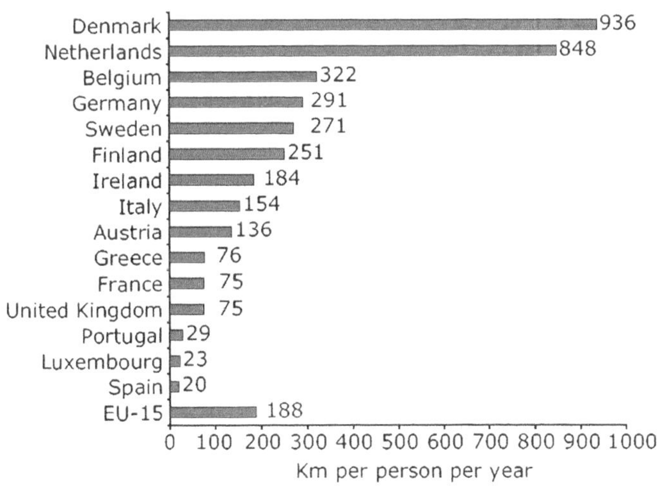

Fig. A.1. Kilometre Covered Per Year by Bike in 2000 in Different European Countries (EEA, 2008).

Method and Figures for Estimating Avoided Emissions

The estimate of avoided emissions is based on the relation

$$E = S * D * FE$$

where
S = Additional bicycle journeys per year
D = Average riding distance avoided (km)
FE = Factor of emission (emission avoided per km covered)

The following data were taken

P = 100 million additional journeys/year compared to 2015 IPPP scenario
D = 4.4 average distance covered by car avoided by a bicycle journey
FE = 177 g CO_2/km

The length of the car journey avoided by means of a bicycle journey is presumed 25% longer than that of the bike journey, inasmuch as the car journey is less direct and also requires parking. Note that the result – 4.4 km – is in any case much below the average distance of the car journey, inasmuch as bicycle journeys are of a greater proximity nature.

The emission factor estimate was made on the basis of the method shown in the chart below, on the basis of the average emission factor in urban cycle (212 gCO_2/km, AMA, 2009) and taking as occupation coefficient 1.2 (AMA, 2009).

Table A.2. Calculation of Emission Factor for Emissions Avoided by Cycling Mobility.

A	212 g/km	Average car emission factor in Milan (urban cycle)
B	1.2 passengers/car	Car occupation coefficient
$c = a/b$	177 g/km avoided	Average km emission factor avoided by a car in Milan

CO_2 Emissions Avoided

The application of the method leads to an estimate of avoided emissions totalling 77 ktonCO_2/year.

The summary picture of the applied method is shown on Chart 5-15.

Table A.3. Summary Picture of the Method Used to Estimate the Emissions Avoided by the Further Development of Cycling Mobility.

A	135 thousand journeys/day	Average bike journeys in 2005
B	320 days/year	Days with bike journeys
$c = a*b/100$	43 million journeys/year	Yearly bike journeys in 2005
D	48 million journeys/year	Yearly bike journeys in 2015 OI
E	148 million journeys/year	Yearly bike journeys in 2020
F	3.5	Average bike journey distance (km)
$g = c*f$	151 million km/year	millions of km covered by bike in 2005
$h = d*f$	168 million km/year	kilometres covered by bike in 2015 OI
$i = e*f$	518 million km/year	kilometres covered by bike in 2020
L	1,308,981 inhab.	Milan inhabitants (2005)
l'	1,450,000 inhab.	Milan inhabitants at 2020 according to PGT
$m = g/l$	116 km/inhab./year	km covered by bike/year in 2005
$n = h/l'$	116 km/inhab./year	km covered by bike/year in 2015 (BaU)
$o = i/l'$	357 km/inhab./year	km covered by bike/year in 2020
P	100 million journeys/year	Increase in bike journeys BAU – 2020
G	4.4 km	Average car journey distance avoided (km)
Q	177 g/km avoided	average emission factor (CO_2 avoided by bike journeys)
$r = o*p*q/1000$	77 ktCO_2/year	CO_2 avoided

Measure Implementation Hypothesis

- Rationalisation of the existing system of cycling lanes, with indication of the low traffic intensity and particularly safe itineraries;
- Widespread cycling jobs dedicated to overcoming cycling mobility barriers;
- Planning and realisation of additional cycling lanes and itineraries;
- Planning and realisation of parking areas for bicycles;
- Extension of bike-sharing service;
- Bicycle purchasing incentives;
- Strengthening of the integration between local public transport system and bicycles (extending the possibility of loading bikes onto public transport means and local railways);
- Activities dedicated to communication and information, such as
  - campaigns for informing the public about existing possibilities tied to this means of transport and to the gradual creation of infrastructures (e.g. cycling charter, like that of Ferrara, with information about repair services, deposits, sales, rental, public tyre inflation facilities);
  - organisation of promotional events, road-use education courses for safe bicycle use, like those already set up in Milan for road and mobility education (e.g. the city of Karditza, Greece, has set up a park for this purpose).

Critical Points of the Estimate and the Need for More In-Depth Investigation
A more precise estimate will only be possible if more precise data become available concerning journeys by bicycle at 2020.

Other critical data include the distance covered by the single bike journey and the distance of the substituted car journey.

APPENDIX B: SHEET RELATING TO THE ENERGY EFFICIENCY OF EXISTING HOMES

R1 – Increase in the Energy Efficiency of Existing Homes

Description of Measure
Improvement in the energy efficiency of homes already in existence in the municipal area, with consequent drop in energy consumption figures for winter heating.

References in Municipal Plans

- General development plan of the Milan City Council 2006–2011
- Reduction in the urbanisation costs for jobs aimed at energy saving
- Incentives for restructuring existing buildings to make them more energy efficient (insulation, window frames, solar panels, thermostatic valves, etc.)
- Draft of new building regulations

To promote energy efficiency in the private residential sector, the General development plan of the Milan City Council for 2006–2011 contemplates the preparation of a plan for the implementation of energy certification of buildings in conjunction with the region and province. The Municipality has also perfected a series of measures to promote the realisation of jobs able to ensure better energy efficiency.

- *Reduction in urbanisation costs for jobs aimed at energy saving*: the reduction is calculated on the basis of the energy performance index for winter climatisation, up to a maximum of 30%, and is applied in the case of new buildings, the extension of existing buildings and building restructuring.

Anyone wanting to obtain a reduction must have the following requirements:

- for new building jobs and restructuring jobs with demolition and reconstruction:
  - adoption of centralised heating systems in buildings with over 5 dwelling units
  - adoption of heat accounting systems for individual dwelling units
  - adoption of consumption recording systems with obligation of remote reading access for the council Authorities, for control and monitoring purposes

- for jobs related to the recovery of attics: obligation of installing ventilated roofs.
  - Incentives for the replacement of heating systems: this consists in the granting of easy-term loans by specific banks, with the support of the Lombardy Region. The incentives apply in the case of upgrade measures being implemented as regards the transformation and use of heating energy for home heating and the production of domestic hot water of the centralised type, for dwelling houses, offices and the like, as long as the energy efficiency of the system is upgraded by at least 15% compared to the previous annual figure. The agreement includes the replacement of heat generators, structural jobs on building-plant systems and single-family accounting system installation jobs. Jobs on new buildings are not included, nor jobs involving the switch from centralised heating systems to independent heating systems.

The proposal for a *new building regulation*, shortly to be approved, also contains specific incentives for the promotion of energy efficiency upgrade jobs on private buildings. The incentives are tied to the observance of a series of obligatory and optional requirements relating to the eco-compatibility and bio-compatibility of buildings.

Council Sectors Involved
Area Development Department (One-stop building counter – Environmental policy implementation sector)

Basic 2005 Figure
The number of homes at 2005 is estimated at 641,534. The figure was obtained by supplementing the ISTAT figure of the 2001 Census with the data relating to the new building licences granted by the Milan City Council imagining that all the granted licences were used.

As regards the number of homes with energy certification, there are no figures available for 2005, inasmuch as, until 2007, there was no obligation to notify certification. It is thought that the number of apartments in the municipal area already subject to energy efficiency jobs is very limited; because such types of jobs are not considered in the BaU scenario, the estimate of reductions is made by assessing the additional jobs with respect to those existing in 2005.

2020 Target

In the absence of a precise figure regarding the number of homes subject in recent years to restructuring jobs with energy requalification, a target has been imagined for 2020 of 150,000 homes to be upgraded in terms of energy efficiency. This goal also takes into account the possible increase in the number of jobs due to the promotion policies implemented at municipal, regional and national level.

In view of the difficulty in forecasting the types of jobs which will be done, the goal of the Climate Plan is that the restructuring jobs done will allow the switch of the restructured apartments from the current energy efficiency class G (with consumptions above 160 kWh/m^2/year) to class E (with consumption figures below 120 kWh/m^2/year), with a specific saving that is estimated at around 60 kWh/m^2/year. Taking into account the average size of an apartment in Milan, which is around 80 m^2 (ISTAT, 2001), a total yearly saving is obtained of 4,800 kWh per apartment.

Table B.1. Estimate of the Total Yearly Saving Per Apartment due to the Carrying Out of Energy Efficiency Upgrade Jobs.

	641,534	no.	Apartments in Milan
A	150,000	no.	Apartments with energy efficiency jobs 2020
B	80	m^2/apartment	Average apartment size
C	60	kWh/m^2/year	Specific saving, from current class G (> 160) to class E (> 120)
$d = b*c$	4,800	kWh/apartment/year	Yearly saving per apartment
$e = a*d/10\hat{}6$	720	GWh/year	Total expected energy saving

Method and Data for Estimating Avoided Emissions

The estimate of the avoided emissions is based on the relation:

$$E = R * A * FE$$

where

R = total yearly saving (kWh/apartment/year)

A = no. additional apartments with upgraded energy efficiency class (difference between 2020 figure and 2005 figure)

FE = Factor of emission (emissions avoided per kWh saved)

The following data have been assumed:

$R = 60*80$ kWh/apartment/year
$A = 150{,}000$
Factor of emission = 241 gCO_2/kWh

The emission factor estimate was made on the basis of the method shown in the following chart.

Table B.2. Calculation of the Emission Factor for Emissions Avoided Thanks to the Upgrading of Energy Efficiency in Homes.

A	60 gCO_2/MJ	Average emission factor natural gas + diesel fuel
B	0.9 MJ supplied/MJ burned	Average efficiency
C	0.277 kWh/MJ	Conversion factor
$d = a/b/c$	**241** gCO_2/kWh	Average emission factor homes

An average efficiency was considered of 90% of the conversion of energy burned in boilers in useful heat for the apartments

CO_2 Emissions Avoided
The application of the method leads to an estimate of avoided emissions totalling **173** ktonCO_2/year.

Hypothesis of Measure Implementation

- Strengthening of communication activities and consultancy services to the public by means of Sustainable Energy Space, to disclose information on available incentives and assist with necessary administrative procedures; preparation of information campaigns diversified according to the recipient (owners, renters…).

Critical Points of the Estimate and the Need for More In-Depth Investigation
The main limit of the survey lies in the lack of a precise picture of homes and buildings, in terms of volumes, energy needs, actual use, whether residential or tertiary, domestic Hot Water (DHW) heating and production plants. This way, it is not possible to estimate the emission reduction potential

deriving from the upgrading of average efficiency. The latest data refer to 2001 and come from an ISTAT Census, the reliability and precision of which is very doubtful as regards this type of survey.

Survey and update activities are really needed to obtain energy data relating to buildings in the Municipality of Milan, useful for estimating the potential tied to the measure.

CHAPTER 6

WIN, LOSE, OR DRAW? ASSESSING THE SUCCESS OF THE ENVIRONMENTAL JUSTICE MOVEMENT IN EMISSIONS TRADING SCHEMES

Krystal Tribbett

ABSTRACT

Purpose – *Emissions trading is often heralded as an efficient approach to environmental regulation. In the mid-90s Communities for a Better Environment (CBE), a Los Angeles-based advocacy organization, raised concerns that emissions trading in the South Coast Air Basin, the most polluted region in Southern California, would result in environmental injustice. The organizations concerns received mixed responses from regulators. Historical analysis is used to assess the clash between emissions trading and environmental justice (EJ).*

Methodology/approach – *Emissions trading and EJ arose side by side between the 1960s and the 1990s, yet they disagree on how to clean the air. Historical analysis of legal documents, presidential addresses, letters, working papers, reports, and the like offers a better understanding of the*

Urban Areas and Global Climate Change
Research in Urban Sociology, Volume 12, 135–167

ISSN: 1047-0042/doi:10.1108/S1047-0042(2012)0000012009

development of emissions trading and EJ, and their intersection in environmental policy.

Findings – *Emissions trading was grafted onto Clean Air Act policies not inherently designed for their incorporation. As a result, emissions trading came into direct philosophical opposition with EJ as political pressures calling for both economically efficient antiregulatory-ism and environmental equity forced their intersection. Formally, regional and national government accepted EJ as part of law. However, in principle, emissions trading undermined this acceptance. As a result, CBE could not easily win or explicitly lose its battle against emissions trading.*

Originality/value of paper – *Previous work on the relationship between emissions trading and EJ tend to focus on legal analysis and normative implications of emissions trading. Putting emissions trading and environment justice into historical perspective helps to illuminate larger questions about EJ activism and policy. Also, as California, the United States, and Europe turn to emissions trading to combat not only air pollution but also climate change, important lessons can be learned from the histories and collision of emissions trading and EJ.*

Keywords: Emissions trading; environmental justice; Regional Clean Air Incentive Market (RECLAIM); communities for a better environment; environmental policy; Southern California

Until the 1990s, command-and-control was the standard means to attempt to abate air pollution in the United States. Traditional command-and-control approaches to air pollution directly regulate *polluting* industries; regulators set air quality standards and prescribe the means by which to meet the standard. Industries must meet regulator directives regardless of cost (Cook, 1988). However, in the 1990s, emissions trading emerged as an alternative to the command-and-control approach. Under an emission trading system, regulators establish the acceptable pollution level for a given area and grant permits to polluting facilities. Polluting facilities can buy and sell permits, or install control technology in order to meet pollution level standards. Ideally, emissions trading controls air pollution to the same extent, but at lower cost and with greater business flexibility, than command-and-control because businesses have an incentive to choose the least cost, most efficient, means available to meet regulatory standards

(Tietenberg, 1985, 2006). Public welfare is protected and economic growth is sustained.

At least, that was the theory (Bryner, 1997; Crocker, 1966; Dales, 1968; Montgomery, 1972; Hahn & Noll, 1982). Environmental justice organizations were doubtful of this claim. In 1997, Communities for a Better Environment (CBE), a Los Angeles-based organization, filed a legal complaint against emissions trading. According to CBE, in order to save money, oil companies had abused emissions trading and citizens in San Pedro and Wilmington, two Port of Los Angeles communities, suffered an uneven distribution of harmful emissions (CBE v. Chevron, 1997a; CBE v. Chevron, 1997b; CBE v. GATX, 1997; CBE v. Tosco, 1997; CBE v. Unocal, 1997a; CBE v. Unocal, 1997b; CBE v. Ultramar, 1997a; CBE v. Ultramar, 1997b). CBE argued that the oil companies, and by extension the South Coast Air Quality Management District[1] (the pollution control agency for the South Coast Air Basin[2]) and the California Air Resources Board violated civil rights and environmental justice (EJ) laws.

This was not the first time that the CBE challenged the integrity and legality of emissions trading. Four years earlier, during the rule development public comment and testimony period of the Basin's emissions trading program, the Regional Clean Air Incentive Market (RECLAIM), CBE submitted comment letters for the District to review, and in the summer of 1993 offered public testimony warning about the potential of RECLAIM to result in environmental injustice, the disproportionate burdening of pollution on the poor and people of color. CBE argued in its letters to the District that RECLAIM was destined to fall short of its goal to protect the people of the Basin. According to CBE, RECLAIM would increase toxic emissions for the first eight years of the program and create hot spots, localized areas of pollution (Jenal, Ramo, & Drury, 1993a, 1993b). CBE asked for a return to command-and-control regulatory regime, arguing that RECLAIM violated the Clean Air Act, the Health and Safety code of California Assembly Bill 1054, and the California Environmental Quality Act (Jenal et al., 1993a, 1993b).

The South Coast Air Quality Management District considered EJ during RECLAIM's development when it welcomed public participation, performed socio-economic assessments, and developed environmental impact reports. The District concluded that the effect of RECLAIM would be legal and equitable, and adopted RECLAIM in October 1993. As a result of CBE's 1997 lawsuit, however, the California Environmental Protection Agency put all district credit rules and trading programs on hold; the United

States Environmental Protection Agency (EPA) considered making EJ concerns part of the approval of any new pollution trading schemes; and the South Coast Air Quality Management District Board of Directors unanimously passed a 10-point program for EJ.

This chapter explores the early history of RECLAIM, California's first emissions trading scheme, with particular focus on the question of EJ. Emissions trading and the EJ movement arose side by side, yet they disagreed on how to clean the air. The goal of emissions trading is regional clean air by Clean Air Act deadlines, achieved at lowest cost and greatest flexibility. The goal of EJ is to avoid short-term or localized increases in emissions that unduly impact particular communities, typically ones that are already disadvantaged, even if clean air goals are achieved in the long run. This early history is significant for two reasons. First, it helps to illuminate larger questions about EJ activism and policy. And second, as California, the United States, and Europe turn to emissions trading to combat not only air pollution but also climate change, important lessons can be learned from the histories and collision of emissions trading and EJ.

THE RISE AND FALL OF COMMAND-AND-CONTROL REGULATION

Situated between the Pacific Ocean on the west and the San Gabriel, San Bernardino, and San Jacinto mountains to the north and east, the South Coast Air Basin has a long history as a simultaneously and paradoxically beautiful and dirty place. Residents of the Basin are exposed to unhealthy levels of air pollutants from factories, cars, and other sources on a daily basis. Toxic pollutants include ozone, volatile organic compounds, nitrogen oxides and sulfur dioxide; these threaten the health and quality of life of Basin residents by causing asthma, headaches, cancer, and other chronic diseases (Ospital, 2006).

California's air pollution policy took shape in the late 1940s and expanded throughout the 1950s and 1960s. In August 1943, Fletcher E. Brown, then mayor of Los Angeles, announced at a press conference that the city's smog would be eliminated in four years. Brown did not keep his promise. The initial failure was due to a misconception that Los Angeles smog was merely a mixture of smoke and fog. In 1945, Los Angeles County officials attempted to abate smog by prohibiting dark smoke from factories. In the 1950s, efforts included banning garbage burning in county dumps and backyard incinerators. These efforts met little success. Research by Southern California scientists, including chemist Arie Haagen-Smit, revealed that Los Angeles urban smog resulted from a mix of tiny particles, carbon monoxide, nitrogen and sulfur oxides, hydrocarbons, and other pollutants produced by industrial processes and automobiles that mixed and when in the presence of sun produced ozone (Haagen-Smit, Darley, Zaitlin, Hull, & Noble, 1952). The Basin's bowl-like morphology is perfect for the perpetuation of this type of smog, which is carried eastward by the ocean breeze, and trapped in Basin cities by the mountains.

For three decades, California worked to bring the problem of smog under control, using a command-and-control approach. Stringent controls and standards of sources of pollution helped to improve the quality of the air despite increases in population, and the number of car miles driven (Carle, 2006). This success set California apart as a state to emulate. When the U.S. Clean Air Act was implemented in 1963, California's programs served as models. The common sentiment was that "If the air could be cleaned up here [in the South Coast Air Basin] (the thinking went), it might be cleaned up anywhere" (Carle, 2006, p. 80).

Until the 1960s in the United States, local and state governments were in charge of enforcing air pollution nuisance laws (Jones, 1975). Federal attempts to address air pollution were prompted by public demand for action after a string of fatal smog incidents in Meuse Valley, Belgium in 1930, Los Angeles in 1945 and 1954, Donora, Pennsylvania in 1948, and London, England in 1952. The first federal legislation focused on supporting smog research and circulating smog information. The Air Pollution Control Act of 1955 did not control pollution at all. Instead, it was "an act to provide research and technical assistance relating to air pollution control." It authorized the United States Surgeon General to conduct research and disseminate information "relating to air pollution and the prevention and abatement thereof" (Air Pollution Control Act of 1955). The Act declared air pollution a danger to public health and welfare, and authorized state and local governments to maintain responsibility for controlling the issue.

The vision for stronger federal involvement in air pollution abatement began in the Kennedy Administration and was sustained through the Johnson and Nixon Administrations. On February 21, 1961 in a special message, President Kennedy referred to air pollution as a problem of "immediacy" and asked Congress to strengthen the federal government's involvement in smog regulation (Kennedy, 1961). According to Kennedy, the air needed help, for "the atmosphere over our growing metropolitan areas… has only limited capacity to dilute and disperse contaminants"

(Kennedy, 1961). Kennedy called for an effective federal air pollution program that would help protect growing metropolitan areas (Kennedy, 1961). Kennedy appointed Assistant Secretary of the Department of Health, Education and Welfare and Mayor of Madison, Wisconsin, Ivan Nestingen summed up the Kennedy Administration's position well when he declared that "[t]here is no reason to wait until every city's problem is as critical as that of Los Angeles before admitting that the problem is increasing elsewhere. We must all intensify our efforts. The hour is late and with the pace of change in today's technology, *tomorrow is already here*" (U.S. Public Health Service 1962, pp. 7–9).

Following Kennedy's assassination in November of 1963, Lyndon B. Johnson signed the legislation that would be known as the Clean Air Act. Johnson's hope was for the Clean Air Act to control and even "halt the trend toward greater contamination of our atmosphere" (Johnson, 1963). Like Kennedy, Johnson believed the issue required "immediate action" (Johnson, 1963). The Clean Air Act (1963) was an "[a]ct to improve, strengthen, and accelerate programs for the prevention and abatement of air pollution." It placed the responsibility for the prevention and control of air pollution "at its source" on states and local governments (Clean Air Act of 1963). Like the Air Pollution Control Act, the Clean Air Act supported federal research and information dissemination regarding techniques for monitoring and controlling air pollution. However, unlike earlier law, the Clean Air Act promoted the creation of emission criteria and gave the Secretary of Health, Education, and Welfare the ability to intervene when control agencies failed to make effective progress toward the abatement of air pollution. It also allowed for technical and financial assistance to state and local governments (Clean Air Act of 1963). According to President Johnson, the Clean Air Act would foster a cooperative effort among state and federal government, as well as industry (Johnson, 1963).

The Clean Air Act of 1963 had widespread support (GovTrack.us, 2012).[3] However, by 1967, the nation still struggled with dirty air; the Clean Air Act of 1963 did not go far enough. Policy makers were ready for a stronger approach (GovTrack.us, 2012).[4] Yet, they recognized the challenge of supporting economic growth while regulating polluting industries.

John W. Gardner, Secretary of the Department of Health, Education, and Welfare argued, "We must now enter a new era in the nation-wide struggle against air pollution ... The scattered hit-or-miss, uncertain control efforts on the part of all levels of government which have characterized the past must give way to a much more rational and scientifically valid national effort" (Air Quality Act, 1967a). Gardner explained during a hearing on the

Air Quality Act of 1967, a Clean Air Act amendment, that the Clean Air Act had been "useful," but it gave the responsibility of setting emission control levels on state and local governments (Air Quality Act, 1967b). Furthermore, "experience has proven that most State and cities will not take the initiative in requiring control measures beyond those required in other places; nor will industries support local or State control action which may place them at a competitive disadvantage" (Air Quality Act, 1967b). Gardner called for setting nationwide emission control levels that would be "fair for everyone, to all industries and all communities" (Air Quality Act, 1967b). John T. Middleton, Director of the National Center for Air Pollution Control clarified,

> the most important of the factors that tend to discourage standard setting at the State and local levels is that such action seems inevitable to bring one major function of State and local governments – the protection of public health and welfare- into direct conflict with another – that of insuring economic growth. No matter how often we remind ourselves that effective control of air pollution is not incompatible with economic progress, the history of air pollution control efforts in this country provides abundant evidence that State and local officials are unable to take decisive action to adopt and enforce effective standards for the control of sources unless the problems have become so obvious, so severe and obnoxious as a nuisance that they cannot be tolerated. (Air Quality Act, 1967c)

According to Middleton, the most equitable solution for the public and polluting industries was to establish emission standards for individual stationary sources of pollution (Air Quality Act 1967c). Essentially, Gardner and Middleton suggested that state and local officials failed to establish emission control levels because it would put them at an economic disadvantage. The federal government could solve this dilemma by creating and enforcing nationwide emissions control levels of stationary sources of pollution to create a level playing field.

The Johnson Administration pushed for national emission standards for stationary sources, yet as enacted, the Air Quality Act of 1967 focused on regional, not national, standards.[5] The amendment required the Secretary of Health, Education, and Welfare to establish air quality control regions. It required states in the control regions to set and enforce pollution control standards through state implementation plans (SIPs) and control agencies. The Air Quality Act gave the Secretary of Health, Education, and Welfare the authority to intervene in cases where air pollution posed a substantial danger to public health. Otherwise, state and local government maintained responsibility of addressing air pollution.

In 1969, Richard Nixon entered the White House and continued a strong dedication to addressing environmental problems. He avowed "we [have] to work on the environment because it is now or never" (Nixon, 1970a). The stated goal of his first environmental legislation, the National Environmental Policy Act of 1969 (NEPA) was to, among other things, "promote efforts which will prevent or eliminate damage to his environment and to promote efforts which will prevent or eliminate damage to the environment and biosphere and stimulate the health and welfare of man" (Clean Air Act of 1970). NEPA also created the EPA, which Nixon hoped would organizationally pull together the research, monitoring, standard-setting and enforcement activities previously distributed among different departments and agencies.

The Clean Air Act of 1970 was a significant departure from previous federal air pollution legislation. Previous federal attempts at smog abatement focused on supporting research, providing technical assistance to control agencies, and leaving enforcement in the hands of state and local governments. The NEPA established the Council of Environmental Quality, which issued guidelines to federal agencies for the preparation of environmental impact statements. These statements were required for every report on proposed legislation connected to the environment (Clean Air Act of 1970). For the first time, federal agencies like the EPA had to consider the environmental impacts of proposed projects, as well as alternative courses of actions. After the passage of NEPA and the creation of the EPA, Nixon argued in his state of the union address that the government's role was "restoring nature to its natural state" (Nixon, 1970b). Furthermore, clean air was not "free" and a debt was finally called after years of "carelessness" (Nixon, 1970b). Nixon proposed what he considered "the most comprehensive and costly program in [pollution control] in American history" (Nixon, 1970b).

With the Clean Air Act of 1970, policy makers embraced direct control of stationary and mobile sources of pollution by authorizing the development of comprehensive federal and state regulations with set deadlines and financial consequences. The premise of the Clean Air Act was the management of air pollution by requiring the achievement of pollutant-specific national ambient air quality standards (NAAQS) by specific dates (Clean Air Act of 1970). The air quality standards specified the maximum allowable concentrations of pollutants in the air. Anything above these concentrations was considered a threat to public health and welfare (Clean Air Act of 1970). The Air Act also required SIPs that the EPA would review, approve, and financially back New Source Performance Standards (NSPS) and National Emission Standards for hazardous air pollutants (Clean Air Act of 1970).

Kennedy, Johnson, and Nixon all shared a sense of urgency and were prepared to implement strict air pollution controls. Nixon followed through with support for the command-and-control approach encompassed in the Clean Air Act of 1970. But almost immediately, there was push-back both inside and outside of the White House. In Washington, the economic branches of the federal government pushed back against command-and-control. The Executive Office of the President, Office of Management and Budget (OMB) expressed concern about the cost of regulations, standards and guidelines pertaining to environmental quality, consumer protection, and occupational and public heath and safety on industry and the federal government (Schultz, 1971). OMB Director George Schultz proposed a formal process known as the "Quality of Life Review" to review social regulatory policy in an effort to raise concern about economic consideration in decision-making processes. Although the Review was supposed to be for all social regulatory agencies, it singled out the EPA (Eads & Fix, 1984, p. 49). Political analyst Brian Cook has observed that the EPA and Clean Air Act of 1970 represented policy change that increased the scope, intrusiveness, and cost of new regulatory apparatus (Cook, 1988).[6] The EPA found itself butting heads with the Quality of Life Review and Commerce Department (Cook, 1988). As a result, the first head of the EPA, William Ruckleshaus, created a unit, the Office of Planning and Evaluation, in the EPA staffed by economists and intended to produce information and analyses the EPA could use to challenge White House and Commerce Department attempts to block EPA regulatory decisions (Cook, 1988).

Early considerations of economic incentive-based environmental regulation focused on waste pollution control. In 1971, economists Charles Schultze and Allen Kneese testified before the Joint Economic Committee that economic incentives were a better way to regulate pollution. During a Senate debate on the Federal Water Pollution Control Act Amendments of 1972, Senator William Proxmire (D-WI) argued that sewage charges were easier to enforce, required less bureaucracy, gave more flexibility to industry, and would be more efficient (Cook, 1988). Nixon proposed economic incentives to address air pollution. He recognized that new comprehensive regulations required that "the price of goods should be made to include the costs of producing and disposing of them without damage to the environment" (Train, 1996, p. 189). The President suggested, "the answer was not to abandon growth, but to redirect it" and in 1971, proposed a "Clean Air Emissions Charge" on emissions of sulfur oxides (Train, 1996, p. 189). Nixon believed the emissions charge incorporated the costs of pollution into the price of the product (Train, 1996, p. 189).

Despite the make-up of the Office of Planning and Evaluation, economic incentives were not well received by the EPA or Congress. Nixon's proposal failed to garner approval. (Train, 1996). Covering many of the main complaints, Senator Edmund Muskie (D-ME) opposed economic incentives arguing that they were not proven; introduced more bureaucratic agents in the regulatory structure; and gave industry the option of polluting for a fee (U.S. Congress Senate, 1977). Political scientist James Q. Wilson has explained the rational behind the initial rejection of economic incentives. "In the precarious early months of the EPA, when environmentalists were expressing skepticism about the Nixon Administration's commitment to environmental programs, any sign that the EPA was even considering pollution charges would have immediately been interpreted as indication that the agency proposed to 'sell licenses to pollute' (Wilson, 1980, p. 376). Such a charge, however misleading, would have dealt a serious blow to the EPA's need to find some political breathing room" (Wilson, 1980, p. 376).

Outside of Washington, polluting industries pushed backed against the deadlines and costs of command-and-control. The EPA published its initial ambient air quality standards for sulfur dioxide, carbon monoxide, nitrogen dioxide, particulates, hydrocarbons, and photochemical oxidants in April 1971.[7] After the EPA published these standards, state governments were required to devise SIPs prescribing how the state would meet the NAAQS within nine months of their circulation. States had until 1975 to attain most primary NAAQS. Designation as a nonattainment area threatened a region's economy. The nonattainment region risked losing some forms of federal financial assistance and faced having to pay annual penalty fees if they did not meet NAAQS goals. Industries in nonattainment regions risked loosing the ability to add to or modify their facilities and could be required by the EPA to install emission control technology. All states, except California, had only limited technological and financial resources from which to create their plans. Most had to estimate the need for reductions in emissions from stationary sources to reduce ambient levels of pollutants (Liroff, 1986). The 1970 Amendments even challenged California. The Act's NAAQS, requirement of SIPs, technology-based standards, and challenging attainment dates resulted in the Basin's failure to meet one or more national air standards by a designated date ("nonattainment") for decades. In fact, Clean Air Act standards were not met in the Basin in five of the thirteen years between 1975 and 1988.

The pressure to meet Clean Air Act deadlines came to a head in 1972 when a smelting company challenged the Clean Air Act's new source performance standards (NSPS). NSPS established national technology-based standards that were stricter for new and modified sources of pollution than for existing sources. Under NSPS, new and modified stationary sources of pollution had to meet emission limits achievable through the use of the best-known technology system (Clean Air Act of 1970). To avoid purchasing potentially expensive technology, the smelting company wanted the Clean Air Act's definition of stationary source broadened from "any building structure, facility or installation" emitting an air pollutant to have stationary source mean "plant" as a whole (Clean Air Act of 1970). If the total amount of emissions coming out of a plant did not increase then the smelting industry could avoid installing new control equipment for plant expansions or improvements. The EPA and the Office of Air Programs and Enforcement, which was responsible for administering the NSPS program, initially opposed the smelting industry's definition. The two regulators viewed the proposal as in direct opposition to the Clean Air Act because it undermined the "technology-forcing" approach of the Clean Air Act (Levin, 1982).

Richard Nixon resigned as president after Watergate in the summer of 1974. The 1973 oil crisis and steel crisis, along with the overall energy shortage of the 70s, the 1970s recession and stagflation challenged U.S. industries (Frum, 2000; Vietor, 1994). The new Ford Administration turned to deregulation in an effort to address the economic pressure associated with rising inflation and unemployment. The Administration embraced economically efficient legislation. This climate allowed the EPA to return to the smelting industry's proposal and in 1976 defined a "stationary source" as "any building, structure, facility, or installation, which emits or may emit any air pollutant" (ASARCO v. EPA, 1978). By applying this new definition, the EPA established a bubble to allow a plant operator who alters an existing facility in a way that increases its net emissions to avoid application of the NSPS by decreasing emissions from other facilities within the plant. The federal embrace of the "bubble policy" was the birth of emissions trading. With the adoption of the bubble policy, the door was opened for the federal adoption of other major aspects of emissions trading schemes: the offset, netting, and banking policies.

The EPA established an offset policy in 1977, only a year after redefining "stationary source." At that time, no major non-attainment area would meet the Clean Air Act 1975 deadline for coming into attainment (i.e., meeting NAAQS). Under the Clean Air Act, regions not in attainment (often urban areas, the locations of the majority of oil, gas, and steel industries) could not construct major new stationary sources or make modifications. The offset policy was the EPA's solution to nonattainment and potentially stifled urban economic growth. It allowed major new and modified sources of emission to site in nonattainment areas provided they used control equipment that produced the "lowest achievable emission rate" (LAER) for the type of industry involved and offset any excess by acquiring greater emission reductions from other sources in the area.

In California, the offset policy seemed a solution to industry initiatives cut short by nonattainment. Standard Oil of Ohio planned to build a major new crude oil receiving terminal for oil from the Alaska pipeline, at Long Beach. The company could not go along with its plans because the South Coast Air Basin, in which the terminal would be located, was in nonattainment. The California Air Resources Board, the South Coast Air Quality Management District, and the oil company made an agreement whereby the oil company bought pollution control equipment for a local power plant and guaranteed delivery of low-sulfur fuel oil to the plant so that the new emissions from the terminal would be offset by reduced emissions form the power plant. The project was abandoned in 1979 based on what the oil company considered regulatory burdens (Liroff, 1980).

One of President Carter's goals was to bring greater efficiency to the design and operation of government programs, especially regulation, and Congress helped him by incorporating the offset policy, including banking, into the Clean Air Act Amendments of 1977. Previously, the EPA rejected the concept of "banking" emissions trading credits for fear of direct conflict with the goal to achieve ambient standards as quickly as possible, and to the concern that banking could increase pollution in regions that met NAAQS (Tietenberg, 1985). The 1977 Amendments made room for banking by protecting significant deterioration (PSD) of regions in attainment through control technology and a cap on the maximum increases emissions allowed (Tietenberg, 1985).

For five years, emissions trading centered on promoting new business, focusing on major new or modified stationary sources of pollution. By 1979, existing sources wanted flexibility in meeting clean air standards as well. In 1980, polluting agencies voiced concern to the EPA about intensive, potential costly, complex new source review process. The EPA proposed "netting" as a solution. Netting increased the flexibility to meet SIP requirements by allowing emission reductions credits earned by a plant to offset the increases expected from new or modified sources in prevention of significant deterioration or nonattainment region. As long as the net increase (counting the emission reduction credits) in plant emissions was insignificant, the plant could net out of what could be a costly review (Tietenberg, 1985).

The Carter Administration subscribed to the idea of regulatory reform by encouraging innovative approaches to regulation. The Reagan

Administration supported regulatory relief by curbing federal involvement in private business. Emissions trading was in line with both approaches. At first, under President Reagan, William Drayton, EPA's Assistant Administrator for Planning and Management during the Carter Administration, and his Regulatory Reform Staff, struggled to develop a "controlled trading" policy statement which would encompass netting, bubbles, offsets, and banking. Michael Levin, then head of the Regulatory Reform Staff, recalled that their work was seen as "...a Democratic smoke screen to divert attention from the 'real issues' of federal intrusion and overly stringent regulation" (Levin, 1982, p. 88). The Reagan Administration worried that reforming air pollution along the lines of controlled trading would undermine the President's objectives for revising the Clean Air Act; "controlled trading" implied too much regulatory intrusiveness (Cook, 1988). However, with a different slant as "emissions trading," an Emissions Trading Policy Statement (ETPS) and technical issues document appeared in the Federal Register in April 1982, and was formalized in December 1986 (U.S. EPA 1982, 1986).[8] Emissions trading implied the deregulation that the Reagan Administration promised. Thus what, began in the mid-70s as the bubble program evolved to include offset policies, netting policies, and banking policies all centered on a unique currency known as emissions reduction credits. With each addition of an emissions trading element, the goal was to protect business, and by extension the economy, by increasing flexibility in meeting standards and thereby lower the cost of compliance. If it worked – if compliance was achieved at a lower cost – then it would be a win-win solution: good for the economy, good for the environment and ultimately good for people on both counts.

Many groups and individuals accepted this "win-win" framework. But not all. Some argued that the offset policy could create hot spots; people in and around the new modified source could suffer from a localized increase in emissions even though regional emissions levels decreased. If that occurred, it would not be a win-win situation at all. It would be a win for industry, and for individuals living in areas where NAAQS were attained, but a loss for individuals living in areas where pollution increased.

THE RISE OF ENVIRONMENTAL JUSTICE: FROM GRASSROOTS TO GOVERNMENT

In 1967, the same year that policy makers discussed how to make the Air Quality Act fair to both industry and the public, an eight year old African

American girl drowned at a garbage dump next to an elementary school and city park in an a predominantly African American neighborhood (Bullard, 1994). This event triggered riots among African American students at Texas Southern University, who, in turn, found allies among civil rights activists (Bullard, 1994). Drawing attention to the issue of hazardous waste in communities of color, the event came to mark the beginning of growing concern for race-based environmental injustice.

While policy makers discussed the Air Quality Act fairness for the public's sake, the racial, ethnic, and socio-economic identities of those in question never came up in early discussions. The mainstream environmental movement focused on the degradation of pristine wild nature through industrialization. The Texas garbage dump drowning drew attention to a need for the preservation and conservation of disempowered and particularly vulnerable people.

EJ had an uneven rise to national prominence, emerging slowly in the 1960s, and framed by the people directly affected by pollution in their communities, and scholars who were also activists. After the Texas incident, and around the same time as the adoption of the Clean Air Act of 1970 and the first publications of studies on emissions trading, scholars including Freeman (1972), Burch (1976), Berry (1977) and Asch and Seneca (1977) reported correlations between socioeconomic status and air pollution in U.S. urban centers. These studies represent some of the earliest empirical evidence of environmental injustice, demonstrating that people of color and low-income communities had been disproportionately located in and around industrial facilities and bore the majority of the environmental burdens.

In the late 1970s and early 1980s, the movement gained strength in communities of color and poor and working class white communities in the United States. In the white working-class community, the movement took the form of a "citizen-worker" or "anti-toxics" movement (Cable & Cable, 1995; Gould, Schnaiberg, & Weinberg, 1996; Levine, 1982). In communities of color, the movement took the form of the "People of Color Environmental Movement." Toxic waste pollution was a particular concern for early EJ groups, but in general they were responding to air, water, and soil pollution in their communities. In 1979, a group of middle-class African American residents in Houston, Texas formed a community action group to block a hazardous waste facility from being built in their neighborhood. The community filed suit to prevent the siting in *Bean v. Southwestern Waste Management*. This lawsuit was the first ever to challenge the siting of a waste facility under civil rights law (Bullard, 1983).

The 1980s were a defining decade for EJ. It was during this time that the EJ movement settled on its first focus of racism. In 1982 EJ drew national attention when an African American community in Warren County North Carolina organized to protest the siting of a polychlorinated biphenyl (PCB) landfill. Warren County was the poorest county in North Carolina and its population was more than 50% Black (Russell, 1989; Szasz and Meuser, 1997). The African American community in Warren argued that the siting of the landfill in a poor, Black community was not coincidental.

The Warren incident's visibility in news reports inspired two key studies, one by the U.S. General Accounting Office (GAO) and one by the United Church of Christ (UCC) Commission for Racial Justice. The GAO examined the demographics of the communities near four large commercial hazardous waste landfills in the South-eastern United States. It found that three of four off-site commercial hazardous waste landfills in Region 4 (which consists of eight states in the South) of the study were located in predominately African American communities (U.S. GAO, 1983). The study Toxic Waste and Race in the United States conducted by the United Church of Christ found, that "three out of every five Black and Hispanic African Americans lived in communities with uncontrolled toxic waste sites" (Chavis & Lee, 1987, p. 14). The study concluded that "indeed, race has been a factor in the location of commercial hazardous waste facilities in the United States" (Chavis & Lee, 1987, p. 15). In addition to these two landmark studies, EJ activist and sociologist Robert Bullard's 1983 study of solid waste sites in a Black Houston communities (Bullard, 1983) found that 21 of Houston's 25 solid waste facilities were located in African American neighborhoods.

In the 1970s and 1980s, envrionmental justice was generally equated with fighting racism. In the words of Reverend Bejamin Chavis, executive director of the UCC Commision for Racial Justice:

> Environmental racism is racial discrimination in environmental policymaking. It is racial discrimination in the enforcement of regulations and laws. It is racial discrimination in the deliberative targeting of communities of color by pollution industries. It is racial discrimination in the official sanctioning of the life-threatening presence of poisons and pollutants in communities of color. And it is racial discrimination in the history of excluding people of color from the mainstream environmental groups, decision making boards, commissions, and regulatory bodies. (Bullard, 1993, p. 3)

Robert Bullard built on Chavis' definition to argue that environmental racism is "any policy, practice, or directive that differentially affects or disadvantages (whether intended or not unintended) individuals, groups, or communities based on race or color." In this definition, Bullard captured the full complexity of the problem and acknowledged that environmental racism may not be intended. Those in positions of authority at the local, regional, state, and national levels may, by virtue of the political structures they maintain, unintentionally oppress the most vulnerable members of society. The challenge for EJ scholars and activists is to uncover the policymaking structures that result in environmental racism, and reform them.

These ideas were emerging and coalescing just as emissions trading found itself, by way of the Clean Air Act Amendments of 1990, newsworthy. Referred to President George H.W. Bush as the "most significant air pollution legislation in the nation's history," the 1990 Amendments authorized, among other things, a national cap-and-trade emission trading program for sulfur dioxide emissions, the precursors of acid rain, and authorized states and regions to develop Economic Incentive Programs, including emissions trading as a strategy for combating urban smog (Associated Press, 1990; Clean Air Act of 1990).[9] Even before entering the White House, President George H.W. Bush was advised by Environmental Defense Fund president Fred Krupp, by way of White House counsel, Boyden Gray, to make good on his promise to be an "environmental president" by finding a solution to acid rain (Conniff, 2009).

In the 1980s, acid rain, an international problem, was caused mostly by sulfur dioxide emitted from coal-fired plants that reacts with water molecules to form acids. This resulted in acid rain that damaged ecosystems, plants, animals and infrastructure. In the years leading up to Bush's election, reports on economic incentives in policy touted the benefits of market mechanisms. Academics Bruce Ackerman and Richard Stewart claimed in a report that market mechanisms were more cost effective than technology-based standards and increased industry's ability to meet Clean Air Act goals (Ackerman & Stewart, 1985). A group called Project 88-Harnessing Market Forces to Protect our Environment – Initiatives for the New President, launched by U.S. Senators Timothy Wirth and John Heinz, produced a series of reports conferences and briefings of White House officials and Congressional members of staff claiming that economic incentives should be at the center of environmental policy (Stavins, 1989). The Bush Administration turned to emissions trading to address acid rain a problem Bush campaigned on and an approach it thought could potentially combat global warming (Conniff, 2009). Previous incorporation of emissions trading policies supported the shift from command-and-control, but the 1990 Clean Air Act Amendments officially opened the door to the use of market mechanisms to control agencies that were struggling to abate air pollution.

In the South Coast Air Basin, the Clean Air Act objectives from the mid-70s through the 1990s forced the South Coast Air Quality Management District to promote the toughest SIPs in its history. According to James Lents, then Executive Director of the South Coast Air Management District from 1986 to 1997, "The unstated position of the [District] between 1977 and 1986 was we will try to improve the air quality, but Los Angeles is never going to have clean air.... In 1987, however we decided that would no longer be the case, that it is our mandate to have air in this basin that is healthy to breathe" (Waldman, 1991, pp. 86–87). With its 1989 SIP, the District released its toughest set of regulations for the South Coast Air Basin or anywhere else. The regulation was in three tiers, comprised of over a hundred controls, among them ride sharing, alternative work hours, new technologies, and extensions to old controls. The plan was to be carried out by the District, the California Air Resources Board (CARB), the California EPA, and local agencies and government. The 1991 Plan was just as vigorous and projected steady declines in air pollution emissions and concentrations. But, like elsewhere, the command-and-control approach embodied in the 1989 and 1990 SIPs had run out of steam. Faced with the worst recession since the great depression of the 1930s and the label of a severe nonattainment area for ozone, the District needed a policy that would balance public health and economic growth. Emissions trading, it was thought, would provide that balance.

At the same time, EJ was changing from a grassroots movement to a national policy. By the 1990s, numerous reports documented evidence of disproportionate environmental impacts in poor and minority communities. Reports by Robert Bullard and Bunyan Bryant and Paul Mohai concentrated on environmental hazard distribution and showed quantitatively and geographically that hazards were dumped on communities of color and poor and working class whites. The reports also began to expand the framework of the EJ movement. In an extensive review of 15 case studies, Bryant and Mohai (Bryant & Mohai, 1992) showed that the distribution of pollution was inequitable by income and in all but one case inequitable by race. Other studies emerged covering a broad range of issues including children's exposure to lead, farm worker exposure to pesticides, contaminated air and drinking water, and placement of transportation thoroughfares.

It was also around the 1990s that EJ activists and scholars replaced the movement's focus of environmental racism with environmental equity. They believed equity broaden the emerging movement by capturing racial inequalities as well as gender and social class environmental inequalities as well (Taylor, 2000). The idea of equity came and went quickly for the movement. EJ proponents decided their goal was to prevent pollution not distribute it, as the notion of equity suggested (Holifield, 2001). However, as the EJ movement gained momentum and legitimacy largely because its existence could be proved quantitatively, mainstream environmental groups as well as state and national agencies adapted movement goals. In its 1990 report on Environmental Equity: Reducing Risk for All Communities, the EPA claimed that environmental equity lent itself to measurement using methods of scientific risk analysis (U.S. EPA, 1990; Holifield, 2001). By doing so, the EPA ignored the EJ movement's rejection of distributing pollution and opened the door to a philosophical clash between EJ goals at the grassroots and EJ goals at the government level.

The 1991 First National People of Color Environmental Leadership Summit saw the movement transforming its framework from racism to justice. During the Summit, delegates, including hundreds of grassroots and national leaders from all fifty U.S. states and around the world, adopted 17 "Principles of Environmental Justice" (First National People of Color Environmental Leadership Summit, 1991). The justice frame deepened the scope of the movement by incorporating the concepts of equity, impartiality, and equality (Taylor, 2000). The Principles are concerned with environmental issues as they relate to humans, nature, and rural and urban environments. They focus on how these issues affect home, community, work, and play environments. The concerns are local, regional, national, and international in scope crossing racial and social class lines (Taylor, 2000). The framework even affected scholars; it required them to not just describe what activists were fighting against but to consider what they were fighting for (Pellow, 2000).[10]

By the time President Bill Clinton entered the White House, market incentives were a mainstay for urban air pollution control. Clinton upheld the commitment to market-based environmental regulation and placed EJ on the national agenda. The Clinton Administration legitimized the importance of the justice frame and the existence of the EJ movement by creating the EPA's Office of Environmental Justice and the National Environmental Justice Advisory Council. The definition of environmental justice adopted by this office further clarifies the federal understanding of the EJ.

> The goal of environmental justice is to ensure that all people, regardless of race, national origin or income, are protected from disproportionate impacts or environmental hazards. To be classified as an environmental justice community, residents must be a minority and/or low income group, excluded from the environmental policy setting and/ or decision-making process; subject to a disproportionate impact from one or more

> environmental hazards; and experience a disparate implementation of environmental regulations, requirements, practices and activities in their communities. (U.S. EPA, 1990b)

Clinton also passed Executive Order 12898, which addressed environmental injustices within federal laws and regulations by reinforcing the Civil Rights Act of 1964, Title VI (Clinton, 1994). Executive Order 12898 called for improved methodologies for accessing and mitigating impacts, health effects from multiple and cumulative exposure, collection of data on low-income and minority populations who may be disproportionately at risk, and impacts on subsistence fishers and wildlife consumers. It also encouraged participation of the impacted populations in the various phases of assessing impacts – including scoping, data gathering, alternatives, analysis, mitigation, and monitoring (Bullard, 2004). Together, the Executive Order 12898 and the EPA's definition of EJ identified the previously unrecognized sufferers of environmental problems, and called for their participation in policy. It is suggested that science could measure injustice. Clinton successfully made EJ a government concern however EJ goals were set that would undermine emissions trading goals.

CBE AND EMISSIONS TRADING IN THE SOUTH COAST AIR BASIN

By the early 1990s, the stage was set for injustice-conscious and economically sensitive emissions trading in the South Coast Air Basin. In 1993, the District adopted its first trading program, Rule 1610, also referred to as the "car scrapping program." Under Rule 1610, "licensed car scrappers" could purchase and destroy old cars (SCAQMD, 2012). The District then gave the scrappers emissions credits based on the projected emissions of the cars (SCAQMD, 2012). The scrappers could sell their credits to stationary sources of pollution like factories (SCAQMD, 2012). The stationary sources could use the credits to avoid emissions reductions that would require them to purchase emissions technology under the command-and-control-based SIPs. Rule 1610 was supposed to reduce pollution to an equal or greater level and at lower cost to industry than command-and-control by scrapping older, higher polluting cars.

In the same year, the District adopted RECLAIM. Unlike Rule 1610, RECLAIM (which is still in effect) uses a cap-and-trade framework whereby stationary polluting facilities, like refineries and power generators, can buy and sell credits to emit nitrogen oxides and sulfur oxides up to a designated level or cap. RECLAIM stood to replace command-and-control regulation by achieving the same environmental ends at lower monetary cost to the District and to polluting facilities (SCAQMD, 1993). At the time of its implementation in January 1994, RECLAIM was the most ambitious of any state or federal government emission-trading program; it later became an exemplar for emission trading (Bryner, 1997; Kamieniecki, Shafie, & Silvers, 1999; Young, 1996).

From its inception, environmental groups questioned RECLAIM's commitment to public health. Of the almost 10 million people that live in Los Angeles County, over 45% are Hispanic, roughly 28% are White, 14% Asian/Pacific Islander, and almost 10% African American (U.S. Census Bureau, 2012). An economically, racially, and ethnically diverse place, the impact of air pollution and air pollution policies raised EJ concerns among EJ advocates. Critics opposed RECLAIM giving industries the right to pollute (Bae, 1997; Bansal, Davis, Buntine, & Piazza, 1998; Drury, Belliveau, Kuhn, & Bansal, 1999; Fowlie, Holland, & Mansur, 2009; Lejano and Hirose, 2005; SCAQMD, 1999).

EJ groups, like CBE, were specifically concerned with RECLAIM's potential to impact local air quality. Because industries could buy emission credits to meet air quality standards instead of reducing emissions, air quality around industries could worsen, resulting in "hot spots" (Bansal et al., 1998; Bansal & Kuhn, 1998; Drury et al., 1997). RECLAIM was thought to risk environmentally unjust outcomes by placing an unequal burden of air pollution in EJ communities near RECLAIM facilities.

James Lents claimed that the development of RECLAIM included "the most extensive public participation process ever initiated by the South Coast Air Quality Management District for the development of any environmental regulation" (SCAQMD, 1994). And it is true that CBE played a major role in public discussions. As it did, CBE was itself transformed from a general environmental organization to one with a core focus on EJ. As a proxy for citizens of the South Coast Air Basin who suffered from high concentrations of pollution from freeways, power plants, oil refineries, seaports, airplanes and chemical manufactures CBE called attention to delayed attainment of air quality goals under RECLAIM, the possible creation of hot spots if Rule 1610 was incorporated into the program, and RECLAIM's failure to provide adequate public participation.

CBE argued in its letters to the District that RECLAIM was destined to fall short of its goal to protect the people of the Basin. According to CBE, RECLAIM would increase emissions by 26,280 tons for nitrogen oxides and 16,790 tons for sulfur oxides for the first eight years of the program (Jenal et al., 1993a, 1993b). The organization argued that such emissions increases put RECLAIM in jeopardy of violating AB 1054. CBE noted that the Health and Safety Code of AB 1054 required any market incentive program for air pollution to provide a level of enforcement and monitoring, and emission reductions, comparable with command-and-control air quality measures that would otherwise have been adopted by the District. Furthermore, the law held that RECLAIM must not "in any manner, delay, postpone, or hinder district compliance with state ambient air quality standards" (Jenal et al., 1993a, 1993b). CBE argued that such delay would occur.

CBE argued that in other ways, RECLAIM failed to adequately invite public participation granted by the California Environmental Quality Act. In the first comment letter sent by CBE to address proposed RECLAIM Rules, Jim Jenal, then Clean Air Program Director for the CBE, was concerned that the omission of three EJ categories – income, education, and ethnicity – from the District's analysis of ozone exposure, could lead to "possible disparate impacts" (Jenal et al., 1993a, 1993b). According to the letter, Jenal was informed by Henry Hugo (who worked with Barry Wallerstein a member of the RECLAIM Socio-Economic Working Group) "the information [about disparate impacts on communities of color] was simply forgotten in the rush to get the [environmental] report out the door last week" (Jenal et al., 1993a, 1993b). How could a public be made aware of adverse impacts if such variables were not included?

A major source of contention involved automobile emissions, a large source of air pollution in the South Coast Air Basin. A mobile source emissions credit program incorporated into RECLAIM would allow RECLAIM facilities to obtain credits from car-scrapping. CBE worried that this would open "a Pandora's box of air pollution problems" (Jenal et al., 1993a, 1993b). Automobile emissions were previously distributed around the Basin, once traded on the RECLAIM market, could be used to permit *increase*d emissions at the stationary sources that purchased the mobile credits. CBE asked for public review of individual trades in addition to public review of the whole program, writing, "[the California Environmental Quality Act] grants affected communities the right to know about, and to comment on [trades] before they occur, so that the public may safeguard its rights, and policy makers can fully consider the environmental effects of their actions" (Jenal et al., 1993a, 1993b). Finally, CBE also raised one seemingly contradictory impact: the potential urban blight that RECLAIM could create. CBE claimed that some industries would not be able to buy emission credits to pollute and as a result close or relocate. This would lead to job loss, further stressing an already-stressed community.

CBE wanted the District to abandon RECLAIM and return to command-and-control. Command-and-control had worked to improve air quality in the Basin; ozone levels had decreased, sulfur and lead standard met, and for the first time in 1992 the annual nitrogen dioxide standard was not exceeded (SCAQMD, 20002). CBE claimed that the participation of the public was only symbolic. Emphasizing this, during a public hearing on September 10, 1993, which took place after the comment letter process, Joel Schwartz, then staff scientist for the Coalition for Clean Air, stood before the South Coast Air Basin Management District governing board to argue that RECLAIM rules needed reevaluation because they had come to believe that the program would produce more air pollution than the SIP it would replace (Schwartz, 1993). The same day, Jim Jenal and Richard Drury, staff attorney for CBE, argued that RECLAIM would give businesses the ability to buy the "right" to pollute and could lead to job losses if companies decided to sell their credits, shutdown, and leave without securing the futures of their workers (Jenal and Drury, 1993). But RECLAIM was accepted that day, unrevised based on the new testimony.

The District disagreed with CBE's allegations; it claimed the organization "misinterpreted" or "confused" information in RECLAIM draft reports (Jenal et al., 1993a, 1993b). The District even suggested that CBE's claims were not "supported by facts" (Jenal et al., 1993a, 1993b). The District admitted that RECLAIM could lead to hot spots. A 1993 District study found that poorer people and black and Latin people were exposed to more pollution than better off white and Asian people (SCAQMD and Fullerton Foundation, 1993). A non-District study found that the majority of toxic chemical releases from manufacturing facilities in Los Angeles County were in Hispanic communities (Burke, 1993). However, the District argued that despite short-term increases in emissions, RECLAIM would meet the requirements of AB 1054, which mandated that RECLAIM emission level endpoints be the same as endpoints that would result from the full implementation of the 1991 Air Quality Management Plan (Jenal et al., 1993a, 1993b). The 1990 Clean Air Act set forth a compliance date of 2010. The District argued that RECLAIM would help the Basin meet the Clean Air Act's by the deadline despite short-term increases in emissions, thus mobile source credits were included in the RECLAIM program despite CBE warnings (Jenal et al., 1993a, 1993b).[11]

The District believed that it did invite public participation into the RECLAIM development and review process. Although it had not included socio-economic information in the first RECLAIM draft report, such information would be included in future reports (Jenal et al., 1993a, 1993b).

The District responded to CBE's concern that RECLAIM could lead to urban blight by noting that CBE had not "provided any credible evidence that the proposed project will contribute to urban blight" (Jenal et al., 1993a, 1993b). From the District's prospective, "urban blight need only be analyzed for a project if the project has the potential to exacerbate the problem" (Jenal et al., 1993a, 1993b). In their view, no evidence had been presented by CBE to support that claim and District analysis suggested RECLAIM's impact on blight would be negligible (Jenal et al., 1993a, 1993b). In the end, the District only made small "revisions and clarifications" to RECLAIM based on the public comment process (Jenal et al., 1997a). The District maintained that RECLAIM abided by state and federal laws.

RECLAIM REVISITED

Four years later, CBE challenged emission trading in the Basin again, this time based on the claim that emissions trading had indeed violated Civil Rights Law. The detail at stake was something called Marine Tank Vessel Operations Rule 1412, which required all marine tank vessels to limit emissions of volatile organic compounds (VOCs) to a predetermined level. CBE claimed that oil companies used Rule 1610- the Old Vehicle Scrapping Rule 1610-instead of the 1142 to evade its legal responsibility to control VOCs. Most VOCs, which are toxic gases, are released when VOC laden oil is emitted into the air after oil is loaded into a tanker. To meet Rule 1142 standards, marine terminal facilities sometimes installed expensive vapor recovery systems. CBE argued that four oil companies, Unocal, Chevron, Ultramar, Tosco and GATX purchased credits obtained under Rule 1610 to meet the requirements of Rule 1142, saving money instead of lowering their emissions (CBE v. Chevron, 1997a, 1997b; CBE v. GATX, 1997; CBE v. Tosco, 1997; CBE v. Ultramar, 1997a, 1997b; CBE v. Unocal, 1997a, 1997b; CBE v. SCAQMD, 1997).

In doing so, CBE argued, the District and oil companies violated Title VI of the Civil Rights Act, as well as EPA implementing regulations and Executive Order 12898. According to the Civil Rights Act Title VI, recipients of government funds must not use their funding in a discriminatory manner. The Civil Rights Act (1964) states, "no person in the United States shall, on the ground of race, color, or national origin be exclude from participation in, be denied the benefit of, or be subjected to discrimination under any program or activity receiving Federal financial assistance." The oil companies saved money by not installing vapor recovery equipment, but they did so at the expense of San Pedro and Wilmington residents, the majority of whom were Latinos, when increased VOC emissions were produced by their facilities. As a federally funded agency, the District could not support programs that impacted members of minority groups even if the harmful impact on a minority community was unintended (Civil Rights Act of 1964; Fisher, 1995).

CBE believed that the issues raised by Rule 1610 were symbolic of the problems with emissions trading programs in general. When CBE challenged RECLAIM in 1993, it used empirical evidence provided by its own experts to argue in letters and public testimony that emissions trading jeopardized the health and welfare of communities (CBE v. SCAQMD, 1997). CBE called for the abandonment of RECLAIM and emissions trading in general. Still backed by its experts, but not arguing to eliminate emissions trading entirely, CBE's lawsuits now specifically challenged loopholes created when command-and-control and market incentives are poorly integrated (Clifford, 1997a; Clifford 1997b).[12] In this case, continuing to pollute instead of installing vapor recovering technology resulted in environmental injustice.

This time CBE was heard. As a result of CBE's 1997 lawsuit, the California EPA put all District credit rules and trading programs on hold; the United States EPA made EJ concerns part of the approval of any new pollution trading schemes, and the District Board of Directors unanimously adopted four guiding principles and ten initiatives to ensure EJ (Frank, 1997a; Frank 1997b; SCAQMD, 2007; SCAQMD, 2005). This incident ended with the oil companies settling with the EPA and CBE for supplemental environmental projects and monetary penalties because of their violation of Rule 1142. CBE withdrew its civil rights complaint (SCAQMD, 2002).

DISCUSSION

From the 1960s to the 1990s, the goal of abating air pollution as quickly as possible regardless of cost evolved into the goal of improving air as cost-effectively as possible. Grafted onto Clean Air Act policies not inherently designed for their incorporation, emissions trading came into direct philosophical opposition with EJ as political pressures calling for both economically efficient anti-regulatory-ism and environmental equity forced their intersection. As policy makers transformed the Clean Air Act to accommodate economic concerns, thinking about how best to clean the air also evolved. When EJ was taken up at the national and regional levels, how to achieve justice became a matter of proving evidence of injustice.

The challenges came particularly to a head in California, where early efforts to implement emissions trading ran into strong local community concerns. In their effort to challenge RECLAIM rules, CBE focused on aspects of the program that would delay meeting air quality standards and negatively impact the communities where RECLAIM facilities were located, drawing on federal and state laws for support. However, these laws set up a Catch 22. The Clean Air Act required RECLAIM meet emission standard deadlines; the District anticipated that the program would do so. The California Air Quality Act required public participation in rule development and evaluation of the RECLAIM program while AB 1054 ensured RECLAIM would not delay compliance with air quality standards. CBE tried to use these to plead its case against RECLAIM. Neither law, however, guarded against short-term, localized increases in emissions or specified the extent to which the District should heed public comments. Each law required evidence, and the burden of proof lay with both the District and CBE. CBE's anticipatory claims of injustice were not sufficient to surpass RECLAIM's promise to clean the air by set deadlines.

CBE argued against emissions trading and for a return to command-and-control. However, in tracing the development of emissions trading, RECLAIM did not abandon command-and-control at all; it was grafted onto it, building atop of the command-and-control basis of the Clean Air Act. As a result, both sides of the argument were present in the structure of RECLAIM and couldn't be pitted against each other. Formally, regional and national government accepted EJ as part of law. However, in principle, emissions trading undermined this acceptance. Emissions trading's economically centered strategy aimed to meet air quality standards by deadlines with little concern if ends justified the mean. Thus the District admitted that that hot spots could occur, but the overall regional effect negligible and targets would be met by their legal deadlines, and RECLAIM legal. Until RECLAIM was implemented its creation of hot spots was purely speculation. Current reviews of the RECLAIM program find that RECLAIM does reduce emissions in the long run and has not created hot spots (Fowlie et al., 2009).

CBE's case against the District and oil companies in the Port of Los Angeles was different in that when CBE plead its case, the people of

Wilmington and San Pedro had experienced short-term increases in VOC emissions proving that environmental injustice can persist under emissions trading systems. CBE received a favorable response from the District and U.S. EPA in the form of stronger EJ policies and procedures. State and federal agencies recognize the reality of pollution, and the reality of inequitable impact. So do EJ advocates. Yet, at the same time, the rise of neo-liberal policies led to the push for market mechanisms. These two impulses clash. CBE tried to protect vulnerable communities by fighting the system within the system when it warned against risks of injustice. It could not have a real impact on how emissions trading and EJ work together until injustice actually occurred.

Today, CBE is once again challenging emissions trading, this time for carbon dioxide. California Assembly Bill 32, the Global Warming Solutions Act, which was passed in 2006, includes a cap-and-trade strategy to help California reduce state green house emissions by 2020 (CARB, 2011). In 2009, CBE sued the California Air Resource Board on the grounds that it had violated the California Environmental Quality Act by passing AB 32. CBE and its constituents argued that, "Allowing the most entrenched polluters to increase pollution... was not the way to stop poisoning our air and slow catastrophic climate change (Communities for a Better Environment, 2011)." Furthermore, CBE claimed CARB was "dogmatic in its focus on cap-and-trade" which "increased pollution in heavily polluted low income communities and communities of color (Communities for a Better Environment, 2011)." On March 17, 2011, Judge Ernest Goldsmith of the San Francisco Superior Court sided with CBE and temporarily halted CARB from going forward with the cap-and-trade program on the grounds that CARB "undermined the California Environmental Quality Act's goal of informed decision-making" (Environmental News Service, 2011). After further review, the cap-and-trade regulation of AB 32 went into effect on January 1, 2012 (CARB, 2011).

How efficacious is EJ in laws if movement advocates must subscribe to practical politics? Seemingly EJ proponents from the grassroots level had to learn to live with emissions trading and take what they could get. In order to win better EJ procedural polices, CBE had to lose its battle to end emissions trading. CBE had to accept reductions in toxic emissions that emissions trading and more equitable distribution of emissions and call a draw on the fundamental EJ goal of eliminating air pollution for everyone. The pro-emissions trading position is a utilitarian one. How you judge it depends on how you judge utilitarian ethics, writ large.

NOTES

1. Hereafter also referred to as the District.

2. The South Coast Air Basin consists of Orange County and the urban portions of Los Angeles, Riverside and San Bernardino Counties. Hereafter, the South Coast Air Basin will also be referred to as the Basin.

3. It passed the House of Representatives with a vote of 273 to 109.

4. The Clean Air Act of 1967 passed the House of Representatives with a vote of 362 to 0 and the Senate with a vote of 88 to 0.

5. For in depth analysis of the Air Quality Act of 1967, see Martin and Symington (1968). For a criticism of the Air Quality Act of 1967, see O'Fallon (1968).

6. Cook (1988) argues that the Clean Air Act of 1970 and EPA approach to federal responsibility to protect the health and welfare of Americans served as the impetus for deeper inquiry and eventual incorporation of market-based incentives into environmental legislation.

7. Later, standards were established for lead, and revised for photochemical oxidants to a standard for ozone – the principal component of smog. For some pollutants, two different standards are set: a primary, which protects public health, and a secondary standard to protect public welfare measured by the effects of pollution on vegetation, materials and visibility.

8. The Trading Statement required that emission reductions be real – the credit must result from a reduction in actual emission levels; surplus – the emission reduction must be surplus to any reductions required under a State Implementation Plan under the Clean Air Act; enforceable – the agency issuing the permit and the EPA must be able to enforce the reduction; permanent – the emission reduction must last for the life of the new or modified source; and quantifiable – the emission reduction must be measurable or calculable using generally accepted procedures, and with an official emission baseline in order to qualify as emission reduction credits. Furthermore, all trades of credits need to be approved by the regulating agency.

9. The 1990 Amendments also authorized programs for Acid Deposition Control, authorized a program to control 189 toxic pollutants, including those previously regulated by the National Emission Standards for Hazardous Air Pollutants, established permit program requirements, expanded and modified provisions concerning the attainment of National Ambient Air Quality Standards, expanded and modified enforcement authority, and established a program to phase out the use of chemicals that deplete the ozone layer.

10. This chapter is in part a response to the call by EJ scholars to go beyond documenting the existence of environmental injustice. This chapter is an effort to exam that what EJ proponents are fighting for is a means to abate pollution without disproportionately affecting any one group of people. In order to do this without merely distributing pollution is to identify the underlining causes of racism, classism, etc. and change them.

11. Mobile source credits were included in RECLAIM's inception, but rarely used (Fowlie et al., 2009, nt 23).

12. For in depth legal analysis of CBE's 1997 administrative and civil suits, see Chinn (1999).

REFERENCES

Ackerman, B. A., & Stewart, R. B. (1985). Reforming environmental law. *Stanford Law Review*, *37*, 1333–1346.

Air Pollution Control Act of 1955. (1955). Pub. L. 84-159, Ch. 360, 69 Stat. 322.

Air Quality Act of 1967. (1967a). Hearing before the House Committee on Interstate and Foreign Commerce, 90th Congress, 1st Session, at 204.

Air Quality Act of 1967. (1967b). Hearing before the Subcommittee on Air and Water Pollution of the Senate Committee on Public Works, 90th Congress, 1st Session, pt. 2, at 762.

Air Quality Act of 1967. (1967c). Hearing before the Subcommittee on Air and Water Pollution of the Senate Committee on Public Works, 90th Congress, 1st Session, pt. 3, at 1155-56.

ASARCO Inc. v. EPA (Sierra Club) v. U.S. EPA, No. 76-1030 (United States Court of Appeals, District of Columbia Circuit Jan. 27, 1978). Retrieved from http://ftp.resource.org/courts.gov/c/F2/578/578.F2d.319.76-1037.76-1030.html#fn19

Asch, P., & Seneca, J. J. (1977). Some evidence of the distribution of air quality. *Land Economics*, *54*(3), 278–297.

Associated Press. (1990). Bush Signs Major Revision of Anti-Pollution Law. *New York Times*, November 16. Retrieved from http://www.nytimes.com/1990/11/16/us/bush-signs-major-revision-of-anti-pollution-law.html

Bae, C. H. C. (1997). The equity impact of Los Angeles' air. *Environment and Planning*, *29*(9), 1563–1584.

Bansal, S., Davis, S., Buntine, C., & Piazza, B. (1998). *Holding our breath*. Los Angeles: Communities for a Better Environment.

Bansal, S., & Kuhn, S. (1998). Stopping an unfair trade: Environmental justice, pollution trade, and cumulative impacts in Los Angeles. *Environmental Law News*, *7*, 16–24.

Berry, B. J. L. (Ed.) (1977). *The social burden of environmental pollution: A comparative metropolitan data source*. Cambridge, MA: Ballinger Publishing Co.

Bryant, B., & Mohai, P. (1992). *Race and the incidence of enviromental hazards: A time for discourse*. Colorado: Westview Press.

Bryner, G. (1997). Market incentives in air pollution control. In S. Kamieniecki, G. A. Gonzales & R. O. Vos (Eds.), *Flashpoints in environmental policymaking: Controversies in achieving sustainability* (pp. 85–107). Albany, NY: State University of New York Press.

Bullard, R. D. (1983). Solid waste sites and the black Houston community. *Sociological Inquiry*, *53*(2–3), 273–288.

Bullard, R. D. (Ed.) (1993). *Confronting environmental racism: Voices from the grassroots*. Cambridge, MA: South End Press.

Bullard, R. D. (1994). Environmental justice for all: It's the right thing to do. *Journal of Environmental Law and Litigation*, *9*, 281–284.

Bullard, R. D. (2004). *Environmental Justice in the twenty-first century*. Environmental Justice Resource Center. Retrieved from http://www.ejrc.cau.edu/ejinthe21century.htm. Accessed on May 6, 2012.

Burch, W. R. (1976). The Peregrine falcon and the urban poor: Some sociological interrelations. In P. Richerson & J. McEvoy (Eds.), *Human ecology: An environmental approach*. North Scituate, MA: Duxbury Press.

Burke, L. (1993). *Environmental equity in Los Angeles*. Technical Report No. 93-6. National Center for Geographic Information and Analysis, Santa Barbara, CA.

Cable, S., & Cable, C. (1995). *Environmental problems grassroots solutions: The politcs of grassroots environmental conflict*. New York, NY: St. Martin's Press.

California Air Resources Board. (2011). *Overview of ARB emissions trading program*. California Air Resources Board. October 20, 2011. Retrieved from www.arb.ca.gov/cc/capandtrade/capandtrade.htm

Carle, D. (2006). *Introduction to air in California*. Los Angeles, CA: University of California Press.

Chavis, B. F., & Lee, C. (1987). *Toxic wastes and race in the United States: A national report on the racial and socio-economic characteristics of communities with hazardous waste sites*. New York, NY: Commission for Racial Justice, United Church of Christ.

Chinn, L. N. (1999). Can the market be fair and efficient? An environmental justice critique of emissions trading. *Ecology Law Quarterly*, *26*(1), 81–125.

Civil Rights Act of 1964. (1964). Pub.L. 88-352, 78 Stat. 241, enacted July 2, 1964.

Clean Air Act of 1963. (1963). Pub.L. 88-206, 77 Stat. 392.

Clean Air Act of 1970. (1970). Pub. L. No. 91-604, 84 Stat. 1676.

Clean Air Act of 1990. (1990). Pub.L. 101-549, 104 Stat. 2399.

Clinton, W. J. (1994). Executive order 12898: Federal actions to address environmental justice in minority populations and low income populations. *Federal Register*, *59*(7629).

Communities for a Better Environment. (2011). Press release: Environmental Justice Groups Win, California Air Resources Board Forced to Revisit Alternatives to Unjust Pollution Trading System. March 21, 2011. Retrieved from http://www.cbecal.org/pdf/Petitioners%2520AB%252032%2520press%2520release%2520doc.pdf+&hl=en&gl=us&pid=bl&srcid=ADGEESgiX0t3ikJK0tMej0ReXuvNfQP1_p85OCePXIWJy-d_gKn3eTBugU2d1xhp4eUsaLnrxCLc5dtppFrER-vODwclQa89a6pm6lhxAdZzaUOzuyR-YmADL4nQSwWBPrz75WgX0med&sig=AHIEtbTC8_ijYZeqnN2zMt2eRE7hAS_aMA

Communities for a Better Environment v. Chevron Corporation, Complaint (No. 97-5412) (C.D. Cal. Filed July 23, 1997a).

Communities for a Better Environment v. Chevron Corporation, Civ. No. 98-5173 DT (BQRx) (C.D. Cal. Filed July 23, 1997b).

Communities for a Better Environment v. GATX Capital Corporation, Complaint (No. 97-5410) C.D. Cal. Filed July 23, 1997).

Communities for a Better Environment v. Tosco Corporation, Complaint (No. 97-5411) (C.D. Cal. Filed July 23, 1997).

Communities for a Better Environment v. Ultramar, Complaint (No. 97-5413) (C.D. Cal. Filed July 23, 1997a).

Communities for a Better Environment v. Ultramar, Civ. No. 98-5174 DT (BQRx) (C.D. Cal. Filed July 23, 1997b).

Communities for a Better Environment v. Unocal, Complaint (No. 97-5414) (C.D. Cal Filed July 23, 1997a).

Communities for a Better Environment v. Unocal, Civ. No. 98-5175 DT (BQRx) (C.D. Cal. Filed July 23, 1997b).

Cook, B. J. (1988). *Bureaucratic politics and regulatory reform: The EPA and emissions trading*. Westport, CT: Greenwood Press.

Crocker, T. D. (1966). The structuring of atmospheric pollution control systems. In H. Wolozin (Ed.), *The economics of air pollution*. New York, NY: W.W. Norton & Company, Inc.

Dales, J. H. (1968). *Pollution, property and prices: An essay in policy-making and economics.* Toronto: University of Toronto Press.

Drury, R. T., Belliveau, M. E., Kuhn, J. S., & Bansal, S. (1997). Pollution trading and environmental injustice: Los Angeles' failed experiment in air quality policy. *Duke Journal of Environmental Law and Policy Forum*, *9*(2), 231–290.

Eads, G. C., & Fix, M. (1984). *Relief or reform? Reagan's regulatory dilemma.* Washington, DC: Urban Institute Press.

Environmental Justice Initiatives. (2007). *South coast air quality management district.* Retrieved from http://www.aqmd.gov/ej/ej_original10.htm. Accessed on May 6, 2012.

Environmental News Service. (2011, March 22). *Judge halts California greenhouse gas cap and trade program. Environmental News Service.* Retrieved from http://www.ens-newswire.com/ens/mar2011/2011-03-22-094.html. May 6, 2012.

Fisher, M. (1995). Environmental racism claims brought under title VI of Civil Rights Act. *Environmental Law*, *25*(2), 285–312.

Fowlie, M., Holland, S. P., & Mansur, E. T. (2009). *What do emissions markets deliver and to whom? Evidence from Southern California's NOx trading program?* Working Paper No. 15082. National Bureau of Economic Research. Retrieved from http://www.nber.org/papers/w15082.pdf

Freeman, A. M. (1972). The distribution of environmental quality. In A. V. Kneese & B. T. Bower (Eds.), *Environmental quality analysis.* Baltimore, MA: Johns Hopkins University Press.

Frum, D. (2000). *How we got here: The 1970s.* New York, NY: Basic Books.

Gould, K. A., Schnaiberg, A., & Weinberg, A. S. (1996). *Local environmental struggles: Citizen activism in the treadmill of production* (1st ed.). Cambridge, MA: Cambridge University Press.

GovTrack.us. *Tracking the U.S. Congress.* Retrieved from http://www.govtrack.us. May 6, 2012.

Haagen-Smit, A. J., Darley, E. F., Zaitlin, M., Hull, H., & Noble, W. (1952). Investigation on injury to plants from air pollution in the Los Angeles area. *Plant Physiology*, *27*(1), 18–34.

Hahn, R., & Noll, R. (1982). Designing a market for tradeable permits. In W. Magat (Ed.), *Reform of environmental regulation.* Cambridge, MA: Ballinger Publishing Co.

Holifield, R. (2001). Defining environmental justice and environmental racism. *Urban Geography*, *22*(1), 78–90.

Jenal, J., Ramo, A., & Drury, R. T. (1993a). Comment Letter #1. RECLAIM: Volume III Socioeconomic and Environmental Assessments (p. Appendix I). Diamond Bar, CA: South Coast Air Quality Management District.

Jenal, J., Ramo, A., & Drury, R. (1993b). Comment Letter #17. RECLAIM: Volume III Socioeconomic and Environmental Assessments (p. Appendix III). Diamond Bar, CA: South Coast Air Quality Management District.

Johnson, L. B. (1963, December 17). Remarks Upon Signing the Clean Air Act. The American Presidents Project. Retrieved from http://www.presidency.ucsb.edu/ws/?pid=26421. Accessed on May 6, 2012.

Jones, C. (1975). *Clean air: The policies and politics of pollution control.* Pittsburg: University of Pittsburg Press.

Kamieniecki, S., Shafie, D., & Silvers, J. (1999). Forming partnerships in environmental policy: The business of emissions trading in clean air management. *American Behavioral Scientist*, *43*(1), 107–123.

Kennedy, J. F. (1961, February 23). *Special message to the Congress on natural resources.* The American Presidents Project. Retrieved from http://www.presidency.ucsb.edu/ws/?pid = 8466. Accessed on May 6, 2012.

Lejano, R., & Hirose, R. (2005). Testing the assumptions behind emissions trading in non-market goods: The RECLAIM program in Southern California. *Environmental Science and Policy*, *8*, 367–377.

Levin, M. H. (1982). Getting there: Implementing the "Bubble" policy. In E. Bardach & R. A. Kagan (Eds.), *Social regulation: Strategies for reform.* San Francisco, CA: Institute for Contemporary Studies.

Levine, A. G. (1982). *Love canal: Science, politics and people.* Lexington, MA: Lexington Books.

Liroff, R. A. (1980). *Air pollution offsets: Trading, selling, and banking.* Washington, DC: The Conservation Foundation.

Liroff, R. A. (1986). *Reforming air pollution regulation: The toil and trouble of EPA's bubble.* Washington, DC: The Conservation Foundation.

Martin, R., & Symington, L. (1968). *A guide to the Air Quality Act of 1967.* Washington, DC: National Coal Policy Conference.

Montgomery, W. D. (1972). Markets in licenses and efficient pollution control programs. *Journal of Economic Theory*, *5*, 395–418.

Nixon, R. (1970a, January 1). *Remarks on signing the National Environmental Policy Act of 1969.* The American Presidents Project. Retrieved from http://www.presidency.ucsb.edu/ws/?pid-2446

Nixon, R. (1970b, January 22). *Annual message to the Congress on the State of the Union.* The American Presidents Project. Retrieved from http://www.presidency.ucsb.edu/ws/?pid=2921

O'Fallon, J. (1968). Deficiencies in the Air Quality Act of 1967. Law and contemporary problems. *Air Pollution Control*, *33*(2).

Ospital, J. (2006, October). *Draft 2007 AQMP: Appendix I. South Coast Air Quality Management District.* Retrieved from http://www.aqmd.gov/smog/historical/smog_and_health.htm

Pellow, D. N. (2000). Environmental inequality formation: Toward a theory of environmental injustice. *American Behavioral Scientist*, *43*(4), 581–601.

Russell, D. (1989). Environmental racism: Minority communities and their battle against toxics. *The Amicus Journal*, *11*, 22–32.

Schultz, G. (1971). *Memorandum for the heads of departments and agencies.* Washington, DC: Executive of the President, Office of Management and Budget. Retrieved from http://www.thecre.com/ombpapers/QualityofLife1.htm

Schwartz, J. (1993). The coalition opposes RECLAIM as currently drafted: Defects must be remedied to design an effective program. Testimony presented to the governing board of the South Coast Air Quality Management District, September 10, Diamond Bar, CA.

South Coast Air Quality Management District. (1993). *RECLAIM: The regional clean air incentives market* (Final Volume 1). Diamond Bar, CA: South Coast Air Quality Management District.

South Coast Air Quality Management District. (1994). *RECLAIM program summary: A market incentive air pollution reduction program for NOx and SOx.* Diamond Bar, CA: South Coast Air Quality Management District.

South Coast Air Quality Management District. (1999). Summary minutes of the South Coast Air Quality Management District August 13, 1999. South Coast Air Quality Management District. Retrieved from www.aqmd.gov/hb/1999/9908min.html

South Coast Air Quality Management District. (2002). An evaluation of the South Coast Air Quality Management District's regional clean air incentives market – Lessons in environmental markets and innovation. Diamond Bar, CA: South Coast Air Quality Management District. Retrieved from http://www.epa.gov/region9/air/reclaim/reclaim-report.pdf

South Coast Air Quality Management District. (2005). Guiding Principles of Environmental Justice. Diamond Bar, CA: South Coast Air Quality Management District. Retrieved May 6, 2012, from http://www.aqmd.gov/ej/EJ_guiding_principles.htm

South Coast Air Quality Management District. (2012). South Coast Air Quality Management District, Regulation XVI, Rule 1610. Retrieved from http://www.aqmd.gov/rules/html/tofc16.html. Accessed on May 6, 2012.

South Coast Air Quality Management District & California State Fullerton Foundation. (1993). The distribution of current and future exposure to ozone, fine particulate matter, carbon monoxide, and nitrogen dioxide among demographic groups in the South Coast Air Basin, Final Report 5. April 1993.

Stavins, R. (1989). Harnessing market forces to protect our environment. *Environment*, *31*(1), 4–7, 28–35.

Szasz, A., & Meuser, M. (1997). Environmental inequalities: Literature review and proposals for new directions in research and theory. *Current Sociology*, *45*, 99–120.

Taylor, D. (2000). The rise of the environmental justice paradigm: Injustice framing and the social construction of environmental discourses. *American Behavioral Scientist*, *43*, 500–580.

Tietenberg, T. H. (1985). *Emissions trading: An exercise in reforming pollution policy*. Washington, DC: RFF Press.

Tietenberg, T. H. (2006). *Emissions trading: Principles and practice* (2nd ed.). Washington, DC: RFF Press.

Train, R. E. (1996). The environmental record of the Nixon administration. Presidential studies quarterly. *The Nixon Presidency*, *26*(1), 185–196.

U.S. Census Bureau. (2012). State and county quickfacts: Los Angeles County, CA. Retrieved from http://quickfacts.census.gov. Accessed on August 15, 2012.

U.S. Congress, Senate. (1977). Status of the programs and policies of the Environmental Protection Agency. Hearing before the Subcommittee on Environmental Pollution of the Committee of Public Works. U.S. Senate, 95th Congress. 18 January 1977.

U.S. Environmental Protection Agency. (1982). Emissions trading policy statement; general principles for creation, banking, and use of emission reduction credits. *Federal Register*, *47*, 15076–15086.

U.S. Environmental Protection Agency. (1986). Emissions trading policy statement; general principles for creation, banking and use of emission reduction credits; final policy statement and accompanying technical issues document. *Federal Register*, *51*, 43814–43860.

U.S. Environmental Protection Agency. (1990). *Environmental equity: Reducing risk for all communities*. Washington, DC: U.S. Environmental Protection Agency.

U.S. General Accounting Office. (1983). *Siting of hazardous waste landfills and their correlation with racial and economic status of surrounding communities* (No. GAO/RCED-83-168). Washington, DC: US Government Printing Office.

U.S. Public Health Service. (1962). *Tomorrow is already here*. Washington, DC: U.S. Government Printing Office.

Vietor, R. H. K. (1994). *Contrived competition: Regulation and deregulation in America*. Cambridge, MA: Harvard University Press.

Waldman, T. (1991). L.A. Air Board Starts a fresh wind blowing. *California Journal, April*, 86–87.

Wilson, J. Q. (1980). The politics of regulation. In J. Q. Wilson (Ed.), *The politics of regulation* (pp. 357–394). New York, NY: Basic Books.

Young, H. N. (1996). An analysis of global CO_2 emissions trading program. *Journal of Land Use and Environmental Law*, *14*(1). Retrieved from http://www.law.fsu.edu/Journals/landuse/Vol141/youn.htm

SECTION 2
THE GLOBAL SOUTH

CHAPTER 7

CLIMATE ADAPTATION IN THE FACE OF RESOURCE CONSTRAINTS: LESSONS FROM A COASTAL SOUTH ASIAN MEGA-CITY

Madhu C. Dutta-Koehler

ABSTRACT

Purpose – *This work offers an investigation of the planning and implementation of climate-adaptation and vulnerability-reduction strategies in coastal mega-cities of the Global South, utilizing Kolkata, India, as a case study. This research is designed to identify factors that aid the implementation of climate-centered action in resource-constrained environments of developing countries and provide a set of policy guidelines reflecting best practices.*

Methodology/approach – *This work draws principally upon analysis of semistructured field interviews conducted in Kolkata, India, during December 2010 and January 2011. The findings are informed by additional data sources as well, including field observations, informal dialogues and meetings, and a review of secondary literature.*

Urban Areas and Global Climate Change
Research in Urban Sociology, Volume 12, 171–195

ISSN: 1047-0042/doi:10.1108/S1047-0042(2012)0000012010

Findings – *This work identifies several key success factors, including organizational restructuring, resource redistribution, technological innovation, use of external consultants, coupling of climate and development projects, and integration of climate approaches into infrastructure projects.*

Research limitations – *This research draws upon Kolkata as a case study; thus the work's broader applicability and utility depend on the similarities between the situation in Kolkata and that of other urban areas. As a local study, this work may also offer fewer insights for regional and national policy.*

Originality and value – *This work fills a timely, unmet need for a greater understanding of climate-adaptation action in the context of cities of the developing world. The extensive use of personal interviews provides unique insights into the minds of planning officials and professionals and draws upon their practical experience to draw lessons for a wide range of similar environments.*

Keywords: Kolkata; South Asia; coastal mega-city; urban planning; climate adaptation

BACKGROUND

Much of the world's rapid urbanization is occurring in South Asia. By the end of the next decade, five of the world's eleven mega-cities – those with a population of at least 10 million – will be in this region. Delhi, Dhaka, and Mumbai will each contain over 20 million people, while Kolkata is projected to grow in population to over 15 million residents and become the world's eighth-largest city. Currently, one-third of the world's population now resides in the urban areas of low- and middle-income countries; these two billion urban dwellers face escalating risks from the environmental, economic, and societal impacts of climate change (Satterthwaite, Huq, Pelling, Reid, & Lankao, 2007).

Numerous factors, including extreme poverty, high population densities, and unstable, worsening climatic conditions are forecasted to exacerbate the environmental dangers confronting mega-cities such as Kolkata and the rest of South Asia.[1] Climate scientists predict markedly higher water levels, increased storm surges, more cyclones, and more frequent inundation events for Kolkata over the next 50 years (see, e.g., Jain, Rao, Jitendra, & Dube, 2010). Three further factors combine to exacerbate the situation in Asia's mega-cities. The first is the rapid rate of urbanization, which has spawned rampant and increasingly unsustainable development. The second is the inherently weak institutional capacities of these cities and the concomitant inability of planning agencies to operationalize and implement effective adaptation plans. The third and perhaps most important aspect is the acute resource constraints that these cities face: the bulk of national and city expenditures is channeled toward the provision of basic necessities such as food, electricity, and water. Climate-adaptation funding is thus not a priority for most cities in this region, largely insofar as pressing immediate needs detract resources and attention away from longer-term threats such as those posed by global warming.

The grave implications from acute climate risks, however, make it imperative to rapidly identify and implement effective climate-change adaptation and vulnerability-reduction measures in South Asia's mega-cities. Moreover, given the ongoing and increasingly intensifying risks of climate change, these cities cannot afford long gestation periods to absorb a range of planning and policy experimentation in order to "get it right." It has never been more necessary to gain a critical understanding of effective climate adaptation strategies that are suitable in this context; the effects of the planning approaches adopted in these Asian mega-cities will have far-reaching and permanent implications for the futures of not only the cities themselves but also for the broader Asian region, which encompasses over 60% of the world's population.

THE STATE OF ADAPTATION PLANNING

Though many of these South Asian cities have formalized climate-change action plans, crucial doubts remain both in the scholarly and scientific community as to whether these plans can tackle complex, interdependent climate-related risks and be effectively implemented.[2] More important, very few of these cities have climate-*adaptation* plans; the primary focus in climate-change planning remains on mitigation or, simply stated, on the reduction in greenhouse gases, particularly as a by-product of economically driven efficiency measures. Climate scholars argue, however, that the interrelationships between mitigation and adaptation have been largely neglected in planning and that an "integrated approach" provides the best platform for sustainable development while augmenting readiness for climate change impacts (Davoudi, Crawford, & Mehmood, 2009; Tanner, Mitchell, Polack, & Guenther, 2007). This current situation further erodes confidence in the ability of the existing strategies and institutions to engender effectual results, especially in the resource-constrained mega-cities of South Asia where the primary focus of local planning authorities remains on economic development. India's limited efforts toward climate-change planning have taken place principally at the national level. The most comprehensive result to date is the *National Action Plan on Climate Change*, or NAPCC (Prime Minister's Council on Climate Change, 2009). The NAPCC exemplifies the limitations, however, of governmental approaches in India. The Plan outlines laudable, crucial broad policy directions, but offers few specifics concerning projects, legislation, or programs to be implemented, and no details whatsoever regarding resource allocation.[3]

In light of these circumstances, scholars and practitioners alike have increasingly come to realize that sustainable urban development needs to be a central component of national agendas, insofar as climate adaptation and sustainable development are inherently linked and play a determinant role in responding to risk (Schipper, Cigaran, & Hedger, 2008). In fact, the new development paradigm for climate adaptation corresponds to the view that "development activities could be concerned with sustainable development and reducing vulnerability ...[and] represent an opportunity to revisit some long-standing problems of environment and development" (Schipper, 2007).

BRIDGING THEORY AND PRACTICE

In the current climate-adaptation discourse, the emergent theoretical dialectic on the normative approaches to climate adaptation, as adaptation relates to development, consists of two primary paradigms: the "adaptation approach" and the "vulnerability reduction approach." The "vulnerability reduction approach" posits, as the name suggests, that increasing the coping capacities of vulnerable populations without causing cultural or economic dislocation can best augment their resilience to environmental and climatic change (Satterthwaite, 2009). This in turn creates enabling conditions for effective adaptation, thereby contributing to the larger agenda of sustainable development (Schipper et al., 2008; Srivastava & Heller, 2003). The "adaptation approach" is simpler to implement from a policy and planning perspective and goes beyond just reducing vulnerability or mainstreaming adaptation: "adequate development will automatically reduce the levels of relative or total risk" (Lavell, 2004). The "adaptation approach" corresponds to the view of climate adaptation as the new development paradigm.

Although the understanding of these theoretical approaches may provide possible insights as to how climate adaptation efforts could be operationalized in cities of the Global South, they at most offer only partial explanations and do not necessarily lend themselves to practical applications. A synthesis, focused on how all municipal-level planning can take account of climate risks in the decision-making process, seems best-suited for the resource-constrained environments of these cities. Research suggests that the specific adaptation strategies will depend on factors unique to the cities and contingent upon various factors ranging from local institutional barriers to regional climatic conditions.

While it is not explored in depth in this chapter, another potentially useful area of the literature explores the linkages between climate adaptation and vulnerability, and Disaster Risk Reduction (DRR) and Management (DRM) planning efforts (Adger, Huq, Brown, Conway, & Hulme, 2003; Hewitt, 1983, 1997; Ribot, Magalhaes, & Panagides, 1996; Sen, 1981). Additionally, recent scholarship has begun to explore the commonalities and synergies between adaptation to climate change and disaster studies (Alam & Rabbani, 2007; Birkmann et al., 2009; Kousky, Rostapshova, Toman, & Zeckhauser, 2009). These scholars have examined the significant reforms that ensue from natural disasters, distinguishing such systemic shifts from immediate impacts to long-term responses, and have determined that major disasters can alter dominant behaviors and attitudes toward long-term climate-related changes. Given that disaster-related impacts bring underlying issues to the fore, the study of changes that not only are catalyzed by disasters but also influence decision-making trajectories beyond the immediate event may provide valuable insights that reduce the complexity and uncertainty of climate-change adaptation in the long term. This line of research should prove especially useful insofar as climate change will increasingly manifest itself in catastrophic weather events, including more frequent periods of intense rainfall, resulting in floods and coastal erosion, and more numerous cyclones. The imprecise distinction between climate-change risk and natural disaster risk will therefore further erode, enabling mutually reinforcing adaptation and mitigation approaches. Given that many cities of the Global South lack specific adaptation plans at the city level, and rely largely on DRMs to combat climate-related threats, the understanding of potential overlap between these areas may provide substantive guidance in shaping effective climate-change strategies, especially in resource-constrained environments.

Building on the linkages between climate-change impacts and climate disasters, the research on response options for mega-catastrophes engendered or exacerbated by climate change also offers useful recommendations for decision making under uncertainty, especially in cases of cascading consequences where resource limitations inhibit effective response. Kousky et al. (2009), note that, while it is impossible to identify all cascading consequences, focusing on those that we can reasonably predict offers significant room for ameliorative action. In the same vein, uncertainty over benefits and costs does not eliminate the advantages that accrue from utilizing reasonable cost–benefit analysis (CBA) methods for evaluating and comparing options (Posner, 2005). Posner suggests a scenario-analysis approach in which the desirable level of effort on risk reduction can be estimated within a CBA decision matrix. Additionally, where the cost of adaptation options varies widely, gradual measures, such as stricter land-use regulations, would be less resource-intensive, thereby enabling a reduction in the overall cost of "circuit-breaker" responses[4] needed to arrest impacts of climate-change mega-catastrophes (Kousky et al., 2009). This framework thus offers significant advantages in identifying and proposing adaptation measures with the greatest return on investment (ROI). In South Asia, where resources are often channeled to deal with climate disasters such as floods and tsunamis and long-term climate planning decisions often contingent upon available resources, less resource-intensive adaptation options that also have the potential to moderate the cost of such circuit-breaker responses to climate catastrophes are worth investigating.

In terms of suggested "adaptation theories," perhaps the most useful and appropriate option when regulatory entities lack resources and/or clarity in climate adaptation planning is the "no regrets" approach (Callaway, 2004), which builds upon the economic development perspective on climate change (Hitz & Smith, 2004). Callaway suggests that "there are potentially many actions, particularly in developing countries, that can be taken today for reasons that are more directly related to a broad variety of other developmental goals (including reduced vulnerability to existing climate variability)… [I]t may be possible to redesign some no regrets actions with a little more climate change protection at relatively low additional costs" (2004). In many cities of South Asia, where municipal decision makers are hampered by institutional inertia, limited resources, and insufficient capacity, this approach can be adapted most easily to modify the existing planning processes in order to maximize climate-risk reduction.

THE CITY OF KOLKATA

Kolkata, the capital of the State of West Bengal, is the economic, political, and cultural center of Eastern India. Kolkata proper, established in 1690 by the British East India Company, lies on the east bank of the Hooghly River, part of the Ganges river system. The city's Metropolitan Area, or KMA, currently encompasses just over 1850 km^2 of low-lying land, with elevations from about 1.5 to 11 m above sea level. Kolkata receives an average of over 180 cm of rainfall annually, concentrated in the summer monsoons, and thus is highly prone to seasonal flooding. Average daily temperatures range from about 19°C in winter to 30°C in summer, and daily temperatures can reach the low 40s (Kolkata Municipal Corporation, 2012). Temperature spikes have been increasing over the last 20 years, in part due to global climate change. Global warming is also resulting in higher sea levels, another environmental and disaster risk factor for the city, which lies about 200 km from the Indian Ocean, just below the Tropic of Cancer.

The broader KMA has an estimated population of 14.7 million, with a third of that, or 4.75 million, in the KMC, or Kolkata Municipal Corporation (World Bank, 2010). Nearly three-quarters of the city's population are Hindus. High rates of migration from the West Bengal countryside, poorer surrounding states, and neighboring Bangladesh have engendered greater crowding and slum conditions, with an estimated 3–4.3 million residents of the KMA living in substandard housing (Kundu, 2003; TCPOMUD, 2009), significantly exacerbating health and weather-event risks. The KMC has a population density of about 23,100 people per square kilometer, more than twice that of New York City, and 10 times that of Los Angeles (World Bank, 2010); density has risen significantly since economic liberalization in 1991. Housing and density issues combine to markedly aggravate climate-event and natural disaster risks.

Kolkata, like India collectively, has certainly benefited economically from liberalization, although less so than have many other regions of India. The mean per-capita GDP for India as a whole is currently estimated at $3500 PPP (CIA Factbook, 2012), slightly higher than that for West Bengal, while in the core area of KMA, the figure is about $4900, about half that of Delhi. GDP growth figures for Kolkata and West Bengal have consistently lagged behind those for India, in large measure because of an unfriendly business climate. Kolkata, which has experienced communist-led state governments and radical labor union politics for most of the past 50 years, ranks last among Indian cities in the ease of doing business (World Bank and International Finance Corporation, 2009), representing another challenge for economic development. West Bengal's weak governmental finance structure further impedes progress on climate-change adaptation action plans. Slower economic growth rates, high population density, and deleterious environmental factors thus collectively raise the degree of climate-related risk in greater Kolkata and West Bengal.

Vulnerabilities

Kolkata is located in the world's largest mega-delta,[5] which the International Panel for Climate Change (IPCC) (Parry, Canziani, Palutikof, van der Linden, & Hanson, 2007) and others have identified as the region at greatest risk for the deleterious impacts of climate change. The OECD (2007) currently ranks Kolkata among the top 10 cities globally for flood risk; by 2070, Kolkata is projected to top the list. Four principal factors – its topographic, developmental, climatic, and demographic situation – render Kolkata highly vulnerable to such adverse climate events. Kolkata is largely settled in a flat terrain with very little natural drainage relief, which causes riverine flooding and overall poor drainage for the city. It does, however, slope in parts toward the eastern part of the city, which consists primarily of marshy land. Its location in the lower coastal region also makes it more directly susceptible to sea-level rise and storm surges. Kolkata also lies in an active seismic zone, and geological experts predict that rising sea levels will result in more frequent earthquake activity as well.

Like many other South Asian mega-cities, Kolkata's population is increasing rather rapidly, leading to unplanned and unregulated urbanization that is unable to keep pace with the city's existing infrastructure, despite ongoing efforts at infrastructure expansion. This is particularly evident in the city's inadequate and at times failing water, drainage, and sewer infrastructure. Given the increased and intense precipitation patterns of the past decade, Kolkata's overdevelopment, coupled with the cities inadequate sewer and storm water drainage,[6] has caused excessive runoff; this runoff coupled with increased and intense periods of precipitation has led to severe and frequent flash flooding, primarily from "drainage congestion." The siltation and poor maintenance of available channels, construction-related disruptions in storm-water flow, and development of the surrounding marshland, which had served traditionally as natural drainage areas have all aggravated the already poor drainage patterns in the city. The city's rapid urbanization is also depleting the city's water supply because of excessive groundwater extraction. Moreover, the physical expansion of the city into adjacent alluvial plains has exacerbated the existing problems of groundwater contamination[7] and soil erosion. Forecasts also indicate that Kolkata will experience significant saltwater intrusion over the next half-century, further degrading groundwater supplies, destroying much of the area's mangrove forest buffer, and halting the production of protective alluvial deposits at the mouth of the Hooghly River. The region around the city will experience similar, if somewhat less drastic drops in agricultural production and aquaculture, but is more prone to winter drought, as precipitation from December through February has declined significantly due to increasing average winter temperatures. The frequency and severity of flooding in Kolkata has increased markedly over the past two decades; significant episodes are now a frequent event. In 2000, 1500 lives were lost in the record floods, but the floods of 2007, 2008, and 2009 also resulted in hundreds of deaths each and in the dislocation of hundreds of thousands of mostly poorer residents.

The city is already experiencing heightened risks from climatic factors in the form of rapid, above-average rates of sea-level rise, periods of intense rainfall, cyclonic activity, and storm surges, the latter aggravated by reduced sedimentation caused by upstream dams, coastal erosion, considerable subsidence,[8] and depletion of the mangrove forests that protect the coast from cyclonic damage. The demographic makeup of the city further adds to its already vulnerable position. Though the natural population growth rate is at 1.8–2.6%, the overall population of the KMA is estimated to reach 20 million by 2021, with the increase resulting primarily from in-migration, concentrated in the city's slum areas. These figures provide a grim picture of the overdevelopment in the most environmentally vulnerable parts of the city, which aggravate the many concerns outlined earlier.

Institutional Arrangements

Kolkata, like many South Asian cities in the region that had developed under British rule (e.g., Dhaka, the capital of Bangladesh), possesses roughly similar demarcations of local and national responsibility for addressing climate-change and DRR issues. In India, the Ministry of Environment and Forests (MOEF) is the central government department accorded overall responsibility for the development, implementation, and oversight of policy measures regarding climate change. MOEF is expected to coordinate with the Ministry of Earth Sciences, which is principally responsible for monitoring and predicting climatic events, the Ministry of Water Resources, which plans for interstate flood control, and, to a lesser extent, the Ministry of New and Renewable Energy. India is largely centralized in governance, especially in terms of policy creation, so the states usually play a very limited role in formulating strategy and instead focus on day-to-day administration. Responsibility for implementation, however, devolves upon state and local authorities; the glaring disconnect between policies and actual practice can be remedied at times only through judicial redress.

At the state level, though Kolkata is the capital of West Bengal, it does not play an important role in climate adaptation. However, Kolkata is one of few Indian cities to exercise strong local governance through the KMC, which serves as the city's municipal "planning agency." However, financial, technical, and jurisdictional constraints, coupled with limited human capital resources, mean that KMC largely concerns itself with day-to-day management and service delivery rather than with design or implementation of climate-adaptation measures.

The authority for issues of environment and climate change as well as overall master planning falls principally under the Kolkata Municipal Development Authority (KMDA) and the overlapping Kolkata Municipal Commission (KMC). Though the ambitions of the KMDA and KMC are somewhat greater than those of the MOEF, the results are not much more significant in practice (Taylor, 2008). The KMC oversees certain operation and maintenance activities related to adaptation efforts, such as sewage treatment and disposal as well as transportation infrastructure. The KMDA in turn (with some jurisdictional overlaps with other local agencies) leads city-planning initiatives, manages disaster planning, identifies new areas for development, including townships, establishes the city's physical infrastructure, and provides essential water and sewer services.

Kolkata faces many of the same challenges as other South Asia megacities in designing and implementing effective adaptation strategies. The rapid rate of urbanization, intensified by the influx of climate refugees and the city's economic upturn, and the need to provide basic services while preparing for exposure to extreme climatic threats, such as cyclones and flooding, remain hard challenges. Addressing these challenges is further complicated by the complex institutional arrangements, which more often than not lack clarity of vision and unclear division of responsibilities.

Adaptation Efforts

Though Kolkata enjoys considerably greater resources than do many of the other cities of the Global South,[9] it has yet to adopt a formal local climate-adaptation plan despite the many climate-related risks and extreme events that the city faces. It does however, have a rudimentary Disaster Management Plan (DMP), a framework that leans more toward centralized relief efforts than toward comprehensive disaster management. The DMP has moreover never actually been applied and would appear at face value to be ineffective. The current City Development Plan for Kolkata, commissioned by the Government of India's JNNURM[10] Program, which in theory guides the overall strategic development of the city, does not take any long-term effects of climate change into consideration, nor does it outline any explicit adaptation efforts. In keeping with the Central government's vision of creating economically productive, efficient, equitable cities, the plan's primary focus is on furthering economic development and social justice. JNNURM's guidelines cover issues of land tenure, provision of basic services, economic infrastructure for urban development, and reforms in city financing and governance. Issues of climate-change adaptation and vulnerability reduction remain outside the Program's principal focus, though it accounts for access to clean water supplies and sewage disposal. Of the 111 projects funded to date nationally by JNNURM, 68 (61%) have been in these two categories with another 35 in roadway construction; only two mass-transit projects have had any ancillary positive climate-change impact (JNNURM, 2012).

The city is currently in the process of reevaluating its development plans as well as creating an explicit climate adaptation plan; a committee has already been convened and has offered preliminary recommendations. These recommendations, however, have not yet been made public. The plans are said to include investment in both soft and hard infrastructure in various sectors, including but not limited to water, drainage and sewage infrastructure, transportation, land use, disaster management and pollution control.[11] The government of the state of West Bengal has already invested in and is in the process of implementing a number of these measures, which address desiltification of existing drainage channels and expansion and repair of water and sewage infrastructure. Four of these projects have been initiated by the JNNURM, while other key projects discussed later in this chapter primarily fall under the purview of the Kolkata Environment Improvement Project (KEIP), which is currently funded by the Asian Development Bank (ADB).

CASE SELECTION AND METHODOLOGY

The climate-related risks being experienced by cities in Asia are already evident in the sea-level rise, altered rainfall patterns, substantial increases in average temperatures, and mounting frequency and intensity of environmental catastrophes in this region. That most of the Indian subcontinent's metropolises[12] are located in highly vulnerable coastal or delta regions further heightens the risks for not only the physical and human environment but also for regional and global political and economic stability, food production, and natural resources. The city of Kolkata in India clearly exemplifies these types of susceptibility. The analysis of this coastal mega-city offers important control variables regarding the nature of climatic threats, resource capacities, institutional structures, and political environments, while some of its innovative and unique adaptation approaches and institutional reconfigurations account for important differences. These factors make Kolkata a valuable case for testing the effectiveness of planning and policy decisions for a larger cross-section of cities in similar developing regions that are likewise severely threatened by the effects of climate change.

The findings presented in this chapter are based on an in-depth and nuanced understanding of the nature of adaptation strategies in Kolkata undertaken by the city-level public agencies. The investigation draws upon national- and state-level adaptation plans and policies, institutional arrangements, and roles of organizations as well as private entities, since municipal adaptation efforts are necessarily nested within and affected by regional and national initiatives and strategies.

The following section summarizes the preliminary findings germane to the overall research question, concerning the factors that facilitate or hinder the formulation and implementation of effective planning strategies for the management of climate risks and impacts. This work draws principally upon analysis of semistructured field interviews conducted in Kolkata, India during December 2010 and January 2011. The findings are informed by additional data sources as well, including field observations, informal dialogues and meetings, and a review of secondary literature. More specifically the qualitative analysis draws liberally upon the adaptation approaches inferred from the innovative management of the East Kolkata Wetland Systems (EKWS) and reconfiguration of the institutional makeup of the KEIP. The EKWS and KEIP organizational frameworks may offer scalable, readily replicable models for other cities not only in South Asia but also across the Global South.

ADAPTATION IN THE FACE OF RESOURCE CONSTRAINTS

Overcoming Institutional Barriers

The Climate Change Action Plan for the state of West Bengal, India, is currently being developed, while a city-level climate action plan for Kolkata is yet to be formulated, though a high-level governmental committee is working on recommendations. However, despite the lack of formal climate action plans both at the state and local level, several studies have been undertaken in collaboration with the State of West Bengal and the Indian MOEF to understand the vulnerabilities of the KMA to climate change. In addition to these studies, the city of Kolkata has also undertaken a number of major projects that directly or indirectly address some of the primary climate-related risks that the city faces. Some of these risks include flooding in the Hooghly and coastal storm surges, as well as increased and intensified precipitation due to climate change. In particular, local precipitation-related flooding, which leads to frequent water-logging in the city, is increasing in frequency and intensity. This flooding results principally, as noted earlier, from a lack of natural drainage caused by "obstructive" overdevelopment, unregulated/unplanned settlement, inadequate drainage system capacity, trunk-sewer siltation, and "combined" storm and sewerage systems.

Some of the most effective climate- and environment-related planning efforts in the city of Kolkata have been initiated by KEIP, which was established as an independent project-based organization in 2001. KEIP is a multiagency endeavor that is involved in the planning, coordination, and implementation of a range of environment-related projects. It also seeks to improve the "governance and service delivery mechanism of the Kolkata Municipal Corporation" (KMC), which is one of the two principal local planning organizations, along with the KMDA.

KEIP is jointly funded and monitored by the Asian Development Bank, the Government of West Bengal, and the Kolkata Metropolitan Corporation. Some of the major projects that are currently underway or have been successfully completed in Kolkata address the following concerns: sewerage and drainage, solid waste management, and canal rehabilitation. Specific measures include desiltation of existing storm and sewer lines, construction and rehabilitation of sewage pumping stations and treatment plants, and planning and construction of new sewage and storm water networks. KEIP also collaborates with the East Kolkata Wetlands Management Authority (EKWMA) and the Centre for Environmental Management and

Participatory Development (CEMPD), an NGO affiliated with KEIP, in the management and preservation efforts for the EKWS. The collaboration entails allocation of human and fiscal resources as well as the safeguarding of various stakeholder interests. Though most of these projects are intended as "environmental improvements," the ones examined for the purposes of this research directly or indirectly address issues related to climate change. Analysis of these projects based on interviews, observation, and secondary data sources reveals that most of the "planning initiatives" undertaken by KEIP meet the majority of the evaluative criteria identified as "effective planning strategies" for climate adaptation for the purposes of this study. Thus, the ancillary benefits of such planning measures extend to a wide variety of environmental and climate concerns.

The successful completion and sustained positive impacts of some of these projects (e.g., the ecosystem management of the East Kolkata wetlands, desiltation of a majority of the canals, and construction, upgrades and rehabilitation of solid waste pumping stations, treatment plants, and secondary and trunk sewer lines), stem primarily from the overall *institutional restructuring and allocation of dedicated financial and human resources* for KEIP projects. KEIP was structured, and functions as, a stand-alone "planning" body that encompasses a number of complementary units, with authority over design and supervision, project accounts, social development, public relations, and project coordination and management, respectively. Each of these units has clear-cut responsibilities and is staffed by KMC officials and engineers, experienced external engineering, technical and environmental consultants, and managerial experts from relevant private-sector enterprises, as well as by representatives of NGOs. These NGOs are engaged by KEIP to ensure the participation of those private citizens directly affected by the project(s) and to serve as mediators as necessary. KEIP is headed by an administrative officer who is appointed on a "transferable" basis from the Indian Administrative Services (IAS) pool of central government officers. This reliance on outside senior management ensures that the administrative head functions as a neutral party who is not personally vested in the politics or projects of the state or city; the case of KEIP demonstrates that this safeguard succeeds in minimizing the politicization of KEIP's plans and policies, particularly useful in societies such as India that struggle with endemic corruption and lack of transparency.

Another factor that has contributed to the success of planning efforts in this area has been the presence of well-functioning and clearly delineated project *management* and *coordination* units within KEIP. These units ensure smooth implementation of projects, accountability of project delivery, and liaison and interagency coordination with multiple state and city agencies. Some of the agencies with which these units must coordinate policy and action include: the West Bengal Pollution Control Board (WBPCB); the Departments of Environment, Forests, Fisheries, Irrigation and Waterways, Transportation, Land and Land Reforms; local and state police departments; the BSNL (Bharat Sanchar Nigham Ltd., the telecommunications giant); CESC (the Calcutta Electric Supply Corp.); WBSEB (the West Bengal State Electricity Board); the District Administration; and the KMC and KMDA.

An additional element that has proven instrumental in facilitating effective planning measures is KEIP's *institutional innovation* in the form of the computerized "project accounting system," which has been tailored to fit the requirements of the types of projects undertaken at KEIP and to meet the organization's level of technical expertise. This fully digital interface functions as an integrated Business Process System, which has built-in controls to ensure accountability and timely and accurate disbursal of payments, in addition to providing "information visibility and workflow-driven approval hierarchy." This technological innovation at KEIP has succeeded in shortening project-delivery time, facilitating communication among different public agencies and private contractors, and minimizing the bureaucratic delays that had long characterized the functioning of India's public-sector entities.

A majority of the interviewees cited the process of "collaborative consultation" with environmental and technical experts who possess extensive knowledge of local needs and conditions as an important element in furthering the success of the KEIP projects. The fact that "experts" were brought in at the beginning of most of the projects and were also "employed" on a full-time basis by KEIP through the duration of each particular project helped in a number of ways. First, the consultants and regular staff had an opportunity to develop the plans and implement them collaboratively, right from the project's inception through its finish, while being able to make necessary adjustments along the way. Second, since some of the experts were members of the KEIP project teams throughout, they also felt more deeply committed to the projects' success, finding new and innovative ways to deal with problems "on the ground" throughout the projects' course. Third, this model of collaborative consultation allowed for continuous dialogue and expert guidance, thereby reducing the opportunities for hierarchical conflicts. Such conflicts are much more likely to occur with one-off expert consultancies, since the regular employees often perceive experts' involvement as an imposition and thus ignore the experts' input during the actual planning stages or reject the outsiders' recommendations as unrealistic. Fourth, having expert consultants employed as part of the project teams helped to fill the inherent gaps in "technical expertise" and thus increase KEIP's overall institutional capacity.

KEIP serves as an exemplary model of institutional innovation. However, it is important to note that the conception and implementation of these institutional innovations did not occur solely because of internal forces. The formation of a separate project-based organization, the hiring of experts to augment institutional capacity, and KEIP's extant implementation and accountability guidelines and processes were endogenously mandated by the contractual conditions imposed by the Asian Development Bank (ADB) as part of its loan agreement with KMC. Though the ADB's conditions certainly served as a catalyst for some of the institutional changes, many of the other innovations stemmed from internal considerations and careful forethought. KEIP's internally generated reforms, which also became major contributing factors in crafting effective strategies, include:

- Implementation of a substantial organizational restructuring while ensuring appropriate and balanced representation of relevant public agencies.
- Development of a realistic planning process that takes into account issues of design, coordination, financing and implementation at a project's onset.
- Adoption of technological innovations, such as the computerized project accounting system, that enhanced transparency and reduced bureaucratic delays.
- Employment of collaborative consultation to decrease technical deficits while increasing overall institutional capacity.

Conversely, analysis of projects of a nature similar to those tackled by KEIP (such as laying out new sewerage and storm drains and water-supply lines), which were independently undertaken by KMC, demonstrates that the KMC-led projects have been far less successful. The principal shortcomings that hindered the success of these measures – bureaucratic delays, financial uncertainties, lack of technical capacity and, perhaps most important, the ill-defined delegation of responsibilities throughout the process among too many public agencies – have dramatically increased costs and often resulted in substandard construction. My research to date indicates that the two pivotal challenges that the city of Kolkata faces in its current institutional forms are, first, lack of interagency coordination and, second, a huge deficit in institutional capacity, arising not from insufficient numbers of personnel, but rather from a pronounced lack of technical expertise and related project-management experience.

In addition, confidential interviews with planners and officials of KEIP and KMC in Kolkata revealed that ADB and other similar agencies are willing to fund similar infrastructural and environmental projects in the city. However, these agreements are in abeyance because outside agencies have determined that the KMC does not have the adequate technical expertise or institutional capacity to plan for, implement, and manage these projects. Moreover, such donor agencies have stringent guidelines for accountability, quality control, and implementation, guidelines that local agencies such as KMC feel unprepared to handle. This kind of situation was also evident in KMDA, where several infrastructure projects initiated and funded through the central government's JNNURM and managed and implemented by the KMDA are running into costly delays and failing to achieve their intended goals. The reasons cited for these problems range from lack of institutional capacity, expertise, and resources to the inflexible and sometimes unrealistic accountability measures that are conditions of the central government funds disbursed via the JNNURM initiative.

Coupling of Climate-Related Issues and Planning Initiatives

Though Kolkata does not have an explicit set of climate-action plans, my research reveals that climate-related planning initiatives have been given greater priority when they are coupled with "regular" development initiatives. In these cases, addressing issues of economic development, infrastructure or social justice generally represents the principal policy justification for the project, while reducing climate-related risks is often viewed as a positive externality of the project. Such coupling of specific socioeconomic objectives with climate-adaptation or vulnerability-reduction measures would seem to offer greater promise in advancing environmental agendas in developing countries. Moreover, planning initiatives that simultaneously satisfy multiple goals and stakeholder groups correspond roughly to the framework of the "no regrets" approach to policymaking, insofar as any uncertainty regarding the actual effects of future regional and global climate changes would not greatly impact the success or failure of the initiative (Callaway, Kašćelan, & Markovic, 2010).

One such pertinent example of "coupling" of planning objectives is evident in the analysis of the case of the EKWS. Apart from the EKWS, Kolkata has a substantial share of urban wetlands, comprised of intertidal and salt marshes, salt meadows, lakes, and seasonal ponds as well as settlement ponds and oxidation basins where similar management approaches can be implemented (Ghosh 1999).

The EKWMA, which was formed in 2006 through the Department of Environment, Government of West Bengal, India, is responsible for conservation and maintenance of the wetlands. It is comprised of representatives of different departments from the State government, other public agencies, and three NGOs. This body (EKWMA) holds the primary responsibility of present and future conservation and maintenance of the wetlands in the KMA, including the East Kolkata Wetlands, and has prepared a comprehensive and integrated Management Plan in keeping with basic guidelines of the Ramsar Protocol,[13] which is currently being implemented.

The confidential interviews further reveal that the "Ramsar" designation was what prompted a renewed and active interest in the wetlands. This interest also engendered further research on the benefits of the wetlands. Though wetlands protection was not part of an explicit "climate adaptation" initiative, it certainly became part of the "environmental" planning agenda for the city. Moreover, once it was understood that the wetlands also helped with the drainage congestion and, for some areas, obviated the need to build additional sewage treatment plants and pumping stations, the city planning agencies, particularly the Kolkata Environment Improvement Project, actively engaged in developing and refurbishing the existing wetlands infrastructure. This external impetus for wetlands protection helped to augment the uses of the wetlands and preserve them.

The East Kolkata Wetlands consists of intertidal and salt marshes, salt meadows, lakes, and seasonal ponds, as well as settlement ponds and oxidation basins. These urban wetlands encompass an area of 12500 hectares and are one of the largest urban wetland systems in Asia. Approximately 46% of this area is comprised of water bodies, 39% has been converted into agricultural land, and the rest consists of urban and semirural settlements as well as landfills. Because of its unique ecological characteristics and the range of critical benefits that it provides for the city's population, the EKWS was deemed as "a wetland of international importance," and designated a "Ramsar" site in 2002, thereby protecting it from future urban encroachment or damage.

Though the city of Kolkata has had a late start in its conservation efforts, it has made important strides in the preservation of its urban wetlands, motivated by three primary advantages that the wetlands offer the city and its residents. First, the wetlands in the city function as natural drainage areas and flood-flow zones, thereby reducing the impact of flooding due to intense precipitation or drainage congestion. When faced with the prospect of planning for additional and costly infrastructure to reduce flooding in the cities, preservation of wetlands and low-lying flood flow plains thus presents itself as an easier and more cost-effective option. Many of the man-made canals have been desilted and refurbished to allow for easy flow of storm and rainwater, again an effort principally targeted at reducing drainage congestion. In addition, the canals and some of the sewer trunks have also been extended and rerouted to form larger networks; in some cases the sewer trunks have been directly connected to the larger ponds and lakes that are part of the existing natural wetlands. This network of connections between existing sewer trunks and canals with the existing wetland system represents an exemplary and innovative planning strategy. Second, apart from reducing the load on existing storm drainage systems in the city, the wetlands also serve as a "natural purifier" for approximately 250 million gallons of sewage per day, obviating the need to build additional sewage treatment plants and pumping stations to service this area. The wetlands act as sewage farms and further purify the water. The third advantage is principally economic in nature; the wetlands support intensive fishing, aquaculture, and farming activity. Water purified through the wetlands is channeled into larger sewage-fed fish farms and ponds, which cover a total area of about 3,500 hectares; some of the individual fish ponds are as large as 70 hectares. In addition, the treated water from the wetlands (particularly those in water bodies that are relatively purified) is used for agriculture in the surrounding areas. These wetlands produce approximately 13,000 tons of fish and 150 tons of vegetables per day, thereby not only serving as an important source of livelihood for many residents of the city, but also helping to reduce the cost of infrastructure and to solve some of the existing drainage and sewer treatment problems. In addition, since most of food is locally sourced in Kolkata, it keeps the cost of vegetables and fish low for the city's residents.

As we see in the case of Kolkata, existing planning initiatives, such as the building of sewer and storm drains, treatment plants, and pumping stations, have occasionally been "coupled with" climate-related action, such as adapting for possible flooding in the city, while preserving the city's existing ecosystems. Although stand-alone, specific measures designed to address climate issues may be ideal, such coupling nonetheless provides significant ancillary benefits for climate adaptation needs. Given the increasing awareness of the importance of the urban wetlands and the range of benefits such areas provide, the city is now actively engaged in prioritizing and implementing such "coupled" planning initiatives that can address several simultaneous developmental concerns and concomitantly serve climate-adaptation priorities.

Integrating Climate-Risk-Management Objectives into Existing Infrastructure Projects

Analysis of the interviews conducted in Kolkata suggest that extant infrastructure projects that lend themselves easily to concomitant management of climate-induced risks as well, are more likely to receive higher priority for authorization, funding, and implementation. For example, flooding due to drainage congestion has been an ongoing and serious problem for the city. To make matters worse, this type of flooding has increased in magnitude and intensity in the past decade or so because of progressively frequent and intensified local rainfall patterns, likely the result of global warming. At the same time, the sewer and storm-drainage systems, which are designed as combined structures, also need to be augmented and refurbished not only because they have reached their capacity but also because a majority of the lines are in various states of disrepair. Since this situation offers an avenue for combining efforts of a number of public agencies (e.g., municipal planning bodies, agencies in charge of the water and sewer sectors, and environmental and flood-control bodies) within the city to solve a number of problems simultaneously, many of these agencies have pooled resources and concentrated their planning initiatives to address this problem collectively. In fact, a large number of sewer- and storm-water-related projects, some funded through JNNURM, are underway, serving both to address long-standing infrastructure deficits and to alleviate the serious problem of flooding due to drainage congestion.

Overall, the research on the East Kolkata Wetlands suggests the following:

- Climate-related issues are often given priority when they can be coupled with existing planning initiatives.
- Climate-adaptation or vulnerability-reduction measures are most likely to be implemented when integrated into broader infrastructure projects.
- Projects that collectively encompass multiple objectives, such as augmenting urban livelihoods, reducing the need for additional infrastructure, and streamlining planning and implementation issues, are more likely to catalyze institutional as well as stakeholder support.

- In resource-constrained environments, where climate adaptation is often neglected, identifying "climate" projects that can also support the city's developmental objectives are essential to bring climate adaptation to the fore.

Reevaluating Adaptation in the Context of South Asian Mega-cities

This research contributes to the broader understanding of how rapidly urbanizing mega-cities of South Asia currently address climate-change adaptation issues under severe resource constraints. More specifically, drawing on the example of Kolkata, the study examines the factors and institutional contexts and behaviors that enable climate adaptation or that promote the conditions under which effective planning strategies can be initiated and implemented. The research completed thus far suggests the following preliminary implications for policy and practice:

- The perception and actual experience of repeated risks and impacts based on previous climate-related events and other systemic threats offer strong motivation for local and national institutions to prioritize the management of those risks.
- City-level public agencies are more likely to overcome their institutional inertia in addressing climate-change issues when initiatives at the central and state government level are aligned with and further local planning agendas. Additionally, when the local agencies are held strictly accountable to departments at the state or central government level or to external donor and/or loan-granting agencies, these local entities are more likely to complete the planning and execution of such projects in an efficacious and timely manner.
- The institutional propensity for addressing climate issues and efficiency in planning for, and implementing, related plans can be facilitated by the following:
  - Restructuring efforts designed to streamline processes and ensure closer interagency collaboration and coordination.
  - The allocation of new resources or reallocation and redistribution of existing resources to prioritize climate-centered or climate-related action.
  - Technological innovations, such as digital interfaces designed to enhance workflow and monitor compliance with the regulations and

requirements of outside organizations and the agencies themselves. KEIP's Business Process System serves as a commendable example of such efforts.
    - Guidance from and consultation with external technical, managerial, and environmental experts will help in bridging the analytical and institutional capacity deficit that most city agencies in the developing world face.
- The coupling of planning objectives related to climate-risk-management and developmental initiatives that produce synergistic outcomes increases the likelihood that such measures will be prioritized, funded, and implemented by city-level public agencies.
- Climate-adaptation strategies and specific measures that can be easily identified and integrated within existing infrastructure projects are generally given precedence and are implemented in a more effective and timely manner than are stand-alone climate initiatives.

NOTES

1. The key climatic risk factors include sea-level rise, saltwater intrusion, floods, and cyclones. By the year 2050, saltwater intrusions caused by rising sea levels are estimated to penetrate over 100 km into the freshwater estuaries of Kolkata and Dhaka, while the projected population dislocations from flood-related events in Bangladesh alone are estimated to affect over 17 million people (Akter, 2009).

2. Financial, technical, and especially political difficulties effectively forestall climate-change planning efforts in India. The city of Chennai, for example, commissioned a comprehensive, well-received Climate Change Action Plan, prepared jointly by researchers from Madras and Kyoto Universities. The city's Corporation Council, however, has yet to ratify the document after municipal elections brought a new mayor to office last year (The New Indian Express, 2011). Delhi established an Energy Efficiency and Renewable Energy Management Centre in 2008, but the NCR did not name a director for over two years. The Centre's accomplishments to date include a series of largely unenforced energy and development guidelines and a three-page set of energy-saving recommendations for residents. The principal driver of efforts in Delhi is power outages, rather than concerns over climate change (Hoornweg, Freire, Lee, Bhada-Tata, & Yuen, 2011).

3. The NAPCC outlines eight "National Missions," namely, for Solar Power, Energy Efficiency, Sustainable Habitat, Water, the Himalayan Ecosystem, Green India, Sustainable Agriculture, and Strategic Knowledge for Climate Change. The Plan, however, is at best an attempt to articulate desirable avenues for future research and planning and to summarize broad, abstract goals.

4. An example of a "circuit-breaker" response in the event of a climate catastrophe, such as a severe flood, is to identify evacuation routes and "flood shelters" beforehand and ascertain the availability of provisions to prevent cascading

consequences, such as additional deaths arising from contaminated water supplies and inadequate shelter among the most vulnerable populations.

5. Kolkata lies at the terminus of the Indo-Gangetic Plain, which includes the wide area linked by the Indus and Ganges river systems and extends from Karachi to Kolkata. The Plain is home to over 1 billion people.

6. Kolkata's "combined" sewer system, which combines waste- and storm-water discharge, is well over 100 years old. Deficient maintenance and insufficient expansion have resulted in regular problems of drainage congestion.

7. Millions of KMA residents regularly ingest arsenic levels well in excess of acceptable levels. See The *Times of India* (2009).

8. Kolkata has experienced dramatic rates of subsidence, ranging from 6.5 to 13 mm annually over the past 50 years (Chatterjee et al., 2006). If such rates of subsidence continue unabated, the dangers of climate-related flooding in Kolkata proper and the broader KMA will increase exponentially (OECD, 2007).

9. Kolkata, for example, boasts 10 times the per-capita GDP of Dhaka (WWF, 2009).

10. The Jawahar Nehru National Urban Renewal Mission, or JNNURM, part of the Ministry of Development, is empowered, as per its mission statement, to "encourage reforms and fast track planned development of identified cities. Focus is to be on efficiency in urban infrastructure and service delivery mechanisms, community participation, and accountability of ULBs/Parastatal agencies towards citizens."

11. Interview with an official of the Ministry of the Environment, Government of West Bengal, December 2010, who asked to remain anonymous.

12. Dhaka, Mumbai, Karachi, and Chennai are also coastal cities at great risk of climate-related disaster.

13. The Ramsar Convention is an intergovernmental treaty that provides the framework for national action and international cooperation for the conservation and wise use of wetlands and their resources. The Ramsar Convention is the only global environmental treaty that deals with a particular ecosystem.

REFERENCES

Adger, N., Huq, S., Brown, K., Conway, D., & Hulme, M. (2003). Adaptation to climate change in the developing world. *Progress in Development Studies*, *3*(3), 179–195.

Akter, T. (2009). Climate change and flow of environmental displacement in Bangladesh. Policy Paper, Environment Unit of Unnayan Onneshan, Center for research and action on development, Bangladesh.

Alam, M., & Rabbani, G. (2007). Vulnerabilities and responses to climate change for Dhaka. *Environment and Urbanization*, *19*(1), 81–97.

Birkmann, J., Buckle, P., Jaeger, J., Pelling, M., Setiadi, N., Garschagen, M., Fernando, N., & Kropp, J. (2009). Extreme events and disasters: A window of opportunity for change? Analysis of organizational, institutional and political changes, formal and informal responses after mega-disasters. *Natural Hazards*. doi:10.1007/s11069-008-9319-2

Callaway, J. (2004). Adaptation benefits and costs: Are they important in the global policy picture and how can we estimate them? *Global Environmental Change*, *14*(3), 273–282.

Callaway, J., Kašćelan, S., & Markovic, M. (2010). *The economic impacts of climate change in Montenegro: A first look*. Montenegro: UNDP.

Central Intelligence Agency. (2012). *In the CIA World Factbook*. Retrieved from http://www.cia.gov/library/publications/the-world-factbook/geos/in.html

Chatterjee, R. S., Fruneau, B., Rudant, J. P., Roy, P. S., Frison, P.-L., Lakhera, R. C., ... Saha, R. (2006). Subsidence of Kolkata (Calcutta) City, India during the 1990s as observed from space by Differential Synthetic Aperture Radar Interferometry (D-InSAR) technique. *Remote Sensing of Environment*, *102*, 176–185.

Davoudi, S., Crawford, J., & Mehmood, A. (Eds.). (2009). *Planning for climate change: Strategies for mitigation and adaptation for spatial planners*. London: Earthscan, James & James.

Hewitt, K. (Ed.) (1983). *Interpretations of calamity from the viewpoint of human ecology*. Boston, MA: Allen and Unwin.

Hewitt, K. (1997). *Regions of risk: A geographical introduction to disasters*. Harlow, United Kingdom: Longman.

Hitz, S., & Smith, J. (2004). Estimating global impacts from climate change. *Global Environmental Change*, *14*(3), 201–218.

Hoornweg, D., Freire, M., Lee, M., Bhada-Tata, P., & Yuen, B. (Eds.). (2011). *Cities and climate change: Responding to an urgent agenda*. Washington, DC: The World Bank/The International Bank for Reconstruction and Development.

Jain, I., Rao, A., Jitendra, V., & Dube, S. (2010). Computation of expected total water levels along the east coast of India. *Journal of Coastal Research*, *26*(4), 681–687.

KMC. (2012). *Basic statistics of Kolkata*. Retrieved from http://www.kmcgov.in/KMCPortal/jsp/KolkataStatistics.jsp

Kousky, C., Rostapshova, O., Toman, M., & Zeckhauser, R. (2009). *Responding to threats of climate change mega-catastrophes*. Background paper for Global Facility for Disaster Reduction and Recovery Unit. Washington D.C.: World Bank.

Kundu, N. (2003). *Urban slum report: The case of Kolkata*. Kolkata, West Bengal: Institute of Wetland Management & Ecological Design.

Lavell, A. (2004). The Lower Lempa River Valley, El Salvador: Risk reduction and development project. In G. Bankoff, G. Frerks & D. Hilhorst (Eds.), *Mapping vulnerability: Disasters, development and people*. London: Earthscan.

OECD (2007). In: R. Nicholls, S. Hanson, C. Herweijer, N. Patmore, S. Hallegatte, J. Corfee-Morlot, J. Chateau, & R. Muir-Wood (Eds.), *Ranking port cities with high exposure and vulnerability to climate extremes: Exposure estimates*. OECD Environment Working Paper 1, ENV/WKP (2007)1. Paris.

Parry, M. L., Canziani, O. F., Palutikof, J. P., van der Linden, P. J., & Hanson, C. E. (Eds). (2007). *Climate change 2007: Impacts, adaptation and vulnerability*. Contribution of Working Group II to the Fourth Assessment Report of the Intergovernmental Panel on Climate Change. Cambridge, UK: Cambridge University Press.

Posner, R. (2005). *Catastrophe: Risk and response*. New York, NY: Oxford University Press.

Prime Minister's Council on Climate Change. (2009). *National action plan on climate change*. Delhi: Government of India.

Ribot, J., Magalhaes, A., & Panagides, S. (Eds.). (1996). *Climate variability, climate change and social vulnerability in the semi-arid tropics*. Cambridge, UK: Cambridge University Press.

Satterthwaite, D. (2009). The implications of population growth and urbanization for climate change. *Environment and Urbanization*, *21*(2), 545–567.

Satterthwaite, D., Huq, S., Pelling, M., Reid, H., & Lankao, P. (2007). *Adapting to climate change in urban areas: The possibilities and constraints in low- and middle-income nations.* Human Settlement Discussion Paper Series, Climate Change and Cities I. IIED, London.

Schipper, L. (2007). *Climate change adaptation and development: Exploring the linkages.* Norwich, UK: Tyndall Centre for Climate Centre Research.

Schipper, L., Cigaran, M., & Hedger, M. (2008). *Adaptation to climate change: The new challenge for development in the developing world.* New York, NY: UNDP.

Sen, A. K. (1981). *Poverty and famines: An essay on entitlement and deprivation.* Oxford: Clarendon.

Srivastava, L., & Heller, T. (2003). *Integrating sustainable development and climate change in AR4.* Technical Paper 10. Munasinghe Institute for Development/IPCC, Colombo, Sri Lanka.

Tanner, T., Mitchell, T., Polack, E., & Guenther, B. (2007). *Urban governance for adaptation: Assessing climate change resilience in ten Asian cities.* Report to the Rockefeller Foundation. Brighton, UK: Institute of Development Studies.

Taylor, K. (2008). *Planning to preserve the East Kolkata wetlands: Recommendations for management planning.* Master's Thesis, Duke University, North Carolina.

The New Indian Express (2011). *Climate change action plan awaits implementation.* Retrieved from http://ibnlive.in.com/news/climate-change-action-plan-awaits-implementation/198234-60.html

The Times of India (2009). *Poisoned water dying on false promises.* Retrieved from http://timesofindia.indiatimes.com/city/kolkata-/Poisoned-Water-Dying-On-False-Promises/articleshow/4298289.cms?intenttarget=no#.TxNx6PcsyFY.email

Town and Country Planning Organization, Ministry of Urban Development (TCPOMUD). (2009). *Urban land price Scenario- Kolkata -2008.* Delhi, India: TCPO

World Bank. (2010). *Vulnerability of Kolkata metropolitan area to increased precipitation in a changing climate.* Washington, DC: The World Bank.

World Bank & International Finance Corporation. (2009). *Doing business in India 2009, Comparing regulation in 17 cities and 181 economies.* Washington, DC: The World Bank, The International Bank for Reconstruction and Development.

World Wildlife Fund. (2009). *Mega-stress for mega-cities: A climate vulnerability ranking of major coastal cities in Asia.* Gland, Switzerland: WWF International.

CHAPTER 8

ADAPTING TO WHAT? CLIMATE CHANGE IMPACTS ON INDIAN MEGACITIES AND THE LOCAL INDIAN CLIMATE CHANGE DISCOURSE

Fritz Reusswig and Lutz Meyer-Ohlendorf

ABSTRACT

Purpose – *Adaptation to climate change requires that the population at risk and decision makers in various sectors become aware of the possible detrimental impacts in order to take whatever action is needed, especially in highly vulnerable countries and regions. In order to assess the climate change and impact awareness in a particularly vulnerable area – the Indian city Hyderabad, located within a semiarid region – we wanted to learn more about the local climate discourse, in particular the daily newspaper coverage of climate change and weather extremes.*

Methodology/approach – *After having looked at the Indian climate change discourse (CCD) in general, based on literature review, we were studying the local public CCD, based on the in-depth analysis of two English language daily newspapers, and three Telugu (the dominant local language) daily newspapers, covering the period of 2008–2009. This*

Urban Areas and Global Climate Change
Research in Urban Sociology, Volume 12, 197–219

ISSN: 1047-0042/doi:10.1108/S1047-0042(2012)0000012011

qualitative and quantitative analysis was completed by two expert interviews with local journalists.

Findings – *We find that the more recent Indian CCD has shifted if compared to the dominant argumentation pattern of the period before, as reported in other analyses. While the former discourse was characterized by the scheme "the poor/developing countries suffer from anthropogenic climate change caused by the industrialized countries," the recent Indian CCD has become more differentiated, taking into account both impacts elsewhere, and, most notably, conceding a (limited) responsibility of countries like India. On a local level, while reports on weather extremes are very common, we find that local newspapers of Hyderabad do not provide a link between these extreme events and (global) climate change.*

Research limitations – *Our discourse analysis could only cover a short time period of a local CCD, leaving open the questions of (a) its further development, and (b) how things might stand in other places in India. Furthermore it would be necessary to complement our study by analyses of the impact of mass media reporting on people's attitudes and behavior.*

Originality/value of paper – *Given the importance of public participation in adaptation measures, it is crucial to know if and how the wider public and the majority of the nonexpert public administration (which needs to be involved) understands the causes, potential impacts, and possible adaptive action in the face of climate change. This chapter provides a necessary (though not sufficient) element for that assessment. The findings can help to identify weaknesses, and thus to give hints how to improve the adaptive capacity in places like Hyderabad (India).*

INTRODUCTION: CLIMATE CHANGE AS A CHALLENGE FOR INDIA

The global community of climate and climate impact scientists is unequivocal by stating some key facts about climate change:

- The global mean temperature (GMT) has been increasing by +0.8°C during the last 100 years.
- Checking for various causes, science can attribute this global warming mainly to human activities that change the atmospheric concentration of greenhouse gases (GHG), mainly carbon dioxide (CO_2) and methane

(NH_4) due to fossil fuel combustion, land use changes, agricultural production, and landfills.

- While a few countries have managed to reduce their GHG emissions lately, the vast majority of countries displays growing GHG emissions, especially from large and rapidly developing countries, such as China and India.
- Given this trend, and taking into account the inertia of the world's climate system, future warming of the planet is unavoidable. The system is already committed to additional +0.6°C, but future emission scenarios may translate into further +4–8°C until 2100, which is a historically unprecedented climate change, and would definitely violate Article 2 of the United Nations Framework Convention on Climate Change (UNFCCC), making the avoiding of dangerous climate change mandatory to the international community. A warming of +2°C against pre-industrial levels is widely accepted in science and politics as an operational definition of Article 2.
- If the world community – states, businesses, and citizens – does not manage to reduce GHG emissions globally from 2020 to 2030 onward, probabilities are very high that we miss the +2°C goal, with really dangerous and in part irreversible climate change being on its way.

These facts have been stated in the last (the fourth) assessment report of the Intergovernmental Panel on Climate Change (IPCC, 2007), and climate and climate impact research since then has confirmed, if not sharpened these messages, as the fifth assessment report, announced for 2013, will clearly show.

While the potential impacts of a change of GMT will be felt globally, there are large differences across regions, depending not only on the concrete natural conditions but also on the vulnerability of these regions due to human activity. One key element to reduce the regional vulnerability is to be informed about potential future impacts, and the concrete mechanisms that lead to a specific vulnerability.

It has been a widely held assumption that people from developing countries will suffer most from climate change, for example, because they depend more on agriculture and natural resources and services, such as the water cycle. While this remains true, climate impact research and vulnerability studies have also revealed that developed countries and regions, including large cities, are vulnerable to climate change. The European Union, for example, has invested a lot in impact and vulnerability studies, and has issued a "White Paper" on climate change adaptation, indicating that without adaptation Europe will experience major (economic) losses and lose competitiveness (EC, 2009).

Nevertheless, cities and regions in the global south will be affected much more seriously, for various reasons. Because of its location, size, and population density, India, for example, will feel most climate change impacts directly, such as the melting of glaciers in the Himalayans, rising sea levels, weather extremes like droughts and floods, and it might even be facing the destabilization of the Indian monsoon, with severe consequences for its food and water supply. Given the enormous size and variation in India's physical and social geography, possible impacts, as well as its limited adaptive capacity, will result in a broad range of vulnerability to climate change (O'Brien et al., 2004; Shukla, Sharma, Ravindranath, Garg, & Bhattacharya, 2003). In particular, major impacts are projected for the agricultural sector, upon which more than 60% of the total population depends upon. Moreover, there are important health implications, as climate change is likely to impose an additional layer on already existing, severe environmental health risks, such as air, water and soil pollution, heat stress, flooding, water logging, and vector-borne diseases. This is in particular the case in urban areas within a context of unprecedented rapid urbanization.

A more recent, Indian science-based study is more precise with respect to climate scenarios, to four major regions (Himalayan Region, Western Ghats, North-Eastern Region, and Coastal Regions) and to four major sectors (agriculture, forests, human health, and water) (INCCA, 2010b). As India's former Minister for Environment and Forest has put it:

> ... no country in the world is as vulnerable, on so many dimensions, to climate change as India. Whether it is our long coastline of 7000kms, our Himalayas with their vast glaciers, our almost 70 million hectares of forests (which incidentally house almost all of our key mineral reserves) – we are exposed to climate change on multiple fronts. (Ramesh, 2010, p. 9)

While both international and Indian experts agree upon the high risks that climate change poses to India – its economy, its ecology, its population – the social perception of climate change and the "felt" necessity to adapt still lack salience features. Many decision makers – especially at local levels, where adaptation mainly has to take place – are often ignorant about the threat, lack of information about possible impacts, and feasible options to avoid damage. This information and awareness deficit is aggravated by lack of funds to, for example, upgrade infrastructure or implement additional warning systems. In urban areas, many informal settlements, habitat of the large part of the urban poor, are located in risky zones, for example, in flood prone areas. Despite being the largest democracy in the world, the poor in India often lack information and rights, so that appropriate adaptation to potential climate impacts is falling far behind what would be necessary.

The growing Indian middle class, on the other hand, does have the educational and informational resources to become aware of climate risks, and they do also have the financial backgrounds to take personal precautionary action. Many members of that class, however, are more engaged in managing their professional and daily lives, and less convinced that they themselves will be vulnerable.

As climate change is nothing that can be perceived directly by the usual means of everyday lay experience, the mass media play a key role as gatekeepers between science on the one hand and the general public (including parts of political decision makers and the public administration) (Carvalho & Burgess, 2005) on the other. The question thus arises which role the mass media do play in reporting about climate change in India. Do they sufficiently communicate the seriousness of the problem? And where do they portray the main impacts to be? Do they offer solutions, both on a political and on a personal level? How do they weigh adaptation versus mitigation measures?

We would like to answer these questions by approaching the climate change discourse (CCD) in India, before we then select a specific part of it, namely the city of Hyderabad (Andhra Pradesh), to concretize our findings.

CLIMATE CHANGE DISCOURSE IN INDIA

Climate change cannot be perceived by the senses of lay people. A key component of climate change – GMT – is a statistical construct, averaging temperatures from different times, places, and seasons across the globe. Without instrumental records and scientific methods, global warming cannot be detected. And even if some impacts can be observed more or less directly (such as the melting of glaciers, or a tropical storm), the attribution problem still remains: is the rise of GMT really the cause of this effect, or is it a "natural" phenomenon, residing well in the range of normal variations, or influenced by other factors?

This is why the public discourse on climate change is mainly science driven (Weart, 2003). Without science, we would not be aware of climate change, and thus there would be no public discourse on it. On the other hand, having a scientific debate alone may be a necessary, but by no means a sufficient condition for a public discourse on an issue. Society (or relevant parts of it) has to become interested in it. Society has to perceive it as relevant social problem that is worth dealing with, and needs to take some kind of action. We know from many examples that it is not necessarily the "objective" or scientifically stated urgency of an issue that translates it into social relevance. Would that be the case, environmental issues would rank much higher in public attention than they actually do. There are many factors that influence the selectivity of the social system with respect to what enters a discourse. The political system, for example, tends to address issues that it can deal with in terms of understanding and capacity to act. Urgent problems that cannot be solved tend to move downward on the agenda.

A CCD is a thematically focused and (more or less) coupled sequence of publicly visible arguments related to climate change issues (existence, causes, consequences, adaptation, and mitigation) in various contexts (or framings) that different social actors are engaged in to influence (1) one another, (2) specific boundary conditions of social action, and (3) the general public so, that the control/power basis, interests, and worldviews of the speaking actors have a higher chance to prevail in the social interpretation and individual or collective decision-making processes (Reusswig, 2010).

The mass media do play a core role in CCD, as they translate scientific findings in social and political meaning, for example, by identifying factors and actors that have created the problem, or by identifying responsible groups or organizations (sometimes characterized as culprits) for solving it, or by reporting about the failures in solving the problem. The mass media tend to link climate change to the political sphere, as the latter can be regarded as the "problem solving agency" of a society, and as climate politics is often taking place at large international conferences, with nation states as the major actors. Various studies have been undertaken in order to analyze the precise role of the mass media for the public understanding of climate change, and its treatment in politics. Most of them have focused on Europe and the USA (cf. Boykoff, 2010; Boykoff & Boykoff, 2004).

In India, climate change has also reached the mass media. We would like to focus here on the print media, as for scientific issues such as climate change the print media are still the most influential channel. With a readership of over 250 million, India is the second largest print market in the world. However, the Indian print market is highly fragmented, with over 60,000 newspapers printed in 22 languages (Datta, 2011).[1]

India has a strong position in international climate policy negotiations, and in the international debate about climate change, its meaning, and its political solution. In a way, the Indian government remains committed to its historic position that "the environment cannot be improved in condition of poverty" (former Indian Prime Minister Indira Gandhi at United Nations Conference on the Human Environment, Stockholm, 1972). While this was stated before climate change was set on the global public agenda, the Indian position remained the same in the CCD. Discussions regarding "common but differentiated responsibility" (as stated in the UNFCCC of 1992) gained traction in the early 1990s through the seminal work of Anil Agarwal and Sunita Narain at the Center for Science and Environment (CSE), where they made distinctions between what constituted "survival" and "luxury" emissions (Agarwal & Narain, 1991).

More recently, Billett (2010) has looked at the Indian CCD by analyzing the top four English-speaking newspapers of the country (The Times of India, The Hindu, Hindustan Times, and The Indian Express). In searching for keywords ("climate change," "global warming," "greenhouse gas," and "IPCC") in headlines between January 2002 and June 2007, he ended up with a sample of 248 articles that were analyzed by means of discourse analysis. In addition, interviews were conducted with 15 of the major environmental writers in the English-language press. The main results of the study may be summarized as follows:

- Unlike other national print media cultures in Europe and, especially, in the USA, the Indian English-speaking press is almost unequivocally convinced that climate change is a scientifically established fact, and that actual weather patterns and extreme events clearly reveal this fact. The anthropogenic causes of climate change are also widely accepted. Climate skepticism is a nonissue in the Indian national press.
- The adverse impacts of global warming that the articles report of are mainly located in the developing world, especially in India itself. India is seen as a primary climate change victim.
- The responsibility for GHG emissions is mainly attributed to developed countries, especially to the USA. For most articles, Indian emissions are not relevant, while the global North is the main responsible. Most articles argue with (a) India's low per capita emissions and (b) the low historical emissions from India.
- Existing evidence for an internal differentiation of Indian emissions (Billett refers to the study of Greenpeace India 2007) is widely ignored in the articles. India is portrayed as a unity in this respect and the elite is hiding behind the poor.

Taken together, Billett interprets India's mass media coverage on climate change, as it is represented in the English-speaking newspapers, as being a reflection of the Indian elite's take on the issue, namely an nationalist, postcolonial discourse with clearly opportunistic undertones, with a neglect of important facts (growing total emissions from India, growing emissions from the middle class), and a negative impact on India's role in international climate policy (namely the Kyoto Protocol and its follower).

Boykoff (2010) has asked the question whether these findings, based on mass media data from 2002 to 2007, still are valid, given some interesting more recent trends, such as

- A further increase in mass media coverage of climate change in India (cf. Fig. 1).
- A shift in Indian climate policy framing from anticolonialism to modernization (Wagner, 2010).
- The popularization of climate change and climate action by Indian celebrities (e.g., from Bollywood).
- Barack Obama following George C. Bush as U.S. president.
- New forms of mass media coverage of climate change, such as mobile phone information.

This is the point where our own study could enter the stage. Before we do so, one should note that the numbers reported by Boykoff (2010) indicate that 2007 was the most important year – in terms of quantity – of climate

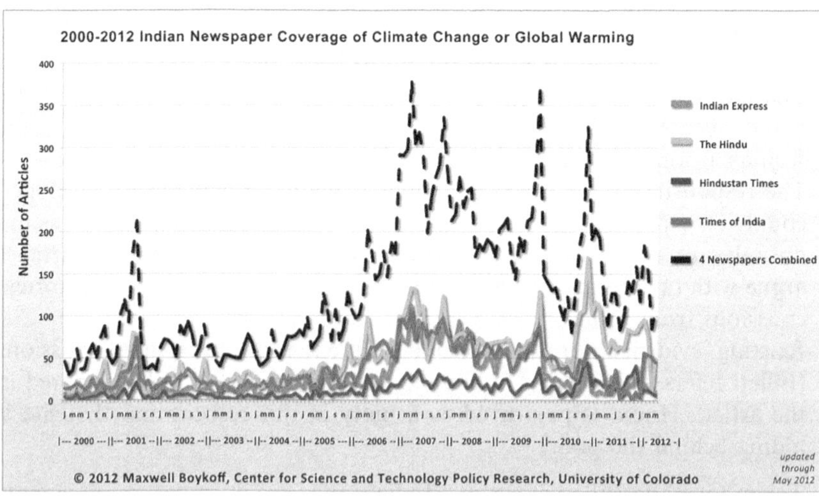

Fig. 1. Mass Media Coverage of Climate Change in India (Boykoff, 2012).

change in the Indian print media in the 2000–2010 period. The exception would be the December 2009, with the Copenhagen climate conference as a major issue of reporting.

This resonates well with the global picture. Despite the reduction in numbers after the 2007 peak, even before Copenhagen, the print media interest did not decline to its pre-2007 levels (mostly below 100), but never fell below 150, indicating a shift in the public discourse on climate change that we have also detected in Germany (Reusswig, 2010).

THE RECENT CLIMATE CHANGE DISCOURSE IN INDIA IN THE 2008–2009 PERIOD

Our own research on India's CCD is situated in the context of a joint research project of German and Indian organizations, termed "Climate and Energy in a complex Transition Process towards Sustainable Hyderabad."[2]

Hyderabad is the capital of the Indian state Andhra Pradesh. Its about 7 million inhabitants grow constantly, especially due to rural-to-urban migration. About one-third of the population lives in absolute poverty, most of them in informal settlements. They are especially exposed to climate change-induced weather extremes, such as flash floods or heat waves.

One of the contributions of the Potsdam Institute is to look at urban lifestyles in Hyderabad, and their potential for a sustainability shift in the city. For that purpose, we were interested in the Indian CCD, and especially in the way it plays out in Hyderabad. Two newspapers were selected for the discourse analysis by taking into account online accessibility, readership, relevance of the paper in Hyderabad, and political affiliation: *The Hindu* and *The Times of India* represent two important English newspapers that cover a quite broad political spectrum with a balanced readership from both sides, left and conservative. Parallel to this exploration, two expert interviews were conducted with staff reporters from *The Eenadu* and *The Hindu*.

By searching for the terms "climate change" and "global warming" in the text corpus, we retrieved 1,100 articles in the 14 months period between January 2008 and February 2009. We excluded those that did not have an explicit reference to India. From the remaining set of articles we then chose 50 that did cover a broad range of issues (impacts, adaptation, mitigation…), all articles that explicitly mentioned consumption or lifestyle, and those that did put climate change in a social and/or political context with India. In a second step, we complemented this sample with another sample of 50 articles generated randomly by choosing every 20th article (or its successor from the list if it had already been chosen by the systematic sampling).

These 100 articles were looked at in a quantitative way, scanning them according to a matrix covering the following (head) categories:

- Author (who is speaking?).
- Causes of climate change (natural/anthropogenic, various sectors/activities).
- Accountability for mitigation action (countries, groups, role of India).
- Impacts of climate change (sectors, human health…).
- Reactions to climate change (mitigation, adaptation, research…).
- Stakeholders (open list of actors mentioned).

For our purposes here, we would like to confine ourselves to the aspects of accountability and of impacts. Only a quarter (27) of the articles did explicitly address the question of accountability, that is, the question of who caused the problem and is (morally) responsible for doing something about it. We know from our own research as well as from Billett (2010) that this question is core to the Indian CCD. The fact that the majority of the articles in our sample did *not* raise the issue may be surprising. However, one can assume that in many contexts the impacts have been in the focus, and that for those impact-oriented articles the question of responsibility was either irrelevant, or implicitly settled. From these 27 articles, 17 clearly held the developed countries being exclusively responsible for global warming, while 5 also included the (other) developing countries in the group of responsible actors (Fig. 2).[3] Eight articles did underline that India had a significant role

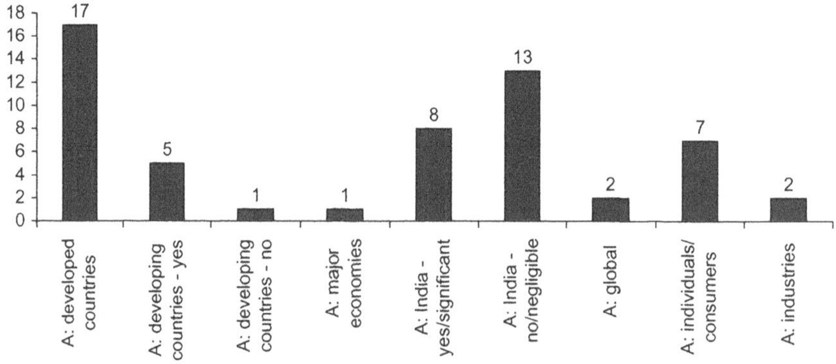

Fig. 2. Accountability for Climate Change (27 articles).

in GHG emissions, while 13 reported that India's contribution was small to negligible. Interestingly, seven articles argued that individual consumers do play a role as drivers, while only two held industry to be the dominant driver, and two only speaking vaguely about a global responsibility.

If we focus on India, the interesting point is the distribution of articles regarding its responsibility. While still a majority of 13 articles claim that India is not responsible, 8 concede such a responsibility. As India's per capita emissions are still very low, this concession refers to the total emissions of India – with 1,727.71 tons of CO_{2eq} in 2007 No. 3 in the world (INCCA, 2010a) – indicating that the per capita emissions do not exclusively form the basis of argumentation any more. This would mean a significant shift of the argumentation of the Indian print media in the 2008–2009 period as compared with the 2002–2007 period, covered by Billett.

The most worrying thing about climate change is not some change in the atmospheric composition of GHGs per se, but the impacts that this change does have on the rest of the Earth system – and on human systems and actors in the first place. Given our baseline of 100 articles, it is slightly surprising to see that "only" 39 articles deal explicitly with impacts. The remaining majority tries to inform its public by speaking about climate change without touching impacts (e.g., by talking about scientific discoveries or policy reactions instead).

Of those that do address impacts, 16 limit themselves to impacts in India while 23 deal with impacts outside India exclusively. If we take again the number of mentioned impacts as a baseline ($n = 140$), we find some interesting clustering (cf. Fig. 3).

This again is a clear shift away from the discourse pattern reported by Billett (2010), in who's case study 54.4% of impacts referred to India. In our case study, only 44.3% of impacts mentioned refer to India, while the majority of 55.7% of reported impacts does not mention India at all. For example, reports on extreme temperature waves refer to other regions exclusively, and reports on the general rise in temperature or melting of ice caps and glaciers see cases from India in a clear minority position. On the other hand, vector-borne diseases and dengue are presented with reference to India exclusively, and reports on adverse impacts on agriculture and rural areas see India in a clear majority position.

A clearer picture emerges if we aggregate the impacts (Fig. 4).

Extreme weather events (mentioned 36 times) play a dominant role in the Indian CCD, followed by health (26) and water issues (17), as well as sea level aspects (14), and agricultural issues (14). However, if we focus on impacts on India exclusively, health impacts (23) are the overriding concern,

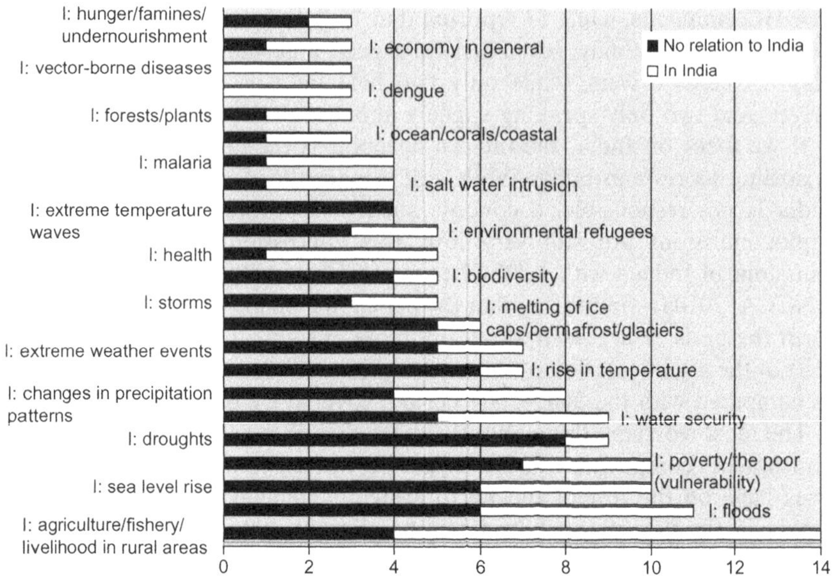

Fig. 3. Impacts of Climate Change (39 articles).

followed by agricultural aspects (10), extreme weather events (10), water issues (8), and sea level issues (7). Impacts on the urban infrastructure are mentioned only twice, with one reference to India. More often development impacts are mentioned (5).

For our project, the "bad news" is that public awareness for direct climate impacts in the urban context seems to be very low, as compared to – say – impacts on agriculture and natural resources in general. However, the "good news" part of our findings is that health, extreme weather and water related issues do play a central role for the Indian CCD, especially with respect to India itself.

In general, we can summarize with respect to impacts that the "self-centeredness" of the Indian CCD Billett (2010) has reported for the 2002–2007 period seems to have given way to a more balanced view on climate change impacts, clearly breaking with the pattern "cause outside, effect inside" (Billett, 2010, p. 11) that has been observed so far.

If we turn to adaptation, we find that differences between reporting on adaptation in general (15) and on mitigation in general (19) are quite small. But while adaptation has only a few, often not explicitly mentioned categories, mitigation contains many, partly very often explicitly discussed

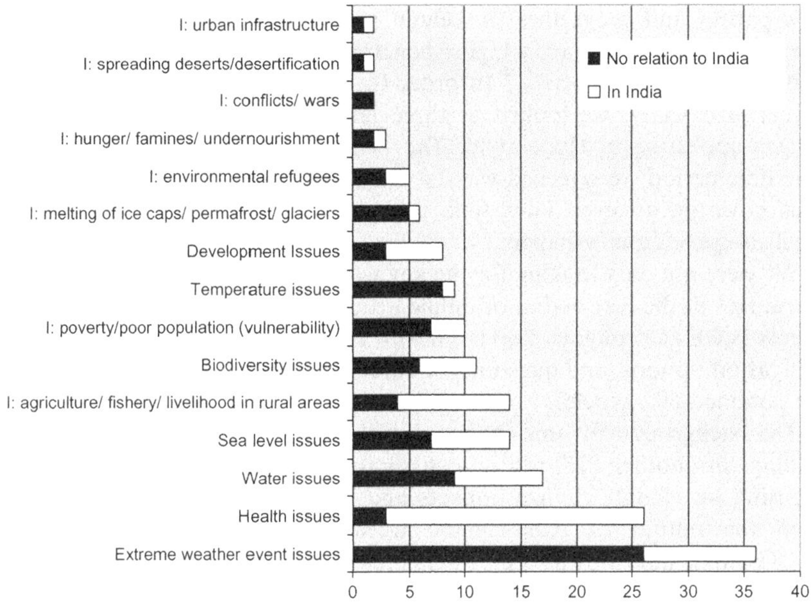

Fig. 4. Aggregated Impacts and Their Relation to India.

sub-categories. With regard to adaptation, only changes in cultivation (6) and conservation of water (4) occurred more than once – reflecting the fact that agricultural impacts play an important role, especially with respect to India. Impacts on the urban infrastructure, for example, play but a minor role – both with respect to India and to the rest of the world (cf. Fig. 4). Taken together, this also means that in the national reporting on climate change, adaptation is less important than mitigation, and within adaptation rural aspects are far more in the focus than urban ones. Given the size, spatial distribution and economic relevance of cities in India, this relative neglect gives rise to many concerns.

THE HYDERABAD CLIMATE CHANGE DISCOURSE IN LATE 2009

While English-speaking newspapers are read in the capital of Andhra Pradesh, Hyderabad has several print-media groups which publish newspapers and magazines in Telugu and Urdu. Especially the Telugu-speaking newspapers have a higher penetration rate, especially in the middle and lower ranks of society.[4] In order to find out more about the specific Hyderabad CCD, we looked at three large Telugu-speaking daily newspapers appearing in Hyderabad: *The Eenadu*, *Vartha*, and *Andhra Jyoti*.[5] The time period we screened was 1st October 2009 to 31st December 2009, thus covering an even later time period than we had covered with the English-speaking newspapers.

We were not only looking for the key word "climate change" and "global warming" in the text bodies of online articles, but also for extreme weather events (such as droughts, floods), health effects, damages to infrastructure, mitigation options, and questions of equity and energy security. As a result, we obtained 605 articles.

The background for this change in the selection criteria is related to the findings of another PIK project contributing to "Sustainable Hyderabad," focusing on climate change impacts and adaptation. Our colleagues have used the output of Atmospheric Ocean General Circulation Models (AOGCMs) and downscaled them to the local weather situation in Hyderabad (Begumpet station data) in order to generate various climate change scenarios (according to IPCC's SRES scenarios) for the city until 2100 (Lüdeke & Budde, 2009). As a result, we can predict with some uncertainties that both heat waves and extreme rainfall events (>80 mm/day).[6] Already by 2050, the frequency of these events will increase by 50–240% under a high emission scenario (A 2). We know from our media analysis – and from interviews with experts – that under the current situation these daily events cause a lot of trouble in the city, including traffic jams due to flooded streets, and flooding of low lying or orographically disadvantaged slum areas (Kit, Lüdeke, & Reckien, 2011, cf. Fig. 5).

Besides to economic losses due to traffic jams especially the poorer sections of the city are thus more exposed to risks of health impacts due to water pollution and an increase in vector-borne diseases. More often and/or more intense heat waves bear the risk of increased cases of admission in hospitals with heat stroke, heat exhaustion, hyperthermia, electrolyte imbalance, particularly children, elderly, poor, socially isolated, those with pre-existing illnesses (Rama Padma, 2011).

It is worth noting that an increased risk of adverse impacts of climate change is a combined function of a changed climate signal and a locally defined sensitivity (exposure), but also of the adaptive capacity of a system or place, which in turn is defined by its socio-economic and institutional capacity. Only in their synergy with the adaptive capacity of a system or

Fig. 5. Flood Impacts on a Poorer Quarter of Hyderabad (Photo: Lutz Meyer-Ohlendorf).

location do potential climate impacts create the climate change vulnerability (cf. Fig. 6).

Against this background, we decided to go for a more detailed and content-oriented approach in looking at Hyderabad's specific CCD. Our analysis is still ongoing, and therefore the following results are preliminary. But we can report some major tendencies on the base of the relatively small sample period of the last quarter of 2009:

- The coverage of climate change (global warming) and related issues (weather extremes) in Hyderabad's Telugu-speaking print media is quite vivid: 605 articles could be retrieved. As we lack a comparing figure from earlier periods, we cannot quantitatively judge any change in coverage. Our interviews with some mass media representatives indicate, however, that there has been a constant rise in attention to the issue.[7]
- As with the English-speaking national press, a lot of articles related to climate change or global warming report on scientific findings and/or international climate negotiations. During the reporting period, a major event took place: the Conference of the Parties of the UNFCCC in

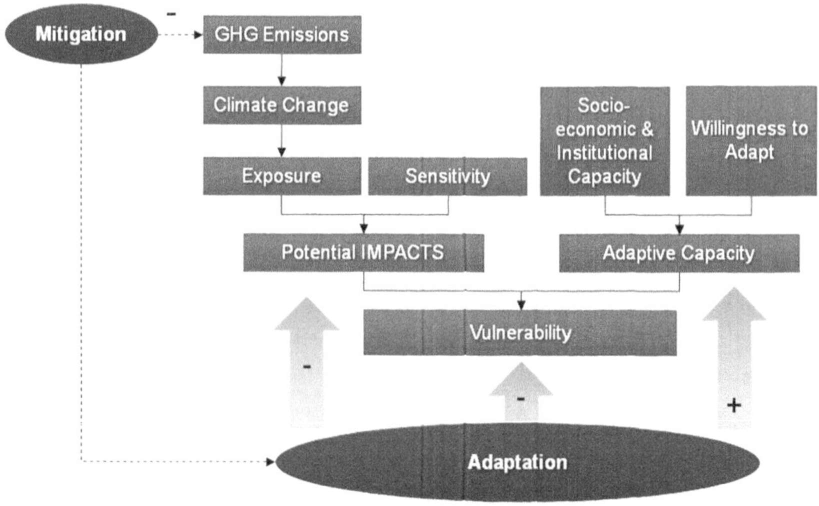

Fig. 6. Relation Between Climate Change Impacts, Vulnerability and Adaptation. In This Model Vulnerability Is Determined by Both Potential Climate Change Impacts and Adaptive Capacity. Adaptation Strategies Can Address Either. *Source:* Isoard (2011).

Copenhagen (December 2009). The Telugu press was reporting in some detail about this event, and of course the Indian position and the behavior of the Government of India was a major point of interest.

- More often than in the national English-speaking newspapers, our three Telugu newspapers did report from local public events in Hyderabad and other places in Andhra Pradesh. In this context, a wide range of voices and stances were reported, such as left leaning protest against the American and/or European "climate imperialism," or pleas for more vegetarian diets as a solution, or spiritual and religious explanations, up to debates about the possible end of the world in 2012 according to the Mayan calendar.
- If we compare mitigation and adaptation as basic options toward climate change, we find – with all caution – a clear bias toward mitigation related articles. We read about the eight National Missions of the NAPCC, we read about solar power, about a vegetarian diet, about celebrities that support environmental campaigns, or about green homes that help protect the global climate. But articles about adaptation to climate change are rare. Only 8 out of 605 articles mention adaptation measures to climate change in India or Hyderabad.

- Reports about weather extremes come often in Hyderabad's Telugu newspapers. Heat waves, lack of rains, but especially heavy rainfalls and their consequences (traffic jams, health impacts, or power breakdowns) are extensively reported on. Seventy-four articles report on rain events and related impacts; none out of these articles give any link to climate change. Thirty articles report on drought, only one out of these relates drought to climate change. Two-hundred fifty seven articles report on flooding, none out of these refers to climate change. Here's a typical example:

> Traffic in several parts of the city came to a stand still and normal life was disrupted because of the heavy rain on Saturday. People in the low-lying areas shifted to safer places as water entered their houses. Power supply was switched off in some areas, as a precautionary measure. Some houses had partially collapsed due to incessant rains for the past three days. Cables and power lines were snapped, as tree branches fell on them. Vehicle drivers found it difficult to go to their destinations, through the water that stagnated on the roads. People were scared when rumours spread that the Himayat Sagar reservoir was full and water was being discharged down the stream. Ministers Anam Ramanarayana Reddy, Danam Nagender, and M. Mukhesh Goud reviewed the steps being taken to shift people living in low-lying areas, in a meeting with higher officials in the GHMC office. Vaartha, 4.10.2009, "Heavy rain disrupts normal life in the capital Hyderabad")

What we read is a short protocol of extreme weather events, of damages occurring, and of the local political and administrative system reacting upon it – or related fears and "rumors". We do not read about any causal or statistical relations between climate change and these local impacts – despite a broad coverage of climate change in the same newspapers.

CONCLUSION: CLIMATE DISCOURSE SHIFTS AND URBAN ACTION

The Indian CCD is special. Given the strong science bias in a topic that is this heavily science dependent, India with its large number of high quality science institutions dealing with climate change has a very good position for a specifically Indian CCD. A look at the authorship of the latest INCCA reports (2010a, 2010b) shows that India has reached the stadium of maturity and self-sufficiency in climate science expertise.

But this is only a necessary, not a sufficient condition for a functioning and effective CCD. Equally important is a mass media landscape that communicates the issue in all the needed accuracy, salience, and understandability. And here our paper ends with some skepticism.

We have seen that the clear-cut disjunction between "cause outside, effect inside" that Billett (2010) has found for the 2002–2007 period no longer holds for the 2008–2009 period. At least in the English-speaking national press (the press that Billett has also analyzed), the picture has become much more differentiated: impacts in other countries are on equal foot with those reported from India, and India as a major emitter of GHG has entered stage. It is no longer the per capita emissions, it is also the total emissions of the country, and the global responsibility that the Indian press is now debating. Even a binding climate policy goal – a nonissue for decades – has become debatable. And with different income segments and lifestyle groups that contribute differently to global warming, India's elite is no longer "hiding behind the poor."[8]

When it comes to impacts and adaptation, things look slightly different. It is striking to see that the traditional CCD in India has portrayed the country as a '"victim" of a changing climate that others, particularly the industrialized countries, have been causing. At the same time there is such a weak coverage of (1) concrete adaptation measures, and (2) such a weak link between reports on weather extremes on the one hand and climate change on the other. This latter point is the preliminary result of a small set of articles (257) from three important Telugu newspapers in Hyderabad, and should thus be treated more like a hypothesis. Nevertheless it gives rise to some concerns:

- On the local level, climate skepticism seems somewhat more articulate than on the national, English-speaking level. But in general, climate change as well as anthropogenic causation is taken for granted in India to a degree that is surprising not only to American, but also to European standards. May be one should think about providing Indian press reporters and editors with some uncertainties – and with strategies to cope with them.
- While climate change is perceived as a reality, and while debates about mitigation options to reduce GHG emissions in general are rather dominant in India, the articles we read lack clarity and explicitness with respect to local situations and people's everyday lives. It is thus difficult for a reader to figure out how a "greener lifestyle" would look like – and where the individual responsibility of the consumer ends, while the political responsibility of the citizen has to step in.

- Climate impacts are reported by the local newspapers in Hyderabad, but either with a clear bias toward the rest of the world, or to India in general (e.g., the Himalayan glaciers). When it comes to local impacts, we observe a vivid reporting activity on weather extremes, but the link to global climate change (including the link to India as a contributor to its own vulnerability) is almost absent.

How can people and organizations prepare for climate change impacts, and how should they adapt, if the information provided is characterized by this obvious gap between broad coverage of everyday "disasters" on the one hand, and the global "disaster" climate change on the other? If successful adaptation requires the active engagement of citizens, civil society organizations, and the business sector, then the mass media coverage of actual weather extremes must not be mute about their possible future change due to global warming. Not because every single actual event is already a consequence of ongoing climate change – this is impossible to proof – but because of the clearly perceivable risk of shifting weather patterns due to climate change. This risk is underlined not only by our own studies, but by many Indian studies too (INCCA, 2010b).

Another facet of the Indian CCD is its state-orientation when it comes to responses. No doubt: policy responses are needed, given the character and size of the problem we are facing. But both mitigation and adaptation require an active engagement of the civil society and the business sector. Otherwise neither the necessary attention nor the required funds will be mobilized. Especially if we take aspects like limited public budgets and competing public missions into account.

A study on the national funding of adaptation to climate change in India looked at the Union's budget in various areas relevant to vulnerable areas and sectors in India, as specified by the NAPCC (Oxfam India, 2010). For the four years examined by the study, union government's expenditure on adaptation to climate change shows an increase from 1.7% of GDP in 2006–2007 to 2.7% of GDP in 2009–2010. While this is in principle good news, the study reveals that more funds are spent on human development than on the stabilization of supporting ecosystems, and that the exact contribution of human development issues, for example, poverty alleviation schemes, to improved adaptive capacity remains unclear:

> While the budgetary measures on adaptation seem to be skewed towards poverty alleviation, the policy statements on existing initiatives on adaptation and the proposed

> national missions are silent on how poverty alleviation should be integrated into the adaptation framework and its linkages with other sectors can be established. Sectors that are crucial to any adaptation intervention such as food security, rural and urban housing for the poor and educational infrastructure have received inadequate attention in the policy response on adaptation. (Oxfam India 2010, p. iv)

The Madras School of Economics (MSE), the M.S. Swaminathan Research Foundation (MSSRF) and the South Asian Network for Development and Environmental Economics (SANDEE) organized a brainstorming workshop on the Economics of Climate Change Adaptation in February 2010 at the MSE, Chennai, to identify policy gaps, research questions and capacity-building needs related to India's need to adapt to climate change (SANDEE, 2010). The summary note concludes that any discussion on climate change adaptation must recognize two sometimes contrasting perspectives on the nature of adaptation: (1) climate change imposes a distinct and additional burden on the society; (2) climate adaptation, is one response among many, to a host of socioeconomic and environmental pressures and cannot necessarily be isolated from regular development activities. While there are remaining gaps in climate and climate impact research as a prerequisite to any adaptation policy, we have to conclude that the design of concrete adaptation measures requires much more knowledge from the social sciences, which is currently lacking in India – but not only there. There is an urgent need for designing short-term training programs for different stakeholders and for careful long-term learning through collaborative research.[9]

One can state rightly so that

> the battle against climate change will be won or lost in our cities. (Sridhar, 2010, p. 347)

This also holds for adaptation to climate change. India is one of the most vulnerable countries to climate change on the planet – mainly due to its wonderful natural richness, and its populated regions, namely cities, at risk. India is endowed with a vivid civil society sector, and an attentive mass media sector. We should all help to improve India's capacity to become aware of the risks and threats it is facing, especially in its rapidly growing urban areas. And we should all support its vivid civil society sector in taking on the issue of climate change more actively, and in critical support of a more targeted public policy. If India manages to prepare for climate change – not only by adapting to it, but also by reducing its emission growth – it will (*ceteris paribus*) prosper and shine. If not, it will suffer and cry. The choice should be clear.

NOTES

1. With a population of more than 1,200 million, and a group of 360 million who can read but do not read any publication, India's print media sector still provides a significant growth opportunity.

2. The project (cf. http://www.sustainable-hyderabad.de/) is funded by the German Ministry for Education and Research (BMBF), and is part of a larger program that addresses environmental problems, especially climate change, in emerging megacities around the world (cf. http://www.future-megacities.org/seiten/programm/forschungsaufgaben/forschungsaufgaben-en.aspx). "Sustainable Hyderabad" is the only project focusing on India, and the German research consortium is cooperating with many Indian partners, such as *The Energy and Resource Institute* (TERI) in New Delhi, or the *Osmania University* in Hyderabad.

3. Note that we count accountability (mentioned more than once), not articles. Total sample size thus exceeds 100. And as we refer to semantic units, not to logical ones, articles that do refer to "developing countries""must not necessarily include India to this category, but might also refer to the developing world excluding India.

4. While more than 90% of individuals from socioeconomic class (SEC) A1 and A2 are reached by print media, this share is only 30% in SEC E 2. But in terms of individuals, the penetration in E2 is about equal to that of SEC A: 21 million (E1, E2) versus 21.6 million (A1, A2) (PWC, 2006, p. 8).

5. Because of our language barriers, we had to let translate the articles. We would like to thank the Centre for Media Studies (CMS) for their support.

6. Heat waves, aggravating the well-known urban heat island effect – and floods are among the major climate change threats to Indian cities (cf. Sharma, 2011, Yehuda, 2011). Coastal cities will face additional problems, as the floods of 2005 in Mumbai have shown (Narulkar, 2011). Future damages under climate change to the megacity of Mumbai could amount to Rs 1,501,725 crore (Kumar, Jawale, & Tandon, 2008). Another study concludes that a severe flood hitting Mumbai under a 2080s changed climate – and no adaptation measures – could result in monetary losses of 2.3 billion USD (against 0.7 billion today). Investments in better drainage systems plus additional measure to increase the adaptive capacity of the city could reduce the potential future damages to 0.3 billion USD (Ranger et al., 2011).

7. We cannot judge whether the seminars with mass media representatives from the Centre for Climate Change at the Engineering Staff College of India in Hyderabad have had an influence here. These seminars are nevertheless very useful. The environment for science reporting in the mass media has become worse in the last 20 years, with journalists from the global south especially lacking the capacity and training to cover the intricacies of climate science and policy, as well as lack access to clear, timely, and understandable climate-related resources and images (Shanahan, 2009).

8. In a German publication of 2010, Sunita Narain has explicitly addressed the internal differences of per capita emissions in India, combined with the suggestion to introduce a personal carbon trading scheme to India (Narain, 2010). In an inaugural speech to members of the International Association for the Study of the Commons (IASC), held in January 2011 in Hyderabad, then Minister for Environment and Forest, Jairam Ramesh, has indicated that Indians are very good at claiming climate

justice in the international context, but rather mute when it comes to address the discrepancies of lifestyles and related emissions within India. These are some small examples for a shift of the Indian CCD. Of course, there are many participants in that discourse still arguing in the same way that Billett has described so vividly.

9. This is one of the reasons why the work of the Centre for Climate Change at the Engineering Staff College of India in Hyderabad is so important.

REFERENCES

Agarwal, A., & Narain, S. (1991). *Global warming in an unequal world: A case of environmental colonialism*. New Delhi: Centre for Science and Environment.

Billett, S. (2010). Dividing climate change: Global warming in the Indian mass media. *Climatic Change*, *99*, 1–16.

Boykoff, M. (2010). Indian media representations of climate change in a threatened journalistic ecosystem. *Climatic Change*, *99*, 17–25.

Boykoff, M. (2012). *2000-2012 India Newspaper Coverage of Climate Change or Global Warming*. University of Colorado at Boulder, Center for Science and Technology Policy Research. Retrieved from http://sciencepolicy.colorado.edu/media_coverage/india

Boykoff, M. T., & Boykoff, J. M. (2004). Balance as bias: Global warming and the US prestige press. *Global Environmental Change*, *14*, 125–136.

Carvalho, A., & Burgess, J. (2005). Cultural circuits of climate change in the UK broadsheet newspapers, 1985–2003. *Risk Analysis*, *25*(6), 14–57.

Datta, P. (Ed.) (2011). *The marketing whitebook 2010–20111. One-stop guide for marketers*. New Delhi: Businessworld.

European Commission (EC) (2009). White Paper. *Adapting to climate change: Towards a European framework for action*. Brussels: Commission of the European Communities. Retrieved from http://eur-lex.europa.eu/LexUriServ/LexUriServ.do?uri = COM:2009:0147:FIN:EN:PDF

Ganguly, K., & Panda, G. R. (2010). Oxfam India. *Adaptation to climate change in India. A study of union budgets*. New Delhi: Oxfam India working papers series, OIWPS – I.

Indian Network for Climate Change Assessment (INCCA) (2010a). India: Greenhouse Gas Emissions 2007. Delhi: Ministry of Environment and Forests, Government of India.

Indian Network for Climate Change Assessment (INCCA) (2010b). Climate Change and India: A 4 × 4 Assessment. A Sectoral and Regional Analysis for 2030s. Delhi: Ministry of Environment and Forests, Government of India.

Intergovernmental Panel on Climate Change (IPCC). (2007). Climate Change 2007: Synthesis Report, Geneva: IPCC.

Isoard, S. (2011). Perspectives on adaptation to climate change in Europe. In J.-D. Ford & L. Berrang-Ford (Eds.), *Climate change adaptation in developed nations: From theory to practice* (pp. 51–68). Dordrecht: Springer.

Kit, O., Lüdeke, M., & Reckien, D. (2011). Assessment of climate change-induced vulnerability to floods in Hyderabad, India, using remote sensing data. In K. Otto-Zimmermann (Ed.), *Resilient cities* (pp. 35–44). Dordrecht: Springer.

Kumar, R., Jawale, P., & Tandon, S. (2008). Economic impact of climate change on Mumbai, India. *Regional Health Forum*, *12*(1), 38–42.

Lüdeke, M. K. B., & Budde, M. (2009). *Evaluating Climate Change Scenarios*. From AOGCMs to Hyderabad. Potsdam: PIK Research Paper No. 1, Sustainable Hyderabad Project.

Narain, S. (2010). Klimawandel: Keine gemeinsame Teilhabe an der Welt. *Aus Politik und Zeitgeschichte*, *32-33*, 3–7.

Narulkar, S. M. (2011). Case study of flooding in Mumbai and possibility of reduction of flood fury by urban hydrologic intervention. *Dikshit*, 2011, pp. 55–82.

O'Brien, K., Leichenko, R., Kelkar, U., Venema, H., Aandahl, G., Tompkins, H., ... West, J. (2004). Mapping vulnerability to multiple stressors: Climate change and globalization in India. *Global Environmental Change*, *14*, 303–313.

Price Waterhouse Coopers (PWC). (2006). *The Indian entertainment and media industry: Unravelling the potential*. New Delhi: PWC.

Rama Padma, Y. (2011). *Climatic change and public health in the context of Hyderabad: Issues of concern & Challenges*. Presentation given at the Young Researchers Day, Sustainable Hyderabad Project, Humboldt University, Berlin, May 18, 2011.

Ramesh, J. (2010). Foreword. In Indian Network for Climate Change Assessment (INCCA) (Ed.), *Climate change and India: A 4 × 4 assessment. A sectoral and regional analysis for 2030s*. New Delhi: Ministry of Environment and Forests, Government of India.

Ranger, N., et al. (2011). An assessment of the potential impact of climate change on flood risk in Mumbai. *Climatic Change*, *104*, 139–167.

Reusswig, F. (2010). The new climate change discourse: A challenge for environmental sociology. In M. Gross & H. Heinrichs (Eds.), *Environmental sociology: European perspectives and interdisciplinary challenges* (pp. 34–61). Dordrecht: Springer.

Shanahan, M. (2009). Time to adapt? Media coverage of climate change in non-industrialised countries. In T. Boyce & J. Lewis (Eds.), *Climate change and the media* (pp. 145–157). London: Lang.

Sharma, R. V. (2011). Climate change and urban floods. In A. K. Dikshit (Ed.), *The cities and the climate change* (pp. 41–53). Mumbai: Indian Institute of Technology, Centre for Environmental Science and Engineering.

Shukla, P. R., Sharma, S. K., Ravindranath, N. H., Garg, A., & Bhattacharya, S. (Eds.). (2003). *Climate change and India: Vulnerability assessment and adaptation*. Hyderabad: Universities Press (India) Pvt. Ltd.

South-Asian Network for Development and Environmental Economics (SANDEE). (2010). *The economics of climate change adaptation in India – Research and policy challenges ahead*. Chennai: Sandee.

Sridhar, K. S. (2010). Carbon emissions, climate change, and impacts in Indian cities. In Infrastructure Development Finance Company (IDFC) (Ed.), *India infrastructure report 2010. Infrastructure development in a low carbon economy* (pp. 345–354). Oxford: Oxford University Press.

Wagner, C. (2010). India: A difficult partner in international climate policy. In: S. Dröge (Ed.), *International climate policy: Priorities of key negotiating parties*. SWP Research Paper No. 2, Stiftung Wissenschaft und Politik, German Institute for International and Security Affairs, Berlin, pp. 76–73.

Weart, S. R. (2003). *The discovery of global warming*. Cambridge, MA: Harvard University Press.

Yehuda, R. U. (2011). Climate change and urban heat island effect. In A. K. Dikshit (Ed.), *The cities and the climate change* (pp. 115–129). Mumbai: Indian Institute of Technology, Centre for Environmental Science and Engineering.

CHAPTER 9

ENVIRONMENTAL STATE IN TRANSFORMATION: THE EMERGENCE OF LOW-CARBON DEVELOPMENT IN URBAN CHINA

Yifei Li

ABSTRACT

Purpose – *The net increase in China's urban population in the last 50 years equals the current total population of the European Union. The scale and speed of urbanization in China requires a sustainable solution to unprecedented energy demands and elevated carbon emissions. As low-carbon development emerges in urban China, it offers a unique vantage point to examine some fundamental theoretical questions of the environmental state. How do structural socioeconomic changes affect the environmental state? Does the rise of the environmental state offer a basis for regulatory reform on a broader scale?*

Methodology/approach – *Case study of five low-carbon cities in China provides the empirical evidence for the analysis. The five cities represent a continuum in their levels of postindustrialization. I compare low-carbon development strategies in postindustrial cities with those*

Urban Areas and Global Climate Change
Research in Urban Sociology, Volume 12, 221–246

ISSN: 1047-0042/doi:10.1108/S1047-0042(2012)0000012012

strategies in industrial cities. Evidence is collected primarily by way of interviews with planning bureau officials, urban design professionals, involved NGOs, academics, and private sector individuals familiar with the matter.

Findings – *First, in cities where the level of postindustrialization is high, state resources support innovative low-carbon development strategies that attempt to achieve emission reductions in a variety of sectors. In industrial cities, however, the environmental state's regulative power is limited to one or two (sub)sectors. Second, and more importantly, a new pattern of governance is emerging in postindustrial cities. Low-carbon development in postindustrial cities is a much less centralized process, having local levels of governments as key players of low-carbon policy making. When the environmental state intersects with the postindustrial city, it gives birth to a new urbanism that has profound implications for political structuring in China.*

Research limitations – *The analysis in this chapter is based on evidence from a purposefully selected set of Chinese cities, which may render the results biased. Future studies should aim for a more systematic analysis of cities in order to establish more generalizable conclusions. In addition, given the increasing availability of quantitative data at the city level in China, future studies should also seek to incorporate quantitative analyses to better substantiate existing knowledge derived from qualitative sources of evidence.*

Originality/value of chapter – *First, this chapter challenges the Western bias in the existing literature on the environmental state. The role of the civil society is far from salient in the Chinese context, and yet the environmental state demonstrates a robust level of activity despite the weak civil society. It therefore seems that a general theory of the environmental state can be built from existing literature, but needs to be sensitive to non-Western social conditions that might falsify parts of the theoretical claims. Second, the environmental state literature can be consolidated and further developed when examined in conjunction with other literatures in the modernity tradition. I have demonstrated the connection between the environmental state and the postindustrial city. More studies are needed to examine other facets of the environmental state, as it intersects with a multitude of (post)modern conditions.*

Keywords: Low-carbon development; postindustrial city; China

INTRODUCTION

China's surge of urbanization is unprecedented. In the last 50 years, China's urban population has grown more than fivefold, from 107 million in 1960 to 586 million in 2009 (McKinsey Global Institute, 2009). The net increase in China's urban population in the last 50 years equals the current total population of the European Union.

At the same time, the rapid expansion of China's urban areas has accelerated the country's energy consumption and corresponding carbon emissions. Generally, energy consumption of urban residents is approximately 3.5 to 4 times that of rural residents, and even higher for large urban centers (Lin, 2009), meaning the bulk of carbon emissions are tied to cities. Research has shown that one of the indirect consequences of urbanization is a marked increase in global warming, reflecting the consequences of what is known as the "urban heat island effect" (Bornstein, 1968).

As the Chinese population becomes increasingly urbanized, the country is determined to define and implement a new pattern of urbanization (Lei, Zhuang, & Zhang, 2011). The Chinese leadership's commitment to low-carbon development was made clear at the UN Climate Change Summit in September 2009, when President Hu Jintao promised to "step up effort to develop green economy, low-carbon economy and circular economy, and enhance research, development and dissemination of climate-friendly technologies" (cited in National Development and Reform Commission, 2009). In the same year, China pledged to reduce carbon emissions per unit of GDP by 40–45% in 2020, compared to 2005 levels (United Nations Development Program, 2010). The scale and speed of China's urbanization presents an opportunity for China to become a global innovator in low-carbon development (Wang, Berrah, Mathur, & Vinuya, 2010).

The emergence of low-carbon development in China is not a methodical process across the board. Cities have followed different pathways to pursue low-carbon development (Niu, Pan, & Cao, 2011). In other words, the environmental state is rendered differently in different local contexts. In this chapter, I try to understand the environmental state through the comparative lenses of five Chinese cities where low-carbon development projects have been introduced and implemented. These five cities have differing levels of postindustrialization, enabling me to understand how postindustrialization has an effect on the environmental state. This chapter is an attempt to understand the environmental as it intersects with postindustrial cities.

In the next section, I draw on existing literatures on the environmental state and the postindustrial city to construct an analytical framework. Comparative study of five cases of low-carbon development in urban China constitutes the empirical evidence. The analysis concludes that the environmental state exhibits signs of devolution in postindustrial cities, which signals a future trend of the environmental state in China. This finding enriches existing scholarship on the environmental state in two ways. First, it supplements evidence from a developing nation, thereby helping to correct for the Western bias of existing accounts. Second, it makes the linkage between the environmental state and the postindustrial city, stressing the organic relationship between the two theoretical constructs.

THE ENVIRONMENTAL STATE IN POSTINDUSTRIAL ERA

As government bodies across the world take increasing responsibility in environmental protection, sociological analyses of the environmental state[1] have also grown rapidly in the past decade. It is generally believed that the environmental state is a recent phenomenon that emerged in North America, parts of Europe and Oceania during the 1960s, and became a global happening during ensuing decades (Mol & Buttel, 2002). Perhaps due to its historical sequencing, much has been said about the global diffusion of the environmental state (Eckersley, 2004; Frank, Hironaka, & Schofer, 2000; Hass, 1994). This perspective understands the environmental state as a product of the global redefinition of nation-states' responsibilities (Goldman, 2001; Janicke, 2006).

Aside from global forces, the social conditions and local contexts through which the environmental state materializes on the ground have also gained much scholarly attention (Buttel, 2000). Dryzek (2003) explains the environmental state from the vantage point of social movements. He attends to interest groups, political parties, protest organizations, and other manifestations of the public sphere in the United States, the United Kingdom, Germany and Norway, and how these groups channel resources to shape the environmental state in each of the national contexts. Dryzek finds that although political processes vary across nations, the environmental state exhibits a consistent tendency to devolve political authority from the central governing body to local authorities. In other words, according to Dryzek, the environmental state signals a refashioning of governance toward local control and a broad base of public input.

Furthermore, the maturing of environmental movements has led to changes in social attitudes about environmentalism. As ideas of environmentalism take root among the public, the governing bodies are increasingly pressured to transform the political structure so as to satisfy public demands for various environmental protection regulations and initiatives. Empirical research of formation of the United States Environmental Protection Agency (EPA) (Geltman & Skroback, 1998; Rosenbaum, 1999) has clearly demonstrated how the EPA has been "reinvented" to advance its duties that the general public increasingly perceives as fundamental.

Frickel & Davidson's (2004) recent research builds on Evans' (1995) model of embedded state theory, arguing that substantive actions of the environmental state are conditioned by the interactions between the state and the civil society. The environmental state, as the authors argue, can effectively exercise its governing power when it is well connected to the civil society, and when the latter prospers. They present compelling evidence of how the environmental state works with environmental movement activists and environmental science professionals in order to effectively fulfill its responsibilities.

Therefore, existing scholarship has greatly advanced out understanding of the environmental state. On the one hand, the environmental state is an institution that tends to be open to public inputs in political decision making on environmental protection (also see Redclift, 1992). On the other hand, the government institution reinvents itself so that local authorities enjoy higher levels of autonomy in making environmental protection decisions.

Given the environmental state literature's emphasis on political change, civil society, and evolving social attitudes, it belongs to the well-established tradition of modernity thoughts in sociology. Specifically, it has family resemblance to a host of sociological theories on modernity, including postindustrialization (Bell, 1973), communicative action (Habermas, 1989), reflexive modernity (Beck, Giddens, & Lash, 1994), and ecological modernization (Hajer, 1995; Mol & Spaargaren, 2000). However, the environmental state as a social theory has underutilized the rich theoretical foundation in modernity thinking. As an emerging intellectual enterprise, the environmental state literature should have a more solid relationship with existing arguments of the implications of modernity, especially given the environmental state literature's original formulation in the ecological modernization tradition (Fisher & Freudenburg, 2004).

As a first step to bridge the environmental state literature to modernity thinking, this chapter is an attempt to understand the environmental as it intersects with postindustrial cities. I study how the environmental state is rendered differently in postindustrial versus industrial cities. In so doing, I seek to juxtapose theoretical insights about both the environmental state and those about the postindustrial city, drawing insights from the latter to enrich the former.

The substantial sociological literature on postindustrialization and post-industrial cities warrants a synoptic overview. In Bell's (1973) classic work, he conceives of governance in the postindustrial society as technocratic. In other words, Bell considers the postindustrial urban political system to be determined by individuals who exercise authority by virtue of technical competence, because political decisions in the postindustrial era involve sophisticated forms of knowledge and rationality (Galbraith, 1967). Consequently, political decision making is dependent on certain set of skills individuals acquire through specialized training.

However, the technocratic "mind-view," according to Bell, is also highly paradoxical. Governance involves more than technical considerations, and requires nontechnical skills to maintain the postindustrial social order. Specifically, urban governance emphasizes "the logical, practical, problem-solving, instrumental, orderly, and disciplined approach" (Bell, 1973, p. 349) in exercising public authority. Furthermore, the postindustrial society is also characterized with the growth of public participation. A multiplicity of social forces enters into the political arena, voicing the demands from all kinds of groups. In fact, in the postindustrial age, as Bell argues, the civil society paradoxically grows in tandem with technocracy. Many more social groups seek to establish their social rights, and desire for participation. The paradox in governance is emblematic of the complexity of the postindustrial society.

Bell's "paradoxical technocracy" is further elaborated in Savitch's (1988) comparative study of three iconic postindustrial cities, New York, Paris, and London. Through his comparative lens, Savitch argues that corporatism is the predominant theme of postindustrial urban governance. Specifically, as a multitude of interests begin to be brought together and satisfied, governments in the postindustrial city become increasingly reliant on organizations from the private sector as "collaborators" of policy making. Corporatist politics "is a way of rationalizing large-scale interests and making them consonant with the larger polity" (p. 301). In other words, although in postindustrial cities, the civil society establishes a stronger foothold in public affairs, corporatist political processes shape civil society interests with that of the government. Empirical evidence from elsewhere (Ley, 1980; Perloff, 1980) also supports this line of argument.

Esping-Andersen (1993) provides further evidence from a slightly different angle. He argues that the class structure has "turned a postindustrial direction" (p. 227), by which he means that class formation exhibits a pattern that is at once broadly based and elitist. While the specific mechanisms of such paradoxical class formation are beyond the scope of this chapter, it is important to note that postindustrial societies are conducive to high levels of class mobility, but entry into the top social strata is more difficult than ever. Esping-Andersen therefore offers a class perspective to understand the paradoxical relationship between the governing body and the civil society in postindustrial cities.

Therefore, decades of theoretical and empirical development in the postindustrial city literature has offered tremendous insights with regard to the nuance of urban governance in the postindustrial age. In postindustrial cities, there is the dual movement of technocratic domination and civil society growth, the tension of which explains a great deal of the conflicts and transformations in postindustrial cities.

Notably, these arguments about environmental state, civil society, and technocracy are based almost entirely on evidence from Western societies. Studies have shown that Eurocentric theories of the environmental state have limited explanatory power for understanding the developing world (see Mol, 2006; Mol & Spaargaren, 2000). However, this does not necessarily mean that existing theories are incapable of explaining what happens in non-Western cultures. In this chapter, I draw on theoretical insights that are potentially biased toward the West, but try to apply the theoretical framework to China. The focus is not placed on where theories fail to explain, but on how theories can explain better. In other words, the theoretical aspiration is to construct a more nuanced theory of the environmental state that builds on knowledge about the West and also attends to differing social and political conditions in China – a first step toward a general theory of the environmental state.

To summarize, both the environmental state and the postindustrial city literatures point to the growing importance of the civil society in shaping politics. In addition, the former highlights the devolution of government as one of the consequences of the environmental state and the latter discusses technocratic domination in urban politics. While most of these theoretical insights are based on studies of the West, they seem relevant to understanding similar political and institutional change in China.

This leads to the focal question of the chapter: *what happens when the environmental state intersects with the postindustrial city in China?* The differences between the environmental state in postindustrial cities and that in industrial cities are of pivotal interest. Low-carbon development projects crystalize the Chinese state's recent efforts to protect the natural environment. As these projects unfold in Chinese cities, they offer a unique vantage point to examine some emerging characteristics of the environmental state in China.

DATA AND METHODS

I use evidence from low-carbon development in five Chinese cities, Shanghai, Hangzhou, Tianjin, Chongqing, and Baoding, to address the research question. These five cities are selected from the universe of 276 Chinese cities that have included low-carbon development in their official planning documents (Chinese Society for Urban Studies, 2011). The five cities are selected because they were among the first Chinese cities to embark on low-carbon development. Shanghai and Tianjin are home to two of the most widely discussed low-carbon city projects, and the other three cities are among the eight cities designated by the National Development and Reform Commission (NDRC) in China's 12th Five Year Plan as national pilots for low-carbon development. To be sure, there are other cities that have comparable experience with low-carbon development, but I target a manageable number of cities that have differing degrees of postindustrialization. Furthermore, these five cities are geographically scattered in Northern, South-eastern, and Central China, thereby ruling out bias toward certain regions in China.

This chapter presents a qualitative[2] comparative analysis of low-carbon development in these five cities. The five cities represent a continuum in their levels of postindustrialization. I compare low-carbon development strategies in postindustrial cities with those strategies in industrial cities. Such comparisons enable me to understand how the environmental state is rendered differently under the postindustrial condition. My approach follows Robinson's (2011) conception of "variation-finding" method, which allows for "higher levels of generalization ... [about] questions of governance" (p. 11). The analytical focus is on the differing paths toward carbon emission reductions, and how these differences can be understood from the perspective of postindustrialization.

Evidence is collected primarily by way of interviews with planning bureau officials, urban design professionals, involved NGOs, academics, and private sector individuals familiar with the matter.[3] I use snowball sampling to identify interviewees. Furthermore, to ensure accuracy and credibility, evidence from interviews is triangulated with other available sources, including public documents, newspaper and journal articles, comprehensive plans, and internal communication documents retrieved from government agencies. Noninterview sources of evidence are referenced where appropriate.

POSTINDUSTRIALIZATION IN URBAN CHINA

The process of postindustrialization in China is fueled by its national policy to gradually restructure urban economies. There is a widely shared consensus among policy makers in China that cities must restructure their economies in order to create better living conditions for their residents. This deindustrialization of Chinese cities happens in tandem with the rise of the modern service sector. China's 12th Five Year Plan (2011–2015) forecasts the number of people employed in the service sector to grow by 5 percentage points to 41% nationwide by 2015 (Liu, 2011). Much of this forecasted growth will realize in the form of postindustrial city growth. Thus, while most Chinese cities exhibit emerging signs of postindustrialization, some of them have become truly postindustrial cities. The five cities selected for study in this chapter represent various levels of postindustrialization in urban China (Fig. 1).

Shanghai's level of postindustrialization is among the highest in mainland China. The share of GDP generated by Shanghai's service sector surpassed that of the industrial sector in 1998, and is almost twice the size of its industrial sector today. The growth of the service sector and the corresponding decline of the industrial sector have been especially dramatic in recent years. The service sector is now the primary driver of Shanghai's economy, and will continue to be so for the foreseeable future.

Hangzhou shows clear trends of postindustrialization. The growth of the city's service sector started in the early 1980s, and has remained strong for the past three decades. The city forecasts the service sector to account for more than 60% of the GDP by 2020 (Chen, 2010). From a macroeconomic perspective, Hangzhou exhibited apparent signs of postindustrialization at a time when most urban economies in China, including that of Shanghai, were still dominated by the industrial sector. However, its process of postindustrialization has stalled in the past two decades, while other cities moved further along in the process. Today, the service sector accounts for roughly half of the economic output in Hangzhou, whereas the service sector in Shanghai now accounts for almost 60% of GDP. Since 2008, Hangzhou's

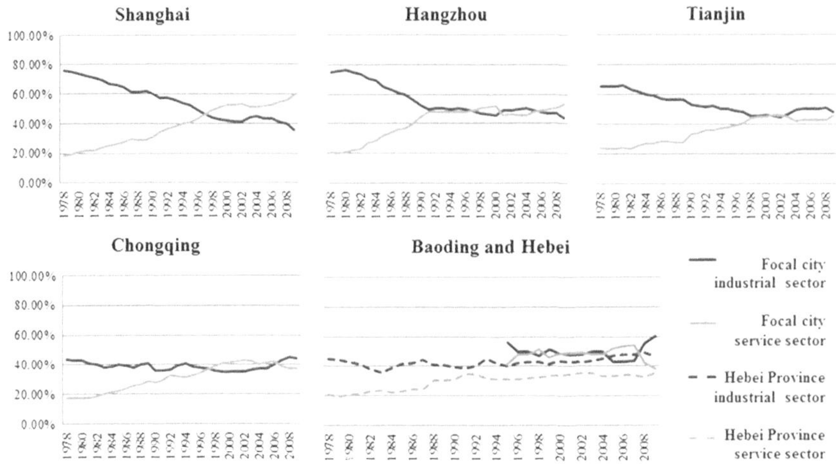

Fig. 1. Industrial and Service Sectors in Five Chinese Cities (% of Total GDP, 1978–2009).

Notes: 1. Percentages do not add up to 100% due to the omission of the agricultural sector. 2. Official definitions of the industrial and service sectors are observed. For details of sectorial definitions, see http://www.shanghai.gov.cn/shanghai/node2314/node4128/node15316/node15317/userobject30ai15041.html. 3. Baoding's economic record keeping did not start until 1995. I have included data at the Hebei (where Baoding is the most populated city) provincial level, as rough reference for Baoding's economic structure prior to 1995. *Source*: National Bureau of Statistics of China.

service sector has again begun to gain momentum, and with the support of local economic restructuring policies, the city's postindustrialization will continue in the years to come.

Tianjin also exhibits signs of postindustrialization. The share of GDP generated by the industrial sector has declined from more than 65% in 1978, the beginning of China's economic reform, to 48% in 2009, with a concurrent rise in the service sector's share. As economic restructuring continues, Tianjin's service sector is projected to contribute a much larger share of GDP, though smaller than that of Shanghai and Hangzhou.

Chongqing's level of postindustrialization is relatively low. The share of economic output generated by the industrial sector has remained almost constant in recent decades. Though the size of the service sector briefly eclipsed that of the industrial sector around the turn of the twenty-first century, Chongqing's industrial sector has witnessed a boom since 2007, while the service sector has experienced relative decline. It is also notable that a large percentage of Chongqing's economy remains agricultural.

The level of postindustrialization in Baoding is similar to that of Chongqing. The gap between the city's industrial and service sectors began to narrow in the mid-1990s. By 1997, the share of the service sector in Baoding was almost equal to the share of industrial sector, and the two remained close until 2004. Interestingly, both Chongqing and Baoding have witnessed decline of the service sector in the last several years, probably reflecting China's efforts to limit industry in major eastern cities while increasing industrial development in central and western areas. Overall, Baoding has not shown clear signs of postindustrialization.

Overall, Shanghai and Hangzhou are truly postindustrial cities. Tianjin has started its process of postindustrialization, but its industries remain robust, whereas Chongqing and Baoding remain industrial. The presentation of the cities in the next section is ordered according to their levels of postindustrialization, with Shanghai being the highest and Baoding the lowest among the five. The next section examines in detail low-carbon development strategies in these five Chinese cities.

THE EMERGENCE OF LOW-CARBON DEVELOPMENT IN URBAN CHINA

An official definition of what constitutes a low-carbon city is lacking. China's Ministry of Environmental Protection and Ministry of Housing and Urban-Rural Development (MoHURD) have both developed standards for "Eco-Cities." Several provinces have also created standards for "Garden Cities," "Green Cities," and other similar designations. All of these standards vary in terms of rigidity and scope. Some large cities have pockets in the urban area under low-carbon development, whereas low-carbon goals in smaller cities typically reach into all areas in the jurisdiction. These differences preclude an overarching definition of low-carbon city in China.

Despite the lack of a standardized definition of what constitutes a low-carbon city, the majority of Chinese cities have set low-carbon goals. According to a study by the Chinese Society for Urban Studies (2011), among the 287 cities in China with municipality status, 276 have proposed low-carbon or eco-city goals. Of these, more than half have begun construction projects in an effort to fulfill these goals, while more than a quarter have specific plans for action in the near future (Yu, 2011). Most of these low-carbon projects did not exist before 2009, demonstrating an emerging national trend of low-carbon development in China. According to China's 12th Five Year Plan, by 2015 China will establish 100 model cities, 200 model counties, 1,000 model districts, and 10,000 model towns of a green and new energy theme to showcase its achievements in low-carbon development. The five cases selected for analysis in this chapter represent some of the most widely discussed low-carbon development projects in Mainland China.

Shanghai Post-Expo Area

The Shanghai Post-Expo Area (SPEA) is located in metropolitan Shanghai, 6 km (3.73 miles) south of People's Square, the center of the city. SPEA covers a total land area of 6.68 km^2 (2.58 square miles), containing residential, commercial, entertainment, and public zones. The Post-Expo Area was the site of the 2010 World Expo, which attracted 70 million visitors. Shanghai has taken the Expo's theme of "Better City, Better Life" as a guiding principle for the city's urban planning since late 2000s. Low-carbon concepts, tools, technologies, and practices were a common thread among the Expo's exhibitions, and the city is determined to use the Expo's legacy to promote further reductions in carbon emissions in the local economy. SPEA is city's pilot area for the introduction of these low-carbon technologies.

SPEA is oriented toward three major uses: conventions and exhibitions, a corporate headquarter zone, and ecotourism. The Shanghai municipal government and a special planning committee are responsible for planning and construction of the project. China's Expo Pavilion and its surrounding area are planned to be developed as the convention and exhibition zone to support Shanghai's strategic role as an international center of finance and commerce.

The corporate headquarter zone in SPEA will supply business premises to domestic and multinational corporations. The zone consists primarily of office buildings, though 10% of the total floor area will be used for entertainment and retail purposes. SPEA's goal is to attract well-known multinational corporations to occupy buildings in this zone. As of August 2011, 13 large-scale state-owned enterprises have signed contracts to move their headquarters to SPEA. Evidently, the development of the headquarter zone in SPEA is heavily supported by economic resources controlled by the central government.

The plan for this zone features a high-density development with low building heights. To achieve this, SPEA will follow the Three Star rating program, an existing rating system administered by Chinese authorities. Through the use of energy saving technologies, all buildings in SPEA are expected to achieve Three Star status, the highest achievable status. However, SPEA exceeds the minimum requirement of the Three Star system. As part of the mission of the World Expo, the SPEA project will put many cutting-edge green building technologies into wide use, including distributed heating systems, geothermal heat pumps, and water source heat pumps. These buildings will showcase new concepts in energy efficiency and are intended to extensions of the Expo's promise to create a better urban life for its residents.

In terms of ecotourism, the project features three parks. A cluster of heavily polluting factories in what is now the SPEA was successfully transformed into Houtan Wetland Park, Expo Park and Bailianjing Eco-Park, in part by using reclaimed materials from the demolished factories. Visitors to the parks, which now form a large green space in SPEA, are exposed to low-carbon technologies, such as a water purification and recycling system that generates significant quantities of potable water.

In addition, one of the key principles in the design of SPEA is its adherence to a traditional spatial pattern iconic of the greater Shanghai region (also known as Jiangnan). Specifically, the overall planning pattern for the area will resemble a traditional Jiangnan garden. Natural ventilation will disperse heat generated by buildings, and paths will be shaded by trees. In addition, water source heat pumps will be installed in buildings as part of the heating, ventilation, and air conditioning (HVAC) systems. The pumps enable heat exchange between the buildings and the Huangpu River, which bisects the SPEA site, and minimize the environmental impact of HVAC operations (Liang, Yang, Meng, Xu, & Zhao, 2010). In addition, the buildings will be connected by public squares, metro stations and corner parks, designed to encourage walking. The intent is to create a comfortable environment combining work spaces and outdoor areas, based on the urban pattern of the Shanghai area's traditional gardens. The entire zone will be receptive to public use while affording privacy to businesses users.

Though endorsed by the central government, low-carbon development in SPEA is effectively determined by local agencies. Planning and implementation is premised on thick knowledge of Shanghai's strategic positioning, climate conditions, and cultural landscape. Local resources are utilized throughout the design stage to ensure that SPEA appropriately matches with its surrounding areas both in the physical and sociocultural senses. Moreover, its low-carbon development encompasses a wide range of areas, aiming to control emissions from a variety of sources, In other words, its low-carbon development strategy is versatile, reflecting the environmental state's flexible capability in a postindustrial context.

Hangzhou's Six-in-One Strategy

Hangzhou is the capital of Eastern China's Zhejiang Province. Hangzhou is one of the earliest in China to have placed the natural environment at the center of its development strategy, and its achievements in sustainability are unparalleled in China.[4] According to Hangzhou's 12th Five Year Plan, the city's comprehensive low-carbon plan comprised of economy, buildings, transportation, lifestyles, environment, and society. This is what the city calls a "six-in-one" strategy of low-carbon development, in which six key elements of low-carbon development create a multifaceted whole.

The economic component of Hangzhou's strategy is exemplified by its industrial park project. Hangzhou has partnered with a Singaporean clean-technology company to build a low-carbon industrial park, with an investment of approximately US$300 million. The industrial park will be tailored to clean energy and renewable energy businesses, taking advantage of the clustering effect of low-carbon industries. The park is expected to create a total of 5 million m^2 (1.93 square miles) of gross floor area, and become the premier low-carbon industrial center of the region. The project broke ground in June 2011, and is expected to be a low-carbon demonstration model for Hangzhou and the region when completed in 2016.

Hangzhou's green transportation programs are among the most advanced in the nation. Public transit is responsible for more than half of the total traffic volume. In 2008, the city introduced a public bicycle program covering most urban areas in its jurisdiction. The local government supplies some 50,000 bicycles, which are available for use at no or very low cost, depending on usage. The city now has 2,150 bicycle rental and return service points, servicing an average of approximately 172,000 users per day (Hangzhou Municipal Government, 2011).

Another element in Hangzhou's low-carbon development strategy is ecotourism. Hangzhou has historically been known as a city of natural beauty. Its most famous tourist attraction is West Lake, located in the center of the city. The vast number of tourists attracted to West Lake has made it difficult to conserve the Lake. The government addressed this challenge by developing and renovating a number of other tourist attractions in order to absorb tourism demand and direct tourist flows away from the lake. These new attractions are scattered across the metropolitan area, and most make a feature of the city's ecological integrity. This strategy is meant to ensure the sustainability of tourism in Hangzhou while providing more attractions for visitors.

The implementation of the six-in-one strategy involves local innovations of the political structure. While Hangzhou's plan and experience in low-carbon development encompasses a wide range of areas, each district in the municipality has only limited resources to invest in low-carbon projects. Had all districts been required to follow a comprehensive low-carbon strategy, that is, to invest in all six elements, very little could have been done in any of the districts. By each district specializing in one or two of the six components, they are able to pick elements that best fit its local conditions. For example, Jianggan District features green office buildings, Qianjiang District specializes in low-carbon industries, and Chun'an County heads ecotourism. Each of the districts has the autonomy to determine its specialization based on existing sectorial composition and public opinions.

Furthermore, many of the implementation projects involve multiple government departments and interest groups. In order to mitigate conflicts and ease coordination, the city selected members from every government department to create a low-carbon leadership committee, which reports to the Secretary-General of Hangzhou's Communist Party, the municipal government's top official. The committee is responsible for coordinating low-carbon city development projects across government agencies.

Similar to the case of Shanghai, Hangzhou's low-carbon development encompasses a wide range of areas. Government resources are channeled into supporting low-carbon development in a variety of sectors. More importantly, the case of Hangzhou demonstrates the environmental state's devolution down to the local level. Success of its overall low-carbon development is conditioned on the collaboration between each of the districts in Hangzhou. Furthermore, the environmental state is effectively "reinvented" in Hangzhou. The municipal government gives ample leeway to each of the district level governments in its jurisdiction to develop their own low-carbon specialties. Its low-carbon leadership committee is also the first of its kind in China that transcends conventional division within the government both vertically and horizontally. The devolution of government is most evident in Hangzhou.

Sino-Singapore Tianjin Eco-City

Like Shanghai, the city of Tianjin is one of four municipalities in China directly administered by the central government.[5] Sino-Singapore Tianjin Eco-City (SSTEC) is located in the northern part of Tianjin's Binhai New Area, a newly established development zone 45 km (28 miles) from the city center and 150 km (93 miles) from Beijing. SSTEC covers a total land area of 30 km^2 (11.58 square miles), and targets a total population of 350,000 people. Construction of SSTEC started in September 2009, and is planned to be completed by 2019.

The SSTEC project has received significant attention because of its aim to become a "model eco and low-carbon city replicable by other cities in China" (World Bank, 2009). The project receives consistent support from the highest levels of leadership in China and Singapore. In 2007, Chinese Premier Wen Jiabao and Singapore's Prime Minister Lee Hsien Loong signed a Framework Agreement to jointly develop the SSTEC. In 2008, Premier Wen and Singapore's Senior Minister Goh Chok Tong both attended its groundbreaking ceremony. This project is the second "flagship bilateral project" between China and Singapore, after the success of Suzhou Industrial Park, established in 1994.

SSTEC uses a set of Key Performance Indicators (KPIs) based on existing eco-city standards in China to monitor and track progress. This innovative system will measure the sustainability of the natural environment and the built environment, and allow officials in others cities to replicate SSTEC's tracking methods. SSTEC's performance on the indicators will be instrumental for the development of future low-carbon projects in China. According to local officials, existing standards in China are vague and not immediately applicable to Tianjin.

Community development is a top priority for SSTEC. The project aims for an ideal mix of work and life for its residents by following the planning standards of the United States Green Building Council (USGBC) Leadership in Energy and Environmental Design for Neighborhood Development (LEED-ND). The LEED-ND standard is used to evaluate environmental sustainability in urban design and community development, and is based on the idea that urban design, land use and the environment are inextricably linked. Buildings in the SSTEC neighborhood are expected to meet the Gold standard of the Tianjin Eco-City's Green Building Evaluation Standard (GBES).

SSTEC's economic development is also geared toward the reduction of carbon emissions in the local economy. Commercial buildings in SSTEC feature energy, water, and lighting efficient technologies. The site's industrial park has ready-built premises as well as customized building solutions that meet SSTEC's KPI criteria. These commercial buildings will showcase ecologically sustainable design features and many aim to achieve LEED Silver certification.

SSTEC relies on quantitative and qualitative measures throughout the planning and construction process. The project-specific GBES has proved to be an effective tool in the development of green buildings, while the established KPIs ensure the consistent implementation of the low-carbon plan. SSTEC demonstrates the successful use of multiple sets of indicators to monitor planning and implementation at both the micro- and macrolevel.

Technocracy is most evident in the case of Tianjin. The use of KPIs throughout the planning and construction stages effectively "blackboxed" much of the discussion of the project details. Its subscription to LEED-ND standard also makes discussion of community development fairly technical. In other words, the evidence suggests that the environmental state has exhibited tendencies toward technocracy only in the case of Tianjin. It is not to be neglected that Tianjin's project marks collaboration between Chinese authorities and their correspondents in Singapore, where the level of postindustrialization is much higher. It seems likely that SSTEC's technocracy is attributable to the involvement of Singaporean planners and policy experts. It is reasonable to infer, then, as Chinese cities become more postindustrial, they will likely witness a growth of technocratic domination of the environmental state.

At the same time, compared to the case of Shanghai and Hangzhou, Tianjin witnessed a stronger presence of the central state in its low-carbon development. Its project-specific GBES was reviewed and approved by the national Green Building Council. The development of the KPIs was also aided by experts from the Ministry of Environmental Protection, let alone the fact that top leaders in China repeatedly voiced their support of SSTEC. Consequently, Tianjin has less autonomy in making low-carbon development choices.

Liangjiang New Area, Chongqing

The city of Chongqing, like Shanghai and Tianjin, is one of the four directly administered municipalities. The Liangjiang New Area is located in northern Chongqing, covering a geographical area of 1,200 km^2 (463 square miles), although 650 km^2 (251 square miles) are uninhabitable mountainous areas. Liangjiang, whose name means "Two Rivers," is located where the Yangtze and Jialing Rivers converge. The establishment of Liangjiang New Area marks a step by China's central government to develop the country's western provinces, which remain relatively poor despite the economic development of the eastern regions.

Chongqing was one of China's major industrial centers during the twentieth century. Even today, industry and construction are two of the biggest drivers of the city's economic growth. Chongqing's history may make it seem like an unlikely case for a low-carbon development demonstration, but the project in Liangjiang represents a notable opportunity for change in traditionally industrial cities. Given the high environmental cost of Chongqing's economic activities in the past decades, the city's leadership views the establishment of Liangjiang New Area as an opportunity for change, and its focus is not only on creating sheer volume of economic activities but also on maintaining the "quality" of economic activities. Liangjiang's low-carbon development carries both symbolic and real meaning for China's historically industrial cities.

Heavy industry accounts for roughly two-thirds of Chongqing's industrial sector by GDP and the city has spent tremendous amounts of resources to enhance industrial energy efficiency in the past. While the energy savings achieved through technological upgrades are impressive, Chongqing has now turned to macroeconomic restructuring for more substantial reductions in carbon emissions in the local economy. Liangjiang's development is oriented toward this goal. The leadership in Chongqing views Liangjiang as a suitable location to experiment with bold economic strategies. If the projects demonstrate favorable results, these strategies may be applied to the rest of Chongqing and beyond. Currently, Liangjiang's strategy for low-carbon development is based on the creation of a financial service center and a renewable energy hub. The first strategy promotes economic activities with relatively low amount of carbon emissions, and the second strategy promotes growth of a low-carbon industry.

According to the plan, by 2015 Liangjiang will construct 6 million m^2 (2.31 square miles) of floor space tailored to the financial service sector, forming an initial base. By 2018, a full-fledged financial center is expected to be functioning, driving the growth of Liangjiang and contributing to the growth of the entire region. Development of Liangjiang's financial service center is viewed as a vehicle for driving restructuring of the local economy toward the service sector. It is important to note that the plan is being implemented only to the extent that infrastructure construction is under way. It remains unclear what specific policies will be enacted to attract financial service providers to Liangjiang to occupy this newly developed financial hub.

The potential for renewable energy is considered one of Chongqing's strategic advantages. The area's climate, existing industries, and macroeconomic conditions have led the city to determine that water source heat pump technology should be the primary solution for energy efficiency in buildings. Branches of both the Yangtze and Jialing Rivers flow through many parts of the city, creating a huge potential for water source heat technology. In addition, the city has studied the possibilities of using lake water, reservoir water, and underground water as sources and continues to look for optimal applications of this technology in buildings with the help of local research institutes and universities.

The national Ministry of Housing and Urban-Rural Development (MoHURD) has recognized Liangjiang's plan for the integrated use of renewable energy as one of ten national demonstration projects of energy efficiency. This recognition from the central government has created local momentum for more renewable energy projects and has led Chongqing to develop plans for a Special Committee on the Application of Renewable Energy in Buildings. However, MoHURD offers more conceptual than practical guidance. Not having specific action plans, the prospect of Liangjiang's low-carbon development remains uncertain. Lack of concerted planning is viewed as the biggest barrier for Liangjiang's low-carbon development (Wang & Zhao, 2010).

Chongqing's low-carbon development is framed as part of China's national policy to develop its western provinces. The positioning of Chongqing as a financial service center has practical meaning only when economic activities in western provinces grow to a certain level, so as to generate substantial needs for financial services. In other words, low-carbon development in Chongqing is conditioned on factors beyond the city's control, and relies on national development policies. Moreover, the city leadership exhibits a vested interest in the financial service and renewable energy sectors for materializing low-carbon development goals, but has not tried to diversify its strategy to incorporate other sectors into low-carbon development. The current foci of financial service and renewable energy evolved from the central state's conceptual framework, and lack localized visions for implementation. Chongqing therefore typifies a top-down approach of low-carbon development.

Baoding's Evolving Low-Carbon Goals

Baoding is the largest city in Northern China's Hebei Province by population. It is an early champion of low-carbon development. In 2006, during the city's 11th Five Year Plan (2006–2010), Baoding proposed a plan to become China's "Electricity Valley," specializing in alternative means of power generation, including wind and solar power. Baoding is expected to become the center of power storage technologies and other related businesses. Baoding is now home to more than 170 new energy companies. These companies contribute 10.2% of the local GDP, making new energy one of the city's "pillar industries," and this percentage is expected to reach 40% by 2050. To ensure the successful creation of the "Electricity Valley," Baoding has established an Electricity Valley Promotion Council, led by Baoding's Secretary-General of the Communist Party, the city's top government official.

In addition to "Electricity Valley," in 2007 the city announced plans to become a "City of Solar Energy." While the former is considered a strategic long-term objective, the latter is a relatively shorter-term component of the overall strategic vision. Specifically, the city will promote the use of photovoltaic lighting equipment in public buildings, major roads, and selected communities. Billboards, road signs, traffic lights, and other public infrastructure will also be reliant on solar energy.

Because of these successful initiatives, Baoding was selected by World Wildlife Fund (WWF) as one of two pilot cities to join an initiative exploring low-carbon development methods (World Wildlife Fund, 2008). This encouraged Baoding to embark on its effort to become a comprehensive low-carbon city. The city government issued a tentative plan for low-carbon city development at the end of 2008, confirming it in 2010 as Baoding's master plan for low-carbon development. In this plan, the government re-emphasized the role of several leading renewable energy companies in Baoding's overall planning, and further promoted the use of renewable energy in rural Baoding.

At the same time, Baoding has been designated by the National Ministry of Science and Technology as a model of renewable energy research, development and application. Resources continue to be channeled into the city to drive advances in renewable energy. Along the way, however, Baoding's success in the renewable energy sector seems to have become a barrier for comprehensive low-carbon development. The gains in investing in renewable energy become so high in Baoding that investing in other sectors for low-carbon development has to bear high opportunity costs. In addition, as investments for renewable energy lands in Baoding from the central state, the city has to further develop its niche in the area. Consequently, Baoding's low-carbon development remains very much tied to renewable energy.

Even though the government wishes to pursue low-carbon development at a more general level, the environmental state has a limited scope in Baoding. Despite available resources from organizations such as the WWF to support comprehensive low-carbon development in Baoding, it remains to be seen how the city will extend its current strategy beyond the renewable energy sector.

Despite the fact that early incarnations of low-carbon development in Baoding, its "Electricity Valley" and "City of Solar Energy" initiatives, were locally created and implemented, the successes of such has brought the city to the central state's attention. As a result, the Chinese state supports the city's renewable energy sector with substantial support, at a time when the city expects to bring its low-carbon development to a more comprehensive level. Not surprisingly, the central state of mind outmaneuvers its local correspondences. Furthermore, both Chongqing and Baoding are cases where low-carbon development explicitly prioritizes a limited number of sectors, contrary to Shanghai, Hangzhou and Tianjin's comprehensive approach.

CONCLUSION

Two general observations can be outlined. First, in cities where the level of postindustrialization is high, state resources support innovative low-carbon development strategies that attempt to achieve emission reductions in a variety of sectors. In industrial cities, however, the environmental state's regulative power is limited to one or two (sub)sectors. Consequently, the environmental state finds it most effective and efficient to seek low-carbon development within the industrial sector. In cities where the postindustrial condition has taken roots, the economic and social structures have become much more complex so as to open up the potential of achieving low-carbon development in a variety of ways. In other words, the environmental state is in a more versatile shape when landed in postindustrial cities. In order to achieve high amounts of carbon emissions, it is crucially importantly for cities to reach into a variety of sectors, due to the simple fact that more sectors, more sources of reduction. This is precisely why city leaderships in Chongqing and Baoding are working hard to try to diversify its low-carbon development.

In a way, postindustrial cities represent the future of China that differs dramatically from the past as represented by industrial cities. Since the late 1990s, as a major part of the country's sustainable development commitment, China has invested considerable resources in research and development of technology to successfully improve industrial energy efficiency (Price, Worrell, Sinton, & Yun, 2001). Such a strategy is insufficient under today and tomorrow's postindustrial conditions. As the empirical evidence suggests, postindustrial cities can and do make substantial gains in carbon emission reductions by way of developing the transportation, building, tourism, and other nonindustrial sectors. Such gains will likely accrue as China furthers its economic restructuring. In other words, as the process of postindustrialization deepens in China, the environmental state will likely prosper. The environmental state is a postindustrial phenomenon.

Second, and more importantly, a new pattern of governance is emerging in postindustrial cities. The environmental state inherits time-tested regulatory tools in China, but also exhibits signs of regulatory innovation. Specifically, initial development is driven by demonstration projects as recognized and supported by the state, a practice characteristic of China's recent political history. On the other hand, low-carbon development in postindustrial cities is a much less centralized process, having local levels of governments as key players of low-carbon policy making. When the environmental state intersects with the postindustrial city, it gives birth to a new urbanism (Wu, 2011) that has profound implications for political structuring in China. This new urbanism is characterized with devolution of state authority. In postindustrial cities, local governments are no longer subject to the absolute totality of the central government. Consequently, the environmental state tailors its strategy to each local context. Environmental state action thus becomes interactively shaped by multiple levels of the government.

The empirical findings from this study allow for an assessment of theoretical insights in the environmental state and postindustrial city literatures. Contrary to both literatures, there is little sign of the civil society shaping environmental politics in China. Given extant knowledge of the authoritarian state in China, this does not come at surprise. More importantly, the environmental state's decentralizing tendency and the postindustrial city's technocratic domination are both supported by empirical evidence. This implies two things. First, a general theory of the environmental state can be built from existing literature, but needs to be sensitive to non-Western social conditions that might falsify parts of the theoretical claims. Those sets of arguments about the civil society seem to have limited scope conditions, but the other theoretical insights have promising applicability to a wide range of contexts. Second, the environmental state literature can be consolidated and further developed when examined in conjunction with other literatures in the modernity tradition. I have demonstrated the connection between the environmental state and the postindustrial city. More studies are needed to examine other facets of the environmental state, as it intersects with a multitude of (post)modern conditions.

NOTES

1. A variety of terms, including environmental state (Frickel & Davidson, 2004; Mol & Buttel, 2002), green state (Dryzek, 2003; Eckersley, 2004), environmental governance (Davidson & Frickel, 2004) and nation-state environmentalization (Frank et al., 2000), have been used to describe the phenomenon. I treat all of these terms as synonymous, and use "environmental state" throughout this chapter for sake of consistency.
2. Readers interested in quantitative approaches to the subject matter can consult Urban China Initiative (2010) and Yip (2008).
3. Because of sensitivity of planning information in China, all interviews were conducted on condition of anonymity.
4. The city's pioneering efforts have been honored nationally and internationally, including the 2001 "UN Habitat Scroll of Honor Award" and the designation as a National Garden City by China's central government, among many others. In 2011, Hangzhou's West Lake was approved as a UNESCO World Heritage Site.
5. Although most cities are subordinate to county and provincial administrations in China, the four directly administered municipalities, Beijing, Tianjin, Shanghai and Chongqing, are directly under the central government.

ACKNOWLEDGMENTS

I gratefully acknowledge the Institute for Building Efficiency at Johnson Controls Inc. for supporting early stages of the project. The chapter is inconceivable without help from Michelle Bai, Tim Griffiths, Chitra Hepburn, Jennivine Kwan, Xun Li, Weiding Long, Jimmy Lu, Haixiao Pan, Tao Pan, Mark Wehling, Matthew Xue, Jianrong Yang, Stanley Yip, and several government officials who spoke on condition of anonymity. Special thanks are due to Julia Currie, Gary Green, Jennifer Layke, Victor Lippit, Jamie Shepherd, Terence Siau, Chris Stcavish, and Elaine Wang for their helpful comments on earlier drafts. All remaining errors are mine.

REFERENCES

Beck, U., Giddens, A., & Lash, S. (1994). *Reflexive modernization: Politics, tradition and aesthetics in the modern social order*. Stanford, CA: Stanford University Press.

Bell, D. (1973). *The coming of post-industrial society*. New York, NY: Basic Books.

Bornstein, R. D. (1968). Observations of the urban heat island effect in New York city. *Journal of Applied Meteorology*, *7*(4), 575–582.

Buttel, F. H. (2000). World society, the nation-state, and environmental protection: Comment on Frank, Hironaka, and Schofer. *American Sociological Review*, *65*(1), 117–121. Looking Forward, Looking Back: Continuity and Change at the Turn of the Millennium.

Chen, Z. (2010). Five billion yuan to kick off low-carbon Hangzhou. *Economics Review*. Retrieved from http://epub.cnki.net/grid2008/brief/detailj.aspx?filename=JJGC201003220150&dbname=CCND2010

Chinese Society for Urban Studies. (2011). *The cases of Chinese eco-city*. Retrieved from http://www.cityup.org/eco_city/download/download/1305532924390.pdf

Davidson, D. J., & Frickel, S. (2004). Understanding environmental governance. *Organization & Environment*, *17*(4), 471–492. doi: 10.1177/1086026603259086

Dryzek, J. S. (2003). *Green states and social movements: Environmentalism in the United States, United Kingdom, Germany, and Norway*. Oxford: Oxford University Press.

Eckersley, R. (2004). *The green state: Rethinking democracy and sovereignty*. Cambridge, MA: MIT Press.

Esping-Andersen, G. (1993). Mobility regimes and class formation. In G. Esping-Andersen (Ed.), *Changing classes: Stratification and mobility in post-industrial societies* (pp. 225–241). London: Sage.

Evans, P. B. (1995). *Embedded autonomy: States and industrial transformation*. Princeton, NJ: Princeton University Press.

Fisher, D. R., & Freudenburg, W. R. (2004). Postindustrialization and environmental quality: An empirical analysis of the environmental state. *Social Forces*, *83*(1), 157–188.

Frank, D. J., Hironaka, A., & Schofer, E. (2000). The nation-state and the natural environment over the twentieth century. *American Sociological Review*, *65*(1), 96–116. Looking Forward, Looking Back: Continuity and Change at the Turn of the Millenium.

Frickel, S., & Davidson, D. J. (2004). Building environmental states. *International Sociology*, *19*(1), 89–110. doi: 10.1177/0268580904040922

Galbraith, J. K. (1967). *The new industrial state*. Boston, MA: Houghton Mifflin.

Geltman, E. G., & Skroback, A. E. (1998). Reinventing the EPA to conform with the new American environmentality. *Columbia Journal of Environmental Law*, *23*(1), 1–56.

Goldman, M. (2001). Constructing an environmental state: Eco-governmentality and other transnational practices of a 'green' world bank. *Social Problems*, *48*(4), 499–523.

Haas, P. M. (1994). *Complex cooperation: Institutions and processes in international resource management*. Oslo: Scandinavian University Press.

Habermas, J. (1989). *The structural transformation of the public sphere: An inquiry into a category of bourgeois society*. Cambridge, MA: MIT Press.

Hajer, M. A. (1995). *The politics of environmental discourse: Ecological modernization and the policy process*. Oxford: Clarendon Press.

Hangzhou Municipal Government. (2011). *Making a livable city*. Retrieved from http://www.hangzhou.gov.cn/main/zjhz/hzlj/2010/tj/T350091.shtml

Janicke, M. (2006). The environmental state and environmental flows. In G. Spaargaren, A. P. J. Mol & F. H. Buttel (Eds.), *Governing environmental flows: Global challenges to social theory* (pp. 83–105). Cambridge, MA: MIT Press.

Lei, H., Zhuang, G., & Zhang, C. (2011). *Understanding low-carbon city development in china: Strategy and method.* Beijing: China Environmental Science Press. Retrieved from http://www.wwfchina.org/wwfpress/publication/climate/lowcity_report.pdf

Ley, D. (1980). Liberal ideology and the postindustrial city. *Annals of the Association of American Geographers*, *70*(2), 238–258.

Liang, B., Yang, Z., Meng, K., Xu, L., & Zhao, D. (2010). Application of water source heat pumps in Shanghai Expo. *Water Use* (05), 96–100.

Lin, B. (2009). *Low-carbon urbanization is way forward for China.* Retrieved from http://www.chinadaily.com.cn/cndy/2009-11/30/content_9073968.htm

Liu, J. (2011). *Service sector: China's emerging powerhouse.* Retrieved from http://www.chinadaily.com.cn/cndy/2011-03/02/content_12098593.htm

McKinsey Global Institute. (2009). *Preparing for China's urban billion.* Retrieved from http://www.mckinsey.com/mgi/reports/pdfs/China_Urban_Billion/China_urban_billion_full_report.pdf

Mol, A. P. J. (2006). Environment and modernity in transitional china: Frontiers of ecological modernization. *Development and Change*, *37*(1), 29–56. doi: 10.1111/j.0012-155X.2006.00468.x

Mol, A. P. J., & Buttel, F. H. (2002). *The environmental state under pressure.* Amsterdam: JAI.

Mol, A. P. J., & Spaargaren, G. (2000). Ecological modernisation around the world: An introduction. In A. P. J. Mol & D. A. Sonnenfeld (Eds.), *Ecological modernisation around the world: Persectives and critical debates* (pp. 3–16). London: Frank Cass.

National Development and Reform Commission. (2009). *China's policies and actions for addressing climate change.* Beijing: Information Office of the State Council of the People's Republic of China.

Niu, D., Pan, T., & Cao, X. (2011). *Low-carbon city development: Concept and practices* (2nd ed.). Unpublished manuscript.

Perloff, H. S. (1980). *Planning the post-industrial city.* Washington, DC: Planners Press.

Price, L., Worrell, E., Sinton, J., & Yun, J. (2001). *Industrial energy efficiency policy in China.*University of California. Retrieved from http://ies.lbl.gov/iespubs/50452.pdf

Redclift, M. (1992). Sustainable development and popular participation: A framework for analysis. In D. P. Ghai & J. M. Vivian (Eds.), *Grassroots environmental action: People's participation in sustainable development* (pp. 23–49). London: Routledge.

Robinson, J. (2011). Cities in a world of cities: The comparative gesture. *International Journal of Urban and Regional Research*, *35*(1), 1–23. doi: 10.1111/j.1468-2427.2010.00982.x

Rosenbaum, W. A. (1999). Escaping the "Battered agency syndrome": EPA's gamble with regulatory reinvention. In M. E. Kraft & N. J. Vig (Eds.), *Environmental policy: New directions for the twenty-first century* (4th ed., pp. 165–189). Washington, DC: Cq Press.

Savitch, H. V. (1988). *Post-industrial cities: Politics and planning in New York, Paris, and London.* Princeton, NJ: Princeton University Press.

United Nations Development Program. (2010). *China and a sustainable future: Towards a low carbon economy and society.* Beijing: China Translation and Publishing Corporation. Retrieved from http://content.undp.org/go/cms-service/stream/asset/?asset_id=2471266

Urban China Initiative. (2010). *The urban sustainability index: A new tool for measuring China's cities.* Retrieved from http://urbanchinainitiative.typepad.com/files/usi.pdf

Wang, H., & Zhao, A. (2010). Path analysis of Chongqing's low-carbon economic development. *Science, Technology and Market* (11), 41–42.

Wang, X., Berrah, N., Mathur, S., & Vinuya, F. (2010). *Winds of change east Asia's sustainable energy future*. World Bank Group.

World Bank. (2009). *Sino-Singapore Tianjin eco-city a case study of an emerging eco-city in china*. Washington, DC: World Bank.

World Wildlife Fund. (2008). Shanghai and Baoding to become China's low carbon city pilots. *WWF China Newsletter*. Retrieved from http://www.wwfchina.org/english/downloads/newsletter/newsletter1-3.08.pdf

Wu, F. (2011). Retreat from a totalitarian society: China's urbanism in the making. In G. Bridge & S. Watson (Eds.), *The new Blackwell companion to the city* (pp. 701–712). Malden, MA: Wiley-Blackwell.

Yip, S. (2008). Eco-city: From concept to implementation with Shanghai Dongtan and Beijing Changxingdian as cases. *City Planning Review*, (08), 15–20 + 27 + 98.

Yu, L. (2011). The current status of low-carbon eco cities in china. Paper presented at the 2011 Annual Conference of the Chinese Society for Urban Studies. Yangzhou, China.

CHAPTER 10

URBAN VULNERABILITY AND ADAPTATION TO THE HEALTH IMPACTS OF AIR POLLUTION AND CLIMATE EXTREMES IN LATIN AMERICAN CITIES

Patricia Romero-Lankao, Hua Qin, Sara Hughes, Melissa Haeffner and Mercy Borbor-Cordova

ABSTRACT

Purpose – *The vulnerability and adaptive capacities of cities in Latin America have received relatively less attention compared to other regions of the world. This chapter seeks to address these gaps by (a) examining vulnerability to the health impacts from air pollution and temperature, and exploring whether socioeconomic factors between neighborhoods differentiate these risks within the cities of Bogota, Buenos Aires, Mexico City, and Santiago and (b) assessing the capacity of urban populations to perceive and respond to vulnerability and risk.*

Design/methodology/approach – *Because of the complex nature of vulnerability, we combined a set of quantitative and quantitative methods and data to determine whether and under what conditions the people in*

Urban Areas and Global Climate Change
Research in Urban Sociology, Volume 12, 247–275

ISSN: 1047-0042/doi:10.1108/S1047-0042(2012)0000012013

these cities are vulnerable (e.g., Time Series Analysis, Generalized Linear Model, and statistical correlations of exposure and human mortality with socioeconomic vulnerability).

Findings – *We found high levels of PM_{10}, ozone, and other criteria air pollutants in three cities for which we had data. However, the pattern of their impacts on health depends on the particulars of pollutant levels and atmospheric and weather conditions of each city. Our results reflect the varied facets of urban vulnerability and shed light on the nature of the associated human health risks. Although wealthy populations have access to education, good quality housing, and health services to mitigate some environmental risks, overall the data show that health impacts from air pollution and temperature in the study cities do not necessarily depend on socioeconomic differentiations.*

Research limitations/implications – *Although we sought to use quantitative and qualitative methods, given the complexity of the research, it has proven difficult to fully explore these issues across scales and with a full accounting of local context.*

Practical implications – *Our findings show that wealthy and educated populations may be equally at risk to the health implications of air pollution. Policies designed to mitigate these risks should not use socioeconomic characteristics as predictors of a population's risk in relation to air pollution.*

Originality/value – *This research contributes valuable insights into the dynamics of vulnerability to air pollution in Latin American cities, a region that has been historically underrepresented in empirical studies of urban risk. We have also combined a range of methods and approaches to improve our understanding of the multifaceted nature of urban vulnerability to global environmental change.*

Keywords: Urban vulnerability and risk; adaptive capacity; socioeconomic differentiation; health risks; air pollutants; climate change

INTRODUCTION

The current context of a changing climate and rapidly growing cities means that urban populations may be at greater risk for the negative health impacts of extreme temperatures and air pollution than ever before. Research has shown, for example, that the exposure of urban populations in Latin America to these hazards is expected to increase with climate change (Magrin et al., 2007). Scholars have previously examined how air pollution and weather patterns affect the risk of negative health impacts including mortality (Bell et al., 2008; O'Neill, Zanobetti, & Schwartz, 2005; Ostro, Sanchez, Aranda, & Eskeland, 1995). However, research on urban environmental risk has yet to develop an understanding of the complexity of factors that shape the vulnerability of cities and urban populations to these hazards and the ways in which differences in this vulnerability are created within and between cities. Further, although a number of studies have attempted to predict how climate change will affect urban areas, cities are also primed to play an important role in efforts to adapt to climate change and to find innovative ways of transitioning to more sustainable economic development pathways. This role of cities as adaptive and innovative actors is generally understudied.

The vulnerability and adaptive capacities of cities in Latin America have received particularly scarce attention compared to other regions of the world. The research project "Adaptation to the Health Impacts of Air Pollution and Climate Extremes in Latin American Cities" (ADAPTE) was initiated to explore the complex nature of urban vulnerability and adaptation to air pollution and climate change in the Latin American cities of Bogota, Buenos Aires, Mexico City, and Santiago. The interdisciplinary and international team seeks to address gaps in existing research by (a) examining vulnerability to the health impacts from air pollution and temperature, and exploring whether socioeconomic factors between neighborhoods differentiate these risks within cities, and (b) assessing the capacity of urban populations to perceive and respond to vulnerability and risk. This chapter highlights the main findings from the research efforts of ADAPTE from 2009 to 2012.

The chapter begins with an overview of three approaches scholars have used to understand and measure the determinants of urban vulnerability to climate change and the development of our own framework (second section). It then presents some background on the environmental and economic conditions of the four Latin American cities included in ADAPTE (third section). The fourth section describes the set of methods and data utilized in the project. Next, in the fifth section results from quantitative assessments of urban vulnerability to health risks to air pollution and temperature variability in three of the four cities are presented. This section also highlights findings from the analysis of a household survey on the capacity of urban populations in the four cities to perceive and respond to vulnerability and risk. We finally conclude with some reflections on how vulnerability is constructed and perceived in the four cities, and the opportunities and challenges of doing interdisciplinary research on urban vulnerability to climate change.

APPROACHING URBAN VULNERABILITY TO CLIMATE CHANGE

Understanding urban vulnerability to the hazards presented by climate change (e.g., more extreme temperatures and higher concentrations of air pollution) requires first understanding risk. While there are many definitions of risk, they have in common the idea that risk is the possibility, probability, or likelihood of harmful consequences and damage (e.g., mortality). For example, when a city is experiencing higher than average summer temperatures (a hazard), there is a risk that this will kill some people. While risk refers to possible harmful consequences, vulnerability relates to the degree to which the system is susceptible to, or able to adapt to, the adverse effects of one or more risks. In a very vulnerable city (or neighborhood), more people would die during a summer heat wave than in a less vulnerable city (or neighborhood). The important role for research then is to understand what makes a given city, or neighborhood within that city, more or less vulnerable, and to gain an understanding of these differences that is relevant for policy and decision making at relevant scales. This is a challenge because of the number and specificity of risks facing each city and the range of often synergistic factors that can contribute to overall levels of vulnerability.

Existing research on vulnerability is fragmented by disciplinary differences in definitions and methods. For example, the Pro Prevention Consortium (2010) identified 25 methodologies for identifying risks and Thywissen (2006) identified 36 and 25 definitions of vulnerability and risk, respectively. The numerous lineages of vulnerability research these methodologies and definitions represent (e.g., disaster risk community, political ecology, and climate change community) have been summarized previously (Birkmann, 2006; Kienberger, 2010; Romero-Lankao & Qin, 2011). Therefore, here we briefly refer to three approaches that have informed the vulnerability framework developed for ADAPTE: urban vulnerability as impact, inherent vulnerability, and risk society theory (Romero-Lankao & Qin, 2011).

Identifying and understanding the factors underlying differences in vulnerability is a central theme for urban vulnerability research. Scholars within the *urban vulnerability as impact* tradition determine the vulnerability of urban populations based on a population's exposure to a hazard and socioeconomic characteristics. Scholarship points to the fact that not all demographic segments of the urban populations are equally affected by temperature, air pollution, and other hazards climate change is predicted to aggravate. The impact of concern is usually a health-related outcome such as hospitalization or mortality and a large proportion of these studies has focused on cities in North America and Western Europe (Curriero et al., 2002; Dockery et al., 1993; Fouillet et al., 2007; Jerrett et al., 2009; Katsouyanni et al., 1993; Knowlton et al., 2007; Medina-Ramón, Zanobetti, Cavanagh, & Schwartz, 2006), whereas only limited research has been conducted in Latin America (Bell et al., 2008; O'Neill et al., 2005; Ostro et al., 1995). Urban vulnerability as impact research has shown that a "very small change in ambient pollution levels can have substantial impacts on large populations, even if that change is associated with only a small increase in risk, since virtually all members of the population are exposed" (Makri & Stilianakis, 2007, p. 2). However, the effects of this exposure do not always follow a linear form. For example, the effects of temperature have a an U- or V-shaped relationship with mortality: deaths increase as temperatures fall below or rise above certain threshold values (McMichael et al., 2008; Michelozzi et al., 2006; Muggeo & Hajat, 2009).

Urban vulnerability as impact research has also shown that socioeconomic characteristics such as age, poverty, and gender can modify or confound the severity of the health effects of hazards such as extreme temperatures and air pollution (Basu, 2009; Naess, Piro, Nafstad, Smith, & Leyland, 2007; Reid et al., 2008). For example, groups such as the elderly, the very young, and people with preexisting medical conditions have been shown to be more sensitive to environmental hazards (Chestnut, Breffle, Smith, & Kalkstein, 1988; Dear, Ranmuthugala, Kjellström, Skinner, & Hanigan, 2005; Pope & Dockery, 1995). Lower levels of education are also associated with higher levels of mortality from extreme temperatures and air pollution (Medina-Ramón and Schwartz, 2007; Smoyer, Kalkstein, Greene, & Ye, 2000). However, the effects of such socioeconomic features as income, poverty, and ethnicity are mixed (D'Ippoliti et al., 2010; Ishigami et al., 2008; O'Neil et al., 2005; Stafoggia et al., 2006).

The influence of socioeconomic factors has also been explored by *inherent urban vulnerability* research, which has a different entry point for analysis: adaptive capacity. Many scholars in this tradition employ a livelihoods approach to explore whether assets and options at the individual, family, or community level (e.g., self-help housing and access to social networks) allow individuals and households to adapt to the hazards they constantly encounter (Moser & Satterthwaite, 2010; Pelling, 2003). Research has shown that in many cases these personal assets are not enough to reduce urban populations' vulnerability because of the role the state plays in shaping adaptive capacity, for instance, by promoting economic growth and poverty reduction. The study of state-centered interventions is lacking in the livelihoods approach despite the fact that they may be a necessary component of vulnerability reduction as the state may be able to construct and maintain such determinants of adaptive capacity as infrastructures and services, land and housing, and public emergency-response systems (Parnell, Simon, & Vogel, 2007; Romero-Lankao & Qin, 2011; Satterthwaite, Huq, Pelling, Reid, & Romero-Lankao, 2007).

Finally, the underlying premise of Ulrich Beck's (1986) risk society theory is that in the current economic era of capitalism, science and technology have become drivers of two processes. First, both become the central mechanism to increase the production of goods, and thus to reduce material needs. Second, they are also both sources of "bads," such as the negative by-products of industrialization, climate change, and air pollution. These bads in turn create risks of uncertain proportions and constrain the possibilities wealthy sectors have to escape from risk or compensate for risks. This constraint on wealthy populations – populations that previously had access to tools and resources for adaptation – increases their vulnerability. Beck refers to this dynamic as the "boomerang effect" and concludes that while "hunger is hierarchical, smog is democratic".[1] The risk society approach to urban vulnerability is therefore distinctive in that it opens options to suggest that differences in vulnerability are not only due to characteristics of the city and its population but also characteristics of the hazard itself.

In ADAPTE, we acknowledge the complex nature of both the health impacts of temperature extremes and air pollution and the determinants of vulnerability. Drawing on these research traditions – and our definition of vulnerability as the degree to which the system is susceptible to, or able to adapt to, the adverse effects of one or more risks – we see the important components of urban vulnerability as exposure to hazards, sensitivity, adaptive capacity, and actual adaptation (Birkmann, 2006; Gallopín, 2006; O'Brien, Eriksen, Nygaard, & Schjolden, 2007; Romero-Lankao & Qin, 2011; UN/ISDR, 2009) (see Fig. 1). *Exposure* is the extent to which urban populations are in contact with or subject to air pollution, temperature, and other hazards. *Sensitivity* is the degree to which subsets of urban populations are susceptible to hazards with patterns of susceptibility often based on such demographic characteristics as age or health. *Adaptive capacity* is the ability to avoid or lessen the negative consequences of

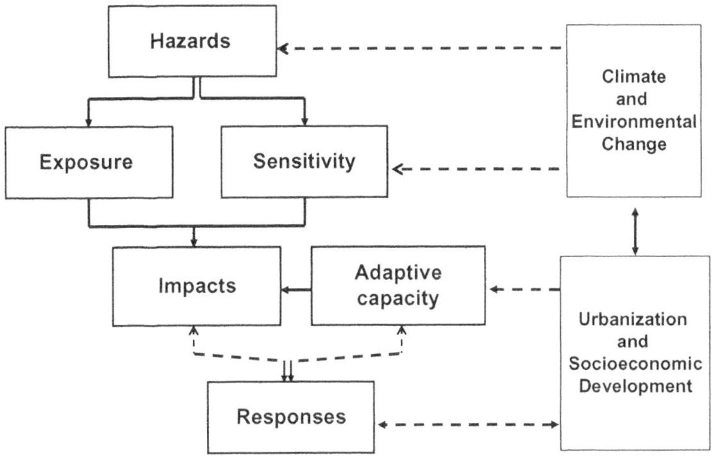

Fig. 1. A Conceptual Framework of Urban Vulnerability to Global Climate and Environmental Change. *Source:* Romero-Lankao and Qin (2011).

hazards based on access to resources, assets, and options. Urban populations and decision makers draw on these resources, assets, and options to moderate potential damages, to cope with the consequences, or to introduce policy changes to expand the range of variability with which it can cope. While actual coping and adaptation actions depend on many determinants of adaptive capacity, the two concepts are also fundamentally different in that adaptive capacity represents a potential for action.

STUDY CITIES: BOGOTA, BUENOS AIRES, MEXICO CITY, AND SANTIAGO

Atmospheric and Climatic Conditions

The climate of the four cities ranges from humid subtropical (Buenos Aires) and Mediterranean (Santiago) to subtropical highland (Bogotá and Mexico City). While no city experiences extreme variations in temperature, there are more seasonal variations in Mexico City, Buenos Aires, and Santiago than in Bogotá (Table 1). Atmospheric conditions can be conducive to air pollutant retention and ozone formation in Bogota, Mexico City, and Santiago, or to the creation of heat waves in Buenos Aires. During the dry

Table 1. Comparative Weather Indicators.

	Bogotá	Buenos Aires	México	Santiago
Latitude	4°32 N	34°35 S	9°26 N	33°28 S
Average annual temperature	14.0	17.96	15.0	14.0
Average & max summer temp (°C)	13.7 & 19.1	22.9 & 27.7	17.8 & 24.5	19.8 & 26.8
Average & min winter temp (°C)	13.6 & 8.2	11.9 & 8.3	16.1 & 9.9	12.8 & 8.0
Annual PM_{10}	68.75	N.D.	51.60	70.18
Annual O_3	11.57	N.D.	32.4	29.84

Source: ADAPTE's own calculations with data from the Meteorological offices of each country.

and winter months, the basin of Mexico City is under the influence of high pressure or anticyclone systems, which lead to the creation of thermal inversions at night and during the first hours of the morning. This explains why atmospheric pollutants remain trapped within the basin. Santiago also experiences thermal inversions during winter months.

Changes in climate at the global and local levels (the latter resulting from land use changes and urban heat-island effects) are expected to alter the existing weather conditions of these cities (Magrin et al., 2007). General Circulation Models suggest increases in mean temperatures of between 1°C and 4 °C, as well as more intense urban heat (Magrin et al., 2007). In Mexico City, for example, mean temperatures have increased by about 1.6°C during the last century as a result of such factors as land-use changes, the heat-island effect, and transformations in the hydrological cycle (Romero-Lankao, 2010). Results from General Circulation Models suggest that Mexico City will register an increase in mean temperatures of up to 4°C together with an up to 20% predicted decrease in mean precipitation by 2080 (INE, 2008). Expected changes in extremes for this city and for Santiago also include the alternation of more intense droughts and a significant decrease in the areas experiencing freezing temperatures (0°C or less). More frequent heat waves are also expected for Santiago and Mexico City (Magrin et al., 2007).

Air pollution and mean regional temperatures in these cities will be intensified with climate change. Such systemic changes will tend to affect larger segments of the population and cut across social and economic boundaries. This means that those responsible for the increase in air pollution and intensity of regional temperatures (i.e., the wealthy) will also begin to feel their effects. Yet, before we can begin to understand whether these cities will be able to adapt to the anticipated impacts of climate change, we still need to understand the vulnerability of urban populations to existing health risks.

Urban Development and Environmental Change

Bogota, Buenos Aires, Mexico City, and Santiago are each the primary centers of their country's economy, generating 25%, 24%, 34%, and 43% of their respective national GDP; each concentrates populations, economic activities, energy, and atmospheric emissions; and last but not least, each is especially affected by hazards climate change is expected to aggravate such as air pollution, changes in average temperatures, and floods. However, rather than always being negatively affected by hazards, demographic, socioeconomic, political and cultural factors give urban populations, infrastructures, and economic activities the ability to bounce back, recover from, and even take advantages of climatic and nonclimatic stresses.

Differentiated urban development is likely to shape the adaptive capacity of these urban populations. With all their dynamism, high levels of integration in the global economy, and presence of a creative middle class, these cities are still faced with high levels of poverty, income inequality, and informality (Table 2). Although the local patterns of population and economic activities have changed in recent decades (with a polycentric urban expansion into second-order localities functionally linked to the city through transportation corridors), spatial segregation within the city remains a key characteristic. Core areas have registered slower growth and in some cases decay; high income, gated communities have grown in suburban and peri-urban areas; and low income, often informal, settlements populate the periphery. Uneven development and inadequate infrastructure and governance structures constrain the ability of urban populations and authorities to adapt to existing and future hazards and stresses. The cities have deficits in such key determinants of adaptive capacity as health (with high infant mortality rates in Mexico and Bogota), education (with socially segregated school systems in all cities), and in housing with inadequate housing stock, informal settlements (exception Santiago), and problems of housing affordability in all cities. Frequently, decaying central areas and peri-urban areas are being inhabited by marginalized populations with inadequate services, a portfolio of precarious livelihood mechanisms, and the absence of appropriate risk-management institutions.

Table 2. Selected Socioeconomic and Demographic Indicators.

Indicator	Bogotá	Buenos Aires	México	Santiago
Population[a]	6,776,009	12,828,669	17,946,313	5,392,804
GDP per capita in US-$ (ppp)[b]	12,692	19,098	17,552	16,826
Elderly (64 and over)[a]	6%	15%	3.9%	6.4%
Persons without high school education[a]	39.50%	22.8%	45.7%	35%
GINI coefficient[b]	0.61	0.44	0.56	0.55
Infant mortality rates (per thousand)[b]	13.5	7.9	17.8	7.6
Informal employment *(% of total workforce)*[b]	n.d.	44	45.7	34
Slum population (% of inhabitants)[b]	16.8	26.2	14.4	5
Homicides per 100 thousand[b]	18.7	6.9	17.6	2.1

Sources:
[a]ADAPTE's own calculations.
[b]Jordán, Rehner, and Samaniego (2010).

Last but not least the levels of crime and violence are high in Mexico and Bogota (Table 2) which might undermine the influence of social capital (i.e., individual levels of social trust and participation in networks) on adaptive capacity.

METHODS AND DATA

We combined a set of quantitative and qualitative methods and data that draw from both the vulnerability as impact and inherent vulnerability approaches to determine whether and under what conditions the people in these cities are vulnerable to temperature and air pollution hazards. Daily temperature data from the meteorological stations of each city provide information on maximum, mean, and minimum temperature. Air pollution data on three primary criteria pollutants, particulate matter (PM_{10}), nitrogen dioxide (NO_2), and ozone, were collected from each city's environmental agency. We obtained the following mortality data from the public health agencies of each city: respiratory mortality (International Classification of Diseases, or ICD 10 cause J) and cardiovascular mortality

(International Classification of Diseases or ICD 10 cause I). In addition, sociodemographic data such as education, poverty, income, age structure, and housing condition from the study cities' census offices were used to construct municipality-level measures of vulnerability based on the livelihoods approach. We developed a multidimensional vulnerability index (MVI) using a multi-criteria model of socioeconomic vulnerability that is based on four different types of livelihood assets: social, human, physical, and financial capitals (Baud, Sridharan, & Pfeffer, 2008).[2]

We used various statistical tools to identify the characteristics of and interactions among such dimension of vulnerability as hazards, exposure, and health impacts. Exploratory Time Series Analysis helped us identify the temporal patterns of air pollution and temperature. A Generalized Linear Model (GLM) with Poisson log-linear distribution was used to explore the relative risks (RR) of mortality from exposure to hazards. We also tested the statistical correlations of exposure to major air pollutants and human mortality with socioeconomic vulnerability to examine whether municipalities with different vulnerability levels can be differentiated with respect to health risks.[3]

Statistical analyses of secondary data are not able to capture the more subtle components of urban vulnerability such as populations' and decision makers' perceptions of risks, whether urban populations actually cope with heat/cold stress and air pollution and, if they do, what mechanisms and adaptations allow them to cope and adapt. Further, these large-scale analyses cannot help determine whether warning systems, health services, and other governmental actions actually help reduce the impacts of hazards on the health of the populations. Therefore, we supplemented the quantitative assessments with household surveys, key informant interviews, and workshops. Our goal was to better gauge such different dimensions of urban vulnerability as exposure, sensitivity, adaptive capacity, and actual adaptations among governmental organizations and populations of selected communities and their perception and knowledge of risks and vulnerabilities.

In this chapter, we will only refer to the results from the surveys, while the analysis of in-depth interviews and workshops are still ongoing. The household surveys were administered to a nonrandom sample of households in communities selected to represent a range of environmental risk and socioeconomic status. We used two main sets of indicators to select the communities where 100 questionnaires per city were conducted: (a) physical and geographical characteristics, that is, degree of environmental risk and (b) socioeconomic status. The surveyed communities include Pudahel and La Florida, in Santiago; Tecano, Las Malvinas, Barrio Caltongo, Canal de

Nativitas, and UH Rinconada del Sur, in Mexico City; San Jose, Club San Fernando, Alvear, Alsina, and Canal in Buenos Aires; and Codito, Fontibon, and Juan Rey, in Bogota. In the survey we further explore how people perceive risks in general and climate relevant risks in particular and how their perceived vulnerability has unfolded over time. The preliminary analysis of the survey data is mainly descriptive. Chi-square test and one factor analysis of variance (ANOVA) with Tukey's post hoc test were also used to examine city variations on key variables in the survey.

FINDINGS

This section presents results on some of the features and linkages between the components of urban vulnerability. It explores the risks of mortality from exposure to hazards as well as whether socioeconomic factors between neighborhoods differentiate these risks within cities. Key findings from the household survey are also included to describe the assets, options, and public support local populations can draw on in response to environmental hazards.

Exposure to Temperature and Air Pollution

Air quality data for Mexico City, Santiago, and Bogotá confirmed that air pollution is one of the most important sources of health risks to populations in urban Latin America (Cifuentes, Borja-Aburto, Gouveia, Thurston, & Davis, 2001). After organizing the air quality data for Mexico City, Santiago and Bogotá, and comparing them with the World Health Organization recommendations for air quality, we found that Santiago has the highest concentration of PM_{10} while Mexico City presents the higher levels of ozone on average (Table 1). Particles suspended in the air of the three cities exceeded WHO standards by about 90%, and levels of nitrogen oxides exceeded WHO standards by up to 73% (Romero-Lankao et al., 2012). PM_{10} temporal trends show that the number of times that Mexico City's records surpass international standards is lower than that of Santiago and Bogotá (Fig. 2).

We used temporal series and GLM to explore the risks of mortality from exposure to hazards and found that these relationships are complex and dependent on several factors such as temperature and unfavorable atmospheric conditions (e.g., high levels of pollution in Mexico City resulting

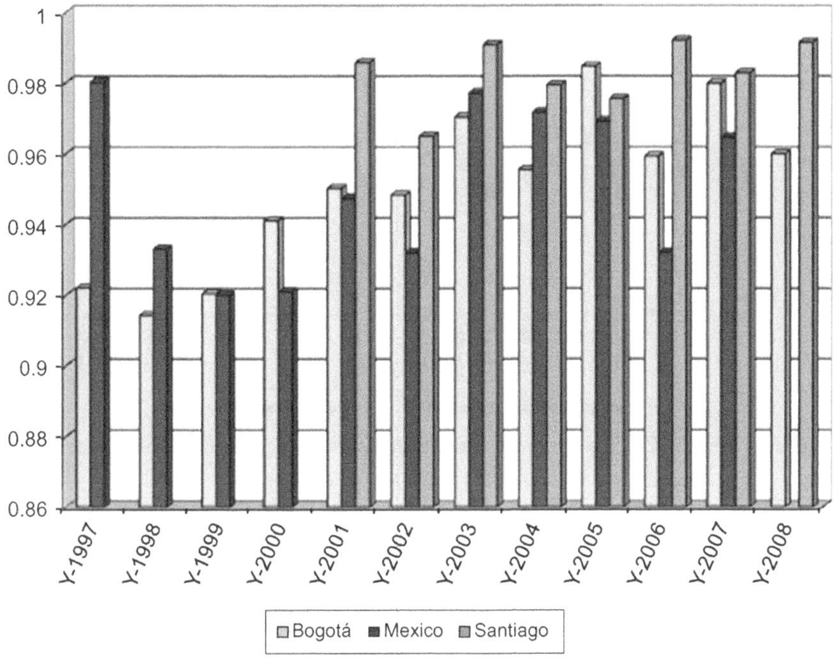

Fig. 2. Nonattainment Levels for PM_{10}, Based on WHO Daily Recommendation ($50\,\mu g\,m^{-3}$).

from high pressure or anticyclone systems, leading to the creation of thermal inversions at night and during the first hours of the morning). For instance, while we found that higher outdoor temperature leads to lower mortality for both respiratory and cardiovascular causes (Fig. 3), higher levels of particulate matter with an aerodynamic diameter of up to 10 μm (PM_{10}) are associated with higher mortality (Fig. 4).

Our quantification of urban populations' likelihood of mortality from exposure to hazards showed there was an increased health risk of cardiovascular and respiratory mortality from higher temperature during the warm season in Mexico City and only of cardiovascular mortality in Bogota (Romero-Lankao et al., 2012). Although some of our numbers are not significant, overall they show a positive correlation between mortality and air pollution (Table 3) meaning that the more air pollution there is in a city the more respiratory and cardiovascular-related deaths there are. However, the pattern of the association differs by city and weather conditions. For instance, the adverse impacts of PM_{10} are especially evident during the

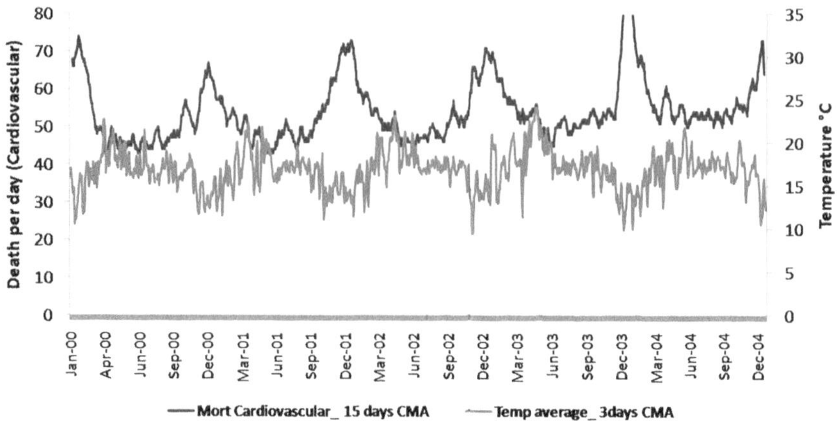

Fig. 3. Mexico City: Negative Correlation Between Temperature and Mortality.

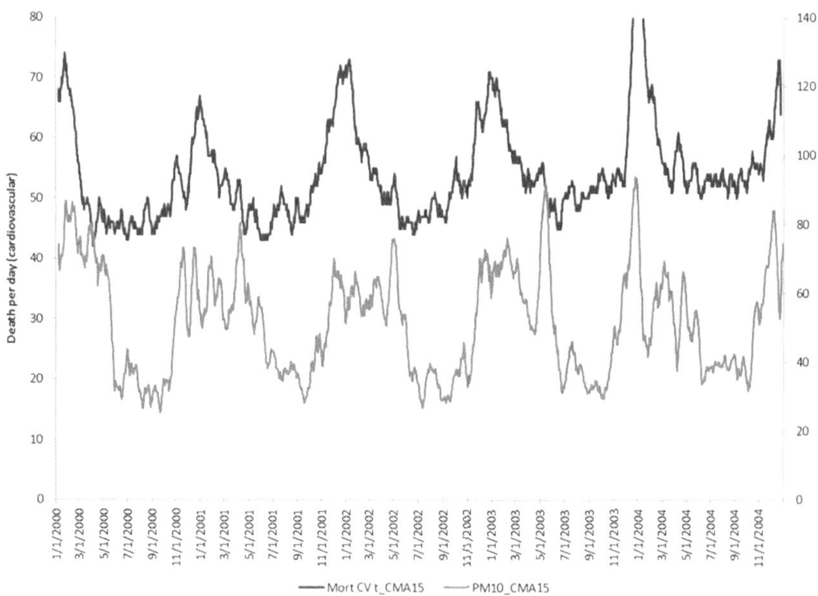

Fig. 4. Mexico City: Positive Correlation Between PM_{10} and Mortality.

Table 3. Relative Risk (RR) of Dying from Exposure to PM_{10} at the City Level by Age.

City	Season	Cardiovascular Deaths	Respiratory Deaths	Cardiovascular Deaths (female)	Respiratory Deaths (female)	Cardiovascular Deaths (elderly)	Respiratory Deaths (elderly)
Bogota (2003–2006)	Warm	**1.001**	0.994	**1.001**	0.998	0.99	0.99
	Cold	**1.001**	**1.005**	**1.001**	**1.001**	**1.00**	**1.00**
Mexico City (2000–2004)	Warm	**1.001**	**1.015**	**1.001**	**1.004**	**1.0**	**1.012**
	Cold	**1.001**	**1.164**	**1.001**	**1.04**	**1.001**	**1.055**
Santiago (2001–2005)	Warm	0.950	0.990	0.978	0.995	0.960	**0.988**
	Cold	**1.012**	**1.014**	1.005	**1.009**	**1.009**	**1.012**

Notes: The numbers in bold represent an increase in relative risk (RR) related to an increase 10 $\mu g/m^3$ in the levels of PM_{10}. RR data were obtained using the Generalized Linear Model (GLM) with Poisson log-linear distribution. The RR fall into three categories defined by $\beta = 1$ (linear), $\beta < 1$ (sublinear), $\beta > 1$ (superlinear). All results included are statistically significant ($p < 0.05$).

cold season in Bogota, Mexico City and Santiago. Additionaly in Mexico City there were positive associations between the levels of PM_{10} in the air and mortality during the warm season (Table 3).

Sensitivity to Temperature and Air Pollution

Although scholarship tends to agree that the elderly, the very young, and people with preexisting health impairments are more *sensitive* to hazards, our measures of the RR of dying from exposure to PM_{10} show mixed and even unexpected results (Table 3). For example, women seem to be more sensitive than the total population only in some cases (e.g., respiratory mortality during the cold season in Mexico City), while they are equally or less sensitive in others (e.g., cardiovascular mortality in Bogota and Mexico City). In other words it does not appear to be the case that women or the elderly have a universal physiological susceptibility to air pollution and temperature, but rather that social conditions (e.g., occupation and equity in access to resources) are responsible for differential effects in some cases and not in others.

Socioeconomic Differentiation and Intra-Urban Differences in Risks

We explored whether the health risks related to air pollution and temperature are equally distributed or spatially and socioeconomically differentiated within cities. We found differences in socioeconomic vulnerabilities, as measured by the MVI, within each study city. There were larger disparities in socioeconomic vulnerability in Bogota (with MVI ranging from 0.06 to 0.72), than in the other two cities. The MVI index ranges from 0.37 to 0.69 and 0.33 to 0.62 between the least and the most vulnerable municipalities respectively in Mexico City and Santiago. This suggests that wealthy populations may have such socioeconomic and political means as education, good quality houses and health services to escape from, or at least to mitigate some environmental risks.

From an environmental justice perspective one may expect that spatial differences in health risks from environmental hazards relate to the socioeconomic characteristics of the population. However, as other scholars before (Szasz & Meuer, 1997; Marshall, 2008) we have found that this is not always the case within Bogota, Mexico City and Santiago. We only found a positive correlation between the level of NO_2 and social vulnerability in

Bogota, while the exposure to PM_{10} was not correlated with vulnerability conditions for all three cities (Romero-Lankao et al., 2012). Furthermore, both some of the most and the least vulnerable districts in the three cities are at similar RR of cardiovascular and respiratory mortality from exposure to PM_{10}. The spatial differences in socioeconomic vulnerabilities within cities also do not necessarily correspond with the spatial distribution of cardiovascular and respiratory mortality rates, and at times the results are quite unexpected. Some of the least vulnerable districts in the three cities had the highest mortality rates, while several of the most deprived communities had the lowest.

Direct Experience with Environmental and Climate Change

To understand the capacity of urban populations to perceive and respond to vulnerability and risk, we first determined what people in the four cities perceive as the primary hazards they face. Across the four study cities, a majority of respondents (65.5%) reported that they had experienced an environmental hazard in the last three years at the time of survey. The most common hazard events identified were "extreme cold weather" (50.0% of respondents) and "extreme precipitation" (46.0%). However, while the majority of the population has experienced an environmental hazard, these experiences do differ between the cities. Respondents from Mexico City and Bogota were less likely than those from Santiago and Buenos Aires to encounter an environmental hazard over the last three years.[4] The type of hazards people experienced were also different in the four cities. For example, in Santiago and Bogota, "extreme cold weather" was the most common hazard, while "heat waves" were the most common hazard in Mexico City. The environmental hazards reported from Buenos Aires were diverse and there was not a dominant one.

In terms of how hazards are impacting health outcomes in the cities, respondents reported that "excessive cold" and "high levels of air pollution" were the top two environmental hazards that caused health problems, both at the aggregate level and for each city individually. In Santiago and Bogota a greater percentage of respondents reported physical problems in relation to excessive cold and air pollution. About a quarter of respondents (26.8%) across the four cities said they or someone in their household had been hospitalized or received emergency care due to an environmental hazard; Santiago and Bogota had much higher percentages of respondents reporting this experience than Mexico City and Buenos Aires.

Multiple Stressors

People do not experience environmental hazards in isolation and other stressors in people's lives may take precedence to responding to environmental hazards. Our survey results showed that the respondents in the four cities are indeed faced with many environmental hazards simultaneously. In addition to the three hazards of relevance for our health analysis (heat waves, extreme cold, and air pollution), respondents also view extreme precipitation, drought, and high winds as important environmental hazards. However, environmental hazards are but one source of stress in people's lives. To better understand the full suite of stressors that people are faced with, we asked respondents in each city what major life events their family experienced in the last five years. This question was open-ended and respondents were encouraged to include anything they thought was relevant. Fortunately, most respondents (51%) could not recall a major stress in the last five years. The others experienced one or more major stresses and they fall into the following general categories: someone in the family lost their job, the entire household moved to a new location, the family suffered financial problems, someone in the family had died, the household was robbed, someone had an illness, or the heads of household separated or divorced. It should be noted that environmentally driven life events like flooding shows up in our results as just as important as socioeconomic life events. These stressors are often very personal but affect the entire family. In general, there are only limited differences between cities when it comes to the most common types of significant life events – household stresses tend to be universal.

Adaptive Capacity

As we did at the city level, to understand the capacity of urban populations to respond to risks in the selected communities the survey adopted a livelihoods approach (Moser & Satterthwaite, 2010) to measure household level assets or capitals. This multi-criteria model is based on four different types of livelihood assets: social, human, physical, and financial capitals (Baud, Pfeffer, Sridharan, & Nainan, 2009). Each type of capital can improve people's ability to cope with and respond to stresses.

To measure social capital we asked respondents about the networks they have and use to respond to environmental risks. The majority of people we surveyed in the four cities appear to have strong social networks to fall back on. When asked if they could rely on a relative or friend to provide support in the event of an environmental disaster, most respondents (64.8%) said that they could. However, there is a segment of the population in each study city that would have no support in the event of a hazard. People without strong social networks represent between 12.0% (Santiago) and 44% (Buenos Aires) of the respondents and will be particularly vulnerable – and in need of government support – when environmental disasters occur. The level of social support was significantly different across the four study cities (Chi-square = 27.073, $p < .001$). Respondents from Santiago were more likely to know someone who would provide temporary housing in emergencies while respondents from Buenos Aires reported the lowest level of social support.

Human capital – education, age, and health – also contributes to people's ability to adopt adaptation strategies and respond to risks. Gender can also help to measure human capital but the relationship between gender and vulnerability to environmental hazards is often context-specific depending on other conditions such as occupation and gender equity broadly in society. The mean age of all sample heads of households in the survey was 49, and the vast majority of them (73.5%) were men; much fewer surveyed households were headed by females in Mexico City than in the other three cities. Around a third (32.8%) of the total surveyed households had someone with a chronic condition; however, only 6.8% had some member(s) with a disability. The majority of the heads of surveyed households (94.3%) knew how to read and write. Most respondents (74.8%) were employed, but only 2.5% have finished college. Some indicators of human capital showed significant variation between the study cities. For example, sample household heads in Buenos Aires were significantly older on average than those in Mexico City and Bogota. Responding households from Santiago were more likely than those from Mexico City to have someone with a chronic condition. In addition, heads of surveyed households in Mexico City had significantly higher education level than those in the other cities.

To identify the level of physical capital in the responding households we asked participants questions about local housing and living conditions. Responding households on average had 4 members and 3 rooms in their home. All respondents reported using electricity for lighting. For cooking, the most common types of energy for the survey sample as a whole were propane gas provided through bottles (56.8% of respondents) and piped gas connected to a public or private provider (36.5%). In Santiago and Bogota nearly all the respondents' cooking fuel comes from these two sources. The most commonly used source of water for cooking was water delivered through a public aqueduct (78.0%). In Mexico City, however, respondents relied most on water from a well with a motor (48%) and water purchased from a water truck (33%). The majority of all surveyed households had a bathroom that connected to a sewage system (66.0%) or a septic well (25.5%). There were some differences between the cities in terms of the level and source of sanitation services. In Santiago and Bogota, all surveyed households had a bathroom with connection to a sewage system, two-thirds of surveyed households in Buenos Aires (68%) had a bathroom with connection to septic tanks, and in Mexico City people relied almost equally on bathrooms connected to a sewage system and on latrines/dumping.

Financial capital broadly refers to those monetary and material resources that are available to people and provide adaptation options. When asked whether their household income covered basic living expenses, nearly two thirds of total respondents (65.3%) said their household income only covered their basic living expenses, while 20.8% reported that it did not cover these basic costs. Sample households in Santiago had higher levels of income than in other cities. Most respondents (75%) reported that their households owned the house or apartment they occupied. Home ownership rates were highest in Santiago and Mexico City, followed by those from Bogota and Buenos Aires in turn.

Responding to Hazards

The survey also allowed us to understand the strategies and options people utilize in response to environmental and climatic hazards. Sample households were asked how they deal with seasonal temperature changes in their homes. Across the study cities, the most common ways people cool their homes were opening windows (39.8% of respondents) and using a swamp or evaporative cooler (38.8%); the most common ways people heat their homes is by using electric stoves (26.5%) and unventilated gas stoves (20.3%). The answers depended on the latitude of the city and how much temperature variability people were exposed to. For example, while most respondents (86.0%) in Buenos Aires used coolers during hot and humid summer weather, very few respondents used coolers in Bogota. Respondents in Bogota also do not use stoves in the winter as much as those in the other three cities, likely because it does not get as cold. People were also creative in their responses to extreme weather. If they could not afford to run the cooler or air conditioner when it was hot, they used other cooling strategies such as opening the windows and wetting the floor. In the winter, those who could not afford a gas stove were more likely to use less expensive options such as electric stoves and space heaters, particularly if they were not formally connected to the electricity network. In informal settlements households are often connected to the electricity network illegally, therefore they do not pay for the service. However, the illegal connections often bring fire risk due to badly installed electrical systems, and our survey found that there have been cases of fires caused by these informal systems.

Another way people can respond to hazards is by changing their house structures. When we asked whether survey respondents have thought about making changes to their home to improve thermal insulation, only 22.0% of all respondents said they have. Moreover, as can be seen in Photo 1, those in San Fernando, Buenos Aires should elevate buildings enough to withstand flood forces such as the storm surges (Sudestadas) occurring throughout the year from the Rio de la Plata estuary. However, because of unstable and inadequate conditions (e.g., job loss, economic hardship, and health problems), many of the respondents have constructed below the flood line. As can be seen in Photo 2, their houses are, therefore, recurrently affected by flooding from the Sudestadas. The only option these households

Photo 1. House in San Fernando, Buenos Aires.

Photo 2. House in San Fernando, Buenos Aires One Day After a Storm Surge.

now have is to cope with floods by moving their belongings to the first floor and using boats. Therefore, efforts should be made to stabilize and increase the asset base in this area to decrease the incentives people have to build houses in a way that makes people vulnerable to floods.

Awareness of public and community response resources plays an important role in people's pursuit of different adaptation strategies. Statistical analysis of the survey data showed that Buenos Aires communities were more likely to have an emergency management program related to environmental hazards. However, despite awareness of the types of hazards common to their region, the majority of respondents (83.3%) were largely unaware of the locally available systems in place to inform them about environmental hazards. Respondents also had very low awareness of emergency systems and the type of events covered by these systems. Of the households that knew of emergency systems, those in Mexico City were aware of heat wave and extreme rain alerts, those in Santiago were aware of extreme rain alerts, but only a few respondents in Bogota and Buenos Aires could think of a particular emergency system.

Our field work in the Buenos Aires communities also found that there was a discrepancy between the ways in which low- and middle-income communities used and relayed information about emergency events. Respondents in the low-income communities reported that they knew an official system existed, but did not know how to use it. They did not know how it operated or which venues it used to disseminate information. Instead, they relied on neighborhood networks and personal knowledge. The middle-income areas of Buenos Aires were more dependent on official announcements.

The survey results showed that respondents in all cities primarily seek out information about climate-related hazards from television, radio, neighbors and family. This suggests that these are good avenues for disseminating information during environmental emergencies. Respondents from Bogota and Santiago had used significantly more media than those living in Mexico City and Buenos Aires to get relevant information during environmental emergencies.

CONCLUDING REMARKS

In this chapter, we assessed some of the features and linkages among the components of vulnerability and risk in the Latin American cities of Bogota, Buenos Aires, Mexico, and Santiago, based on the assumption that the health impacts of temperature and air pollution as well as the determinants of vulnerability are of a nuanced nature. We found high pollution levels of PM_{10}, ozone, and other criteria pollutants in three of the cities. However, the nature of their impacts depends on the particulars of pollutant levels, atmospheric, and weather conditions of each city. For instance, while lower temperatures relate to higher cardiovascular and respiratory mortality (pointing to a negative mortality/temperature relationship), higher levels of particulate matter (PM_{10}) relate to higher mortality (thus pointing to a positive mortality/PM_{10}).

Our findings on the risk of mortality among sensitive groups do not support existing literature, which tends to emphasize a universal physiological susceptibility to air pollution and temperature within vulnerable populations such as the elderly or women. On the contrary, we found inconsistencies in the data on how age and gender related to mortality risk, but determined that more empirical work is needed on the influence, on risk, of occupation, education, equity in access to resources, cultural roles, and other factors.

Even considering these additional determinants of mortality risk, however, our study found that threats such as air pollution reach threshold levels at which the importance of socioeconomic factors falls away. Although wealthy populations in the three cities have access to education, good quality housing, and health services to mitigate some environmental risks, overall health risks from air pollution and temperature in the study cities do not necessarily depend on socioeconomic differentiations. We suggest therefore that health risks from atmospheric conditions and pollutants act without boundaries or social distinctions and show characteristics of a boomerang effect at least at the relatively small spatial scales of urban areas. Furthermore, if we keep in mind that in a plausible future, with increases in the levels of air pollution interacting with more intense urban heat islands, heat waves and other effects resulting from climate change what goes around will certainly come around and pose risks to both rich and poor alike.

Although household vulnerability to the health impacts of air quality and temperature was the defining theme of our analysis at the inception of the project, we could not overlook that urban residents do not experience environmental hazards in isolation. Rather, people operate in contexts of multiple environmental and socioeconomic hazards and stresses. The way people respond to one hazard is related to the way they deal with other hazards. The strategies people develop to manage unemployment or floods go into their toolbox and do reduce (or enhance) their ability to respond to other hazards. Last but not least, socioeconomic stressors in people's lives may take precedence over responding to environmental hazards. Components of adaptive capacity such as network building or the ability to make demands on government agencies influence their resilience to multiple stresses. Therefore it is important to explore the assets, options and governmental policies that urban populations can draw on to actually respond to multiple hazards.

Beyond the disparities in socioeconomic vulnerability within and across cities, our results on the capacity of urban populations to perceive risks and respond to them are mixed. In some cases assets seem to be high (e.g., there are high levels of home ownership and literacy). However, in many other ways, household assets are low (see Box 1). Given that the availability of income and such resources as infrastructures, health and water services, and the quality of dwellings determines people's ability to protect themselves from the consequences of such hazards as air pollution and temperature, the fact that many of these are absent or inadequate compromises the adaptive capacity of studied populations. Furthermore, these populations lack some of the options that can empower them such as access to social networks and awareness of and information about public resources to respond to emergencies. Efforts should be made to stabilize and increase the asset base of these populations; to take advantage of tools to disseminate information such as neighborhood networks and to increase personal

Box 1. A Vignette of a Mexico City Respondent Household.

A family of six lives in a two room house. To make space, their kitchen doubles as a sleeping space. They have electricity, propane gas, and running water but use an outdoor latrine. They have lived the barrio called Paraje Las Malvinas in Mexico City for 36 years, but they do not consider it safe. The husband, 38, works full-time to take care of his wife, 36, three sons, 20, 10, and 7, and their nephew, 5. He has worked for 40 years, but he lost a limb in a work accident. The family feels that his income does not cover the basics. They have trouble affording appliances like a washing machine or a sewing machine. They do not own any personal transportation such as a car, a motorcycle, or even a bicycle. One of the family members has a chronic illness due to extreme cold temperatures, but does not see a doctor. In the summer, they suffer from heat waves but they cannot afford air conditioning or house modifications. Instead, all they can do is open the windows. If a disaster happened today, the family does not feel that they have anyone who would take them in or lend them money. They do not belong to any community groups and are not aware of any community resources. They primarily get their news from the television, and are worried about the alternating extreme rains and droughts that they hear about.

knowledge and awareness of official information or emergency response systems and how to use them.

We would like to close with some reflections on the opportunities and challenges of doing interdisciplinary research on urban vulnerability and risk. The use of quantitative and qualitative methods allowed us to get a different take on the nature and interconnections between the different components of urban vulnerability, mortality and risk. However, our efforts to integrate concepts, methods and data from different disciplines have not been exempted from challenges. It has proven difficult for us to fully explore the dynamics of our subject as well as the issues of scale and context. A set of cultural and communication challenges has arisen, not only from the diverse conceptualization approaches, methods, differing terminologies and mechanisms for analyzing and presenting results that we have attempted to integrate, but also from interpersonal issues affecting team interactions. From its inception, ADAPTE has sought to be an issue-driven interdisciplinary project (Romero-Lankao et al., 2012). The integrating research question and the use of quantitative and qualitative methods allowed ADAPTE to shed a slightly different light on the nature and interconnections between the different dimensions of urban vulnerability to climate and environmental change.

NOTES

1. "Not ist hierarchish, Smog ist democratisch" Beck (1986, p. 48).
2. The MVI ranges between 0 and 1 as all indicators used in the calculation were first normalized based on the method for the UNDP's Human Development Index (Osborne & Difei, 2010).
3. We conducted these quantitative assessments for only three of the study cities (Bogota, Mexico City, and Santiago), because we could not collect sufficient air pollution data for Buenos Aires.
4. Based on results from Tukey's post hoc test.

REFERENCES

Basu, R. (2009). High ambient temperature and mortality: A review of epidemiologic studies from 2001 to 2008. *Environmental Health*, *8*(1), 40.

Baud, I., Sridharan, N., & Pfeffer, K. (2008). Mapping urban poverty for local governance in an Indian mega-city: The case of Delhi. *Urban Studies*, *45*, 1385–1412.

Baud, I. S. A., Pfeffer, K., Sridharan, N., & Nainan, N. (2009). Matching deprivation mapping to urban governance in three Indian mega-cities. *Habitat International*, *33*, 365–377.

Beck, U. (1986). *Risikogesellschaft: Auf dem Weg in eine andere Moderne*. Suhrkamp.

Bell, M. L., O'Neill, M. S., Ranjit, N., Borja-Aburto, V. H., Cifuentes, L. A., & Gouveia, N. C. (2008). Vulnerability to heat-related mortality in Latin America: A case-crossover study in São Paulo, Brazil, Santiago, Chile and Mexico City, Mexico. *International Journal of Epidemiology*, *37*, 796–804.

Birkmann, J. (2006). Measuring vulnerability to promote disaster-resilient societies: Conceptual frameworks and definitions. In J. Birkmann (Ed.), *Measuring vulnerability to natural hazards* (pp. 9–54). New Delhi: TERI Press.

Chestnut, L. G., Breffle, W. S., Smith, J. B., & Kalkstein, L. S. (1988). Analysis of differences in hot-weather-related mortality across 44 U.S. metropolitan areas. *Environmental Science & Policy*, *1*, 59–70.

Cifuentes, L., Borja-Aburto, V. H., Gouveia, N., Thurston, G., & Davis, D. L. (2001). Assessing the health benefits of urban air pollution reductions associated with climate change mitigation (2000–2020): Santiago, Sao Paulo, Mexico City, and New York City. *Environmental Health Perspectives*, *109*(Suppl. 3), 419–425.

Curriero, F. C., Heiner, K. S., Samet, J. M., Zeger, S. L., Strug, L., & Patz, J. A. (2002). Temperature and mortality in 11 cities of the eastern United States. *American Journal of Epidemiology*, *155*, 80–87.

Dear, K., Ranmuthugala, G., Kjellström, T., Skinner, C., & Hanigan, I. (2005). Effects of temperature and ozone on daily mortality during the august 2003 heat wave in France. *Archives of Environmental & Occupational Health*, *60*, 205–212.

D'Ippoliti, D., Michelozzi, P., Marino, C., de'Donato, F., Menne, B., Katsouyanni, K., ..., Perucci1, C. A. (2010). The impact of heat waves on mortality in 9 European cities: Results from the EuroHEAT project. *Environmental Health*, 9.

Dockery, D. W., Pope, C. A., Xu, X., Spengler, J. D., Ware, J. H., Fay, M. E., et al. (1993). An association between air pollution and mortality in six U.S. cities. *New England Journal of Medicine*, *329*, 1753–1759.

Fouillet, A., Rey, G., Jougla, E., Frayssinet, P., Bessemoulin, P., & Hemon, D. (2007). A predictive model relating daily fluctuations in summer temperatures and mortality rates. *BMC Public Health*, *7*, 114.

Gallopín, G. C. (2006). Linkages between vulnerability, resilience, and adaptive capacity. *Global Environmental Change*, *16*(3), 293–303.

INE (2008). *Climate change in Mexico: Information by State and Sector*. Retrieved from http://www.ine.gob.mx/cclimatico/edo_sector/estados/futuro_mexico.html. Accessed on January 12, 2009.

Ishigami, A., Hajat, S., Kovats, R. S., Bisanti, L., Rognoni, M., & Russo, A., et al. (2008). An ecological time-series study of heat-related mortality in three European cities. *Environmental Health: A Global Access Science Source*, 7.

Jerrett, M., Burnett, R. T., Pope, C. A., Ito, K., Thurston, G., Krewski, D., et al. (2009). Long-term ozone exposure and mortality. *New England Journal of Medicine*, *360*, 1085–1095.

Jordán, R., Rehner, J., & Samaniego, J. (2010). *Regional Panorama Latin America: Magacities and Sustainability*. United Nations, Santiago de Chile: Economic Commission for Latin America and the Caribbean.

Katsouyanni, K., Pantazopoulou, A., Touloumi, G., Tselepidaki, I., Moustris, K., Asimakopoulos, D., et al. (1993). Evidence for interaction between air pollution and high temperature in the causation of excess mortality. *Archives of Environmental Health: An International Journal*, *48*, 235–242.

Kienberger. (2010). *Spatial vulnerability assessment: Methodology for the community and district level applied to floods in Buzim Mozambique*. PhD Thesis, University of Salzburg, Salzburg, Austria. 305pp.

Knowlton, K., Lynn, B., Goldberg, R. A., Rosenzweig, C., Hogrefe, C., Rosenthal, J. K., et al. (2007). Projecting heat-related mortality impacts under a changing climate in the New York city region. *American Journal of Public Health*, *97*, 2028–2034.

Magrin, G., Gay García, C., Cruz Choque, D., Giménez, J. C., Moreno, A. R., Nagy, G. J., et al. (2007). Latin America. Climate change 2007: Impacts, adaptation and vulnerability. In M. L. Parry, O. F. Canziani, J. P. Palutikof, P. J. van der Linden & C. E. Hanson (Eds.), *Contribution of working group ii to the fourth assessment report of the Intergovernmental Panel on Climate Change* (pp. 581–615). Cambridge, UK: Cambridge University Press.

Makri, A., & Stilianakis, N. I. (2008). Vulnerability to air pollution health effects. *International Journal of Hygiene and Environmental Health*, *211*, 326–336.

McMichael, A. J., Wilkinson, P., Kovats, R. S., Pattenden, S., Hajat, S., Armstrong, B., ... Kingkeow, C. (2008). International study of temperature, heat and urban mortality: The 'ISOTHURM' project. *International Journal of Epidemiology*, *37*(5), 1121–1131.

Medina-Ramón, M., & Schwartz, J. (2007). Temperature, temperature extremes, and mortality: A study of acclimatisation and effect modification in 50 US cities. *Occupational and Environmental Medicine*, *64*, 827–833.

Medina-Ramón, M., Zanobetti, A., Cavanagh, D. P., & Schwartz, J. (2006). Extreme temperatures and mortality: Assessing effect modification by personal characteristics and specific cause of death in a multi-city case-only analysis. *Environmental Health Perspectives*, *114*, 1331.

Michelozzi, P., De Sario, M., Accetta, G., de'Donato, F., Kirchmayer, U., D'Ovidio, M., & Perucci, C. A. (2006). Temperature and summer mortality: Geographical and temporal variations in four Italian cities. *Journal of Epidemiology and Community Health*, *60*(5), 417–423.

Moser, C., & Satterthwaite, D. (2010). Toward pro-poor adaptation to climate change in the urban centers of low- and middle-income countries. In R. Mearns & A. Norton (Eds.), *Social dimensions of climate change: Equity and vulnerability in a warming world* (pp. 231–258). Washington, DC: The International Bank for Reconstruction and Development/The World Bank.

Muggeo, V. M., & Hajat, S. (2009). Modelling the non-linear multiple-lag effects of ambient temperature on mortality in Santiago and Palermo: A constrained segmented distributed lag approach. *Occupational and Environmental Medicine*, *66*(9), 584–591.

Naess, O., Piro, F. N., Nafstad, P., Smith, G. D., & Leyland, A. H. (2007). Air pollution, social deprivation, and mortality a multilevel cohort study. *Epidemiology*, *18*(6), 686–694.

O'Brien, K., Eriksen, S., Nygaard, L. P., & Schjolden, A. (2007). Why different interpretations of vulnerability matter in climate change discourses. *Climate Policy*, *7*, 73–88.

O'Neill, M., Zanobetti, A., & Schwartz, J. (2005). Disparities by race in heat-related mortality in four US cities: The role of air conditioning prevalence. *Journal of Urban Health*, *82*, 191–197.

Osborne, D. B., & Difei, H. (2010, April). *Technical note on re-calculating the HDI, using projections of components of the HDI*. United Nations Development Programme, Human Development Report Office.

Ostro, B., Sanchez, J. M., Aranda, C., & Eskeland, G. S. (1995). *Air pollution and mortality: Results from Santiago, Chile*. Policy Research Working Paper 1453. World Bank Policy Research Department, Washington, DC.

Parnell, S., Simon, D., & Vogel, C. (2007). Global environmental change: Conceptualising the growing challenge for cities in poor countries. *Area*, *39*(3), 357–369.

Pelling, M. (2003). *The vulnerability of cities: Natural disasters and social resilience*. London: Earthscan.

Pope, C. A., & Dockery, D. W. (1995). Health effects of fine particulate air pollution: Lines that connect. *Journal of the Air & Waste Management Association*, *56*, 709–742.

Pro Prevention Consortium. (2010). Community risk assessment methodologies and case studies. Retrieved from http://www.proventionconsortium.org/

Reid, C. E., O'Neill, M. S., Brines, S. J., Gronlund, C., Diez-Roux, A. V., Brown, D. G., & Schwartz, J. (2008). Mapping community determinants of heat vulnerability. *Epidemiology*, *19*(6), S229.

Romero-Lankao, P. (2010). Water in Mexico City: What will climate change bring to its history of water-related hazards and vulnerabilities? *Environment and Urbanization*, *22*, 157–178.

Romero-Lankao, P., Borbor-Cordova, M., Abrutsky, R., Günther, G., Behrenz, E., & Dawidowsky, L. (2012). ADAPTE: A tale of diverse teams coming together to do issue-driven interdisciplinary research. *Environmental Science & Policy*.

Romero-Lankao, P., & Qin, H. (2011). Conceptualizing urban vulnerability to global climate and environmental change. *Current Opinion in Environmental Sustainability*, *3*, 142–149.

Satterthwaite, D., Huq, S., Pelling, M., Reid, H., & Romero-Lankao, P. (2007). *Adapting to climate change in urban areas: The possibilities and constraints in low- and middle-income nations*. IIED, Human Settlements Discussion Paper: Climate Change and Cities 1; 2007. International Institute for Environment and Development (IIED), London.

Smoyer, K. E., Kalkstein, L. S., Greene, J. S., & Ye, H. (2000). The impacts of weather and pollution on human mortality in Birmingham, Alabama and Philadelphia, Pennsylvania. *International Journal of Climatology*, *20*, 881–897.

Stafoggia, M., Forastiere, F., Agostini, D., Biggeri, A., Bisanti, L., Cadum, E., et al. (2006). Vulnerability to heat-related mortality: A multicity, population-based, case-crossover analysis. *Epidemiology*, *17*, 315–323.

Thywissen, K. (2006). *Components of risk: A comparative glossary*. United Nations University Institute for Environment and Human Security, Bonn, Germany. Available at https://www.ehs.unu.edu/file/get/8335

UN/ISDR. (2009). *Terminology on disaster risk reduction*. United Nations International Strategy for Disaster Reduction, Geneva, Switzerland.

CHAPTER 11

CITIES IN THE FLOOD: VULNERABILITY AND DISASTER RISK MANAGEMENT: EVIDENCE FROM IBADAN, NIGERIA

Andrew Onwuemele

ABSTRACT

Purpose – *Flooding has become a recurring phenomenal in most cities in Nigeria. The 26 August 2011 flooding disaster which occurred in Ibadan is only an indication of magnitude of flooding problem in Nigerian cities. This chapter examines the impacts, vulnerability factors and disaster risk management framework in Ibadan metropolis.*

Methodology/Approach – *The survey design was used for the study and covers eleven local government areas (LGAs) affected by the flood. The study utilized both primary and secondary data. The primary data were obtained by physical observation and in-depth interview of affected households. In-depth interview was also carried out with key officials of State Ministry of Environment and Housing. The study also relies on the data from the Oyo State Government Task Force on Flood Prevention and Management report.*

Findings – *The chapter shows that the 26 August 2011 flood disaster in Ibadan metropolis caused monumental destructions in the city.*

Urban Areas and Global Climate Change
Research in Urban Sociology, Volume 12, 277–299

ISSN: 1047-0042/doi:10.1108/S1047-0042(2012)0000012014

The continuous construction on flood plains, indiscriminate dumping of refuse, excessive rainfall and deforestation were identified as the main vulnerability factors. The chapter shows that there is no adequate framework for disaster risk management in the city.

Research limitations – *About 250 affected households in 11 LGAs were interviewed for the study due to time and budget constraints. This figure is considered meagre considering the number of affected households by the flood disaster. However, the random selection of affected households and key government officials helped to address this limitation.*

Originality/value of chapter – *The simultaneous identification of impacts, vulnerability factors and disaster risk management framework in the city provides an opportunity for the development of a holistic and proactive disaster risk management strategy in Ibadan metropolis.*

Keywords: Cities; flood; vulnerability; disaster risk; management

INTRODUCTION

Flooding has become a recurring phenomenal in most cities in the world. According to Jha et al. (2011), two major global themes point to the fact that the number and scale of flood events will continue and possibly accelerate in the next 50 years. First is the global trend in urbanization and the second is the global changes in climatic conditions. These two powerful human-induced forces are converging in dangerous ways with severe negative impacts upon quality of life and economic and social stability. By the end of the last decade, the world reached a milestone when, for the first time in human history, half of the world's population lived in urban areas. The pace of urbanization in the world today is unprecedented, with a near quintupling of the urban population between 1950 and 2011 (UN, 2010).

Concomitantly, the number of large cities and the size of the world's largest cities are increasing. The number of cities in the world with populations greater than 1 million increased from 75 in 1950 to 447 in 2011, while during the same period, the average size of the world's 100 largest cities increased from 2.0 to 7.6 million. By 2020, it is projected that there will be 527 cities with a population of more than 1 million, while the average size of the world's 100 largest cities will have reached 8.5 million (UN, 2010). One of the negative consequences of this rapid urbanization is the increased vulnerability of cities to extreme weather events such as flooding. Cities and climate change are inextricably linked. Urban enterprises, vehicles and populations are key sources of greenhouse gases, which contributes significantly to the global changes in climatic condition. On the other hand, cities are highly vulnerable to climate change due to several reasons. Firstly, cities host large numbers of the poor who are especially vulnerable to climate change. Poor city residents tend to locate in the most vulnerable locations and housing construction materials are not robust. The consequences of surging seas, wind storms, and flooding are much more dramatic in these areas (The World Bank, 2010a). Secondly, cities are vulnerable due to their immobility. Such infrastructure as bridges, subway systems, buildings, and roads, the historic sense of place, and rootedness of residents which are critical attributes of cities can become liabilities if the local ecosystems that they are based on are unable to adapt to the climate-induced changes (World Bank, 2010b). Thirdly, the geographic locations of most world cities make them to be highly vulnerable to climate change. Traditionally, cities were located near rivers and oceans for transportation and connectivity purposes. This natural geographic advantage is now increasing vulnerability of cities as sea levels rise and wind storms increase in severity and frequency (World Bank, 2010a).

These vulnerabilities are increasing due to continued settlement along coastlines and in flood plains, and may be exacerbated in future by climate change (Johnstone & Lence, 2009). Given the above scenario, it is necessary to pay attention to the worsening global problem of climate change in relation to cities. Flooding is one of the most common of all environmental hazards in cities and it regularly claims over 20,000 lives per year and adversely affects around 75 million people worldwide (Smith, 1996). Across the globe, floods have posed tremendous danger to people's lives and properties. Floods cause about one third of all deaths, one third of all injuries and one third of all damage from natural disasters (Askew, 1999). Since 1900, floods have claimed more than 10,000 lives in the United States alone. In China, some of the world's most disastrous floods have been caused by the unstable Huang He (Yellow River). In 1970, 1985 and 1991, hundreds of thousands of people in Bangladesh were killed when the combination of high tides and a tropical cyclone storm surge caused widespread flooding of the low-lying delta of the Ganges and Brahmaputra rivers (Askew, 1999).

In Nigeria, the pattern is similar with the rest of the world. Excluding droughts, almost 90 per cent of damages relating to natural disasters are caused directly or indirectly by floods (Adeoye, Ayanlade, & Babatimehin,

2009). Though the Nigerian environment today is faced with myriads of ecological problems arising from the impacts of human–environment interactions and natural phenomenon, however, the common ecological problem is urban flooding which now occurs in all the ecological zones in Nigeria which is attributed to global warming. Urban flooding has been identified as one of the major factors that prevents Africa's growing population of city dwellers from escaping poverty and stands in the way of United Nations 2020 goal of achieving significant improvement in the lives of urban slum dwellers. Nigeria is Africa' most populous nation. However, the majority of this population is absolutely poor. The high population coupled with increasing poverty puts increasingly severe demands upon the natural environment, the institutional structures and the resources available to manage them. This scenario has continued to exacerbate the magnitude, intensity and disaster risks associated with urban flooding in Nigeria. The 26 August 2011 flooding incidence which occurred in Ibadan, leaving hundreds dead, is only an indication of magnitude of flooding problem in Nigerian cities.

The devastation arising from urban flooding in different cities in Nigeria have resulted in the loss of lives, livelihoods, infrastructures and property. What is even more worrisome is the increasing effect of climate change on the pattern of rainfall, whose intensity has increased in recent years. Over the years, various studies in Nigeria looked at urban flooding as exceptional events, not related to the ongoing social and developmental processes. Gradually, this attitude changed to an emphasis on after response measures, such as the provision of humanitarian aspects of relief, providing medical care, food and water, search and rescue, and containing the secondary disasters after the disaster had occurred and a growing role for relief agencies such as the Nigerian National Emergency Management Agency (NEMA). This "contingency planning" approach certainly improved the efficiency of relief agencies but left a lot to be desired in terms of appropriateness and effectiveness of relief. The perpetual persistence of urban flooding in many cities in Nigeria is only an indication of the vulnerability of the people to disaster risks posed by flooding. The people's vulnerability is a key factor determining the impact of flooding on them. Thus, a more comprehensive approach of disaster risk management, comprising of hazard assessment, vulnerability analysis, and enhancement of management capacity, which is more closely integrated with the ongoing development processes should be an integral part in any effort towards addressing urban flooding in Nigeria. Regrettably, this has not been the interest of past studies in the region over the years. It is against this backdrop that this study analyzes flooding impacts, vulnerability and disaster risk management within an integrated assessment framework in Ibadan, Nigeria.

THE CONCEPTUAL FRAMEWORK

In this study, cities are interchangeably used with urban areas. The definition of urban areas has generated controversy among scholars over the years to the extent that some scholars refer to it as areas outside the rural areas. This then raises the question of what is rural. In this respect, Obot (1986) defined rural as areas outside the urban areas. To circumvent this conceptual confusion, criteria have been established to classify settlements either as urban or rural. Three of these have been identified as more commonly used (Okpala, 1986). These are:

1. legal and administrative definition/classification in which urban places are defined by law;
2. use of minimum population thresholds; and
3. use of functional definition which employs certain universally accepted urban characteristics.

In spite of these criteria, there are still conceptual issues to be addressed especially at the level of comparability of cities at international levels. For instance, the use of minimum population thresholds in the definition of urban areas is difficult as population thresholds tend to vary from country to country. While Greece, Canada and Japan use 20,000, 1,000 and 30,000 respectively for urban population thresholds, it is only 200 for Denmark and 20,000 for Nigeria respectively. One of the most conscious efforts to defined urban areas was made by Wirth in 1938 when he defined urban areas as relatively in the context of large, dense and permanent settlement of socially heterogeneous individuals (Wirth, 1938). Fadare and Oduwaye (2009) defined urban areas or cities as 'a physical ecological organism evolving through interplay of regulations and entrepreneurial activities'. In this study, however, cities are identified on the basis of the three criteria as listed by Okpala (1986). Thus, a city or urban area in Nigeria is an area legally recognized, with the minimum population thresholds and with the minimum universally accepted urban characteristics. In this respect, local government areas such as headquarters which are legally recognized as urban areas by law but without the population thresholds and some urban characteristics such as the availability of some social amenities such as electricity are outside urban areas in this study.

Flood is an overflowing or irruption of a great body of water over land not usually submerged (*Oxford English Dictionary*). It is an extreme weather event naturally caused by rising global temperature which results in heavy downpour, thermal expansion of the ocean and glacier melt, which in turn result in rise in sea level, thereby causing salt water to inundate coastal lands (Adeoye et al., 2009). Floods emanates from a combination of meteorological and hydrological extremes. In most cases, floods are additionally influenced by human factors. Although these influences are very diverse, they generally tend to aggravate flood hazards by accentuating flood peaks. Thus, flood hazards in built environments have to be seen as the consequence of natural *and* man-made factors (World Meteorological Station/Global Water Partnership, 2008).

Based on multifarious causes of urban flooding, four major types of urban flooding have been identified. These are:

1. local floods
2. riverine floods
3. coastal floods
4. flash floods

Local floods occur due to high surface run-off that is in excess of local drainage capacity. It occurs many times a year in slum areas because there are few drains and these few ones are poorly maintained resulting to blockage, thus reducing the drainage capacity and leading to increased surface run-off and back-up effects causing local floods.

Riverine floods occur when the river run-off volume exceeds local flow capacities which is triggered by heavy rainfall or snow melt in upstream areas, or tidal influence from the downstream. Most cities are traditionally located in coastal areas or flood plains. Urban areas situated on flood plains are particularly exposed to extensive riverine floods.

Flash floods occur as a result of the rapid accumulation and release of run-off waters from upstream mountainous areas, which can be caused by very heavy rainfall, cloud bursts, landslides, the sudden break up of an ice jam or failure of flood control works. They are characterized by a sharp rise followed by relatively rapid recession causing high flow velocities (World Meteorological Station/Global Water Partnership, 2008).

In order to fully understand the concept of vulnerability, it is crucial to be familiar with the different elements that constitute disaster risks. Often risk is conceptualized by equating it with the occurrence of an extreme event or hazard (flood, drought, earthquake, storm, landslide etc.) caused by natural forces or by a combination of natural forces and human influences.

Although the occurrence of these hazards is a primary precondition, it is only one element in the creation of risk. The second element in the creation of risk is the fact that somebody or something has to be at risk (World Meteorological Station/Global Water Partnership, 2008). This widespread definition makes the basic structure of risks very clear. Hence, this chapter is based on this extended definition of risk:

> Risk is the probability of a loss, and this depends on three elements: hazard, vulnerability, and exposure. If any of these three elements in risk increases or decreases, then the risk increases or decreases respectively. (Crichton, 1999)

Risk is thus defined as a function of hazard, exposure and vulnerability as shown below:

$$\text{Risk} = \text{function (Hazard} \times \text{Exposure} \times \text{Vulnerability)}$$

Disaster in this chapter is defined as 'a serious disruptions of the functioning of a community or society causing widespread human, material, economic or environmental losses which exceed the ability of the affected community or society to cope using its own resources' (ISDR, 2004). The Centre for Research on the Epidemiology of Disasters (CRED) classifies an event disaster if at least one of the following has occurred: (1) 10 or more people killed, (2) 100 or more people reported affected, (3) a call for international assistance and/or (4) a declaration of state of emergency.

The impact of flooding is driven by a combination of natural and human-made factors. In describing flood impacts and discussing flood risk management solutions within this chapter, two models of flooding, which are chosen from the multitude of risk frameworks available, are particularly pertinent: the *source pathway receptor model* associated with Fleming (2001), and the *flood risk triangle* (Clark, Priest, Treby, & Crichton, 2002; Crichton, 1999). Fleming breaks down the process of flooding into the identification of a source of flood water, the pathway which is taken by flood water and the receptor for the flood water which is the human settlement, building, or field or other structure or environment.

This definition is complemented by a commonly accepted definition of flood risk which defines the risk of flooding as being a function of the probability of the flood *hazard*, of *exposure* to the flood hazard, and of the *vulnerability* of receptors to the flood hazard. Many versions of such models exist for disasters generally (Thywissen, 2006). In the flood context, Crichton (1999) formalizes this definition of risk into the risk triangle where, importantly, hazard is a function not only of natural processes but also of anthropogenic environmental changes which alter the natural flow patterns and pathways to generate increased flood hazard from a similar magnitude of weather event. In an urban context, this implies that upstream conveyance becomes part of the hazard experienced by cities. This definition also encompasses the important notion of the differences between the relative damage suffered by receptors from the hazard, with this described either as vulnerability to the hazard or alternatively as resilience against the hazard. Vulnerability and resilience are negative and positive measures of the same characteristic. Increases in past impacts and future risks from flooding can therefore result from increases in hazard, exposure of populations and their assets, or the vulnerability of exposed populations and assets to flooding.

Impact from flooding in the urban environment is caused by the action of hazard on exposed and vulnerable receptors. Increases in impacts from flooding can result from increases in hazard, in the exposure of populations and their assets, or in the vulnerability of these exposed populations and assets to flooding. Recent increases in impacts observed by sources such as the CRED database are seen to be fuelled by changes in all three underlying factors. Hazard, exposure and vulnerability all appear to be increasing and are likely to continue to do so in the future unless active steps are taken to prevent this.

Exposure

Exposure refers to the question of whether or not people or values are in range of flood waters (ISDR, 2004). One of the major factors for the rise in urban flood damages is simply the increasing number of people and assets that are physically exposed to floods in cities. The fast and unplanned growth of cities results in a larger number of people living in areas potentially liable to flooding. In many developing countries, cities are witnessing rapid urbanization and urban growth which have been attributed to unprecedented migration from rural areas to cities. This has led to urban sprawl with increasing human settlement development in hazard prone areas such as riversides, wetlands, land below the river, sea or reservoir level.

However, with proper land use planning, urban growth need not necessarily lead to the intensification of risks. The significant factor in this respect is whether flood risks are mainstreamed in the urban land use planning and in the entire development process. Many a times the commitment to flood risk sensitive urban planning depends strongly on the flood frequency. After years or decades without major flood events, it becomes more and more difficult to maintain the flood awareness of both people and authorities. This is particularly the case with urban settlements allowed to develop behind the flood levees. Unfortunately, many urbanization processes take place either without any planning or with plans that ignore or underestimate flood risks as it is with many cities in Nigeria including the study area. Often the construction and land use regulations, the underlying legal basis, as well as a set of concrete plans do exist but are not enforced (World Meteorological Station/Global Water Partnership, 2008).

Vulnerability

Vulnerability is the most crucial component of risk in that it determines whether or not exposure to a hazard constitutes a risk that may actually result in a disaster. If the potential exposure to floods becomes reality, then the vulnerability of people and infrastructure is decisive for the degree of harm and damage. Three types of vulnerability can be distinguished on this premise:

1. physical vulnerability of people and infrastructure;
2. unfavourable organizational and economic conditions; and
3. attitudes and motivations.

Physical vulnerability of people and infrastructure emanates from socio-spatial segregation with reference to the hazard exposure of settlement. Socio-economic differentials allow people in higher income groups to avoid or bear flood risks, while those with low incomes cope with them to their detriment. Consequently, those who cannot afford to purchase or to rent space in secure environments are forced to move to cheaper places. Such locations may be found at the outskirts of town or in areas inside town where the wealthier do not want to live because, e.g., these areas are prone to floods or other hazards (World Meteorological Station/Global Water Partnership, 2008). Given the fact that the livelihood of the urban poor often depends on the proximity to informal economies in the central areas of big cities, many prefer to inhabit hazard areas inside town. Two more factors aggravate this spatial marginalization. On one hand, hazard prone areas are often not privately owned, and thus informal dwellers are less likely to get displaced. On the other hand, however, many urban poors are migrants from rural areas who are not familiar with the respective hazards and therefore tend to underestimate the risk of living in such exposed areas (World Meteorological Station/Global Water Partnership, 2008).

The physical vulnerability of urban populations tends to increase as a result of the dense concentration of potentially dangerous infrastructure and substances in urban areas (bridges, solid and liquid waste, chemicals, electric facilities etc.). Existence of health threatening infrastructure such as sewage treatment plants (usually located at very low spots), waste dumps or dangerous industries at such locations increase additionally the risk of secondary hazards and damages. Special attention in the context of human settlement locations has to be paid to socio-economic factors.

The vulnerability to flood risks in urban settlements, particularly in developing countries in informal developments, can be attributed to the following factors:

- Risk prone areas are the only areas that the poor migrants are able to afford
- Failure to perceive flood risks due to lack of knowledge till a flood strikes
- Infrastructure to reduce risk is not economically viable

Unfavourable organizational and economic condition vulnerability occurs due to the inability of informal settlement dwellers to act effectively together. The corollary of this is that they face difficulties in getting support from government in the event of disaster risks. This lack of support often lead to chaotic circumstances in times of stress. This tends to accentuate the vulnerability of informal settlement dwellers to disaster risks. Attitudes and motivational vulnerability are the result of reluctance towards flood preparedness and mitigation measures on the part of informal settlement dwellers. This may be attributed to lack of hazard knowledge or its impact on human lives and property.

THE STUDY CONTEXT

Background Information

Ibadan is located in south-western Nigeria. It is the capital of Oyo State, and is reputed to be the largest indigenous city in Africa, south of the Sahara. The presence of hills makes the site of the city easily defensible while its location close to the boundary between forest and grassland makes it a melting point for people and products of the forests as well as those of the grassland areas. However, Ibadan was resettled about 1820 as a camp by the soldiers of the Ife, Ijebu and Oyo after they had successfully destroyed the neighbouring kingdom of Owu (Tomori, n.d.). The basic characteristics of Ibadan are discussed below. They include the physical features, locational attributes, historical background, economic activities, demographic structure, settlement pattern and political development.

Location

The city of Ibadan is located approximately on longitude 3°5′ East of the Greenwich Meridian and latitude 7°23′ North of the Equator at a distance of some 145 km west-east of Lagos. Ibadan is directly connected to many towns in Nigeria as its rural hinterland by a system of roads, railways and air routes.

Physical Background

Relief

Ibadan is located on lowlands which are punctuated by rocky outcrops and series of hills. Most of the rock outcrops are located in areas around Iseyin, Shaki, Sepeteri, Okeho and Igbeti. These consist mainly of schist and quartzite inselbergs. The boundary of the lowlands is marked in the north by the edge of the basement complex of the Western Highlands and in the south by the coastal deposits. The major rivers and their tributaries which flow through the state have created some notable river basins. These rivers include the Ogun, Osun, Oyan, Oba and Ona. Thus, the metropolis is characterized by three landform units – hills, plains and river valleys. There are two main types of hills and these are the quartzite ridges and gneiss inselbergs. The most prominent is the Oke-Badan Hill.

Drainage

Ibadan is drained by three important rivers. These are Ogunpa, Ona river and Ogbere rivers. The former drained the eastern while the latter drains the western parts. The Ogunpa flows south-eastwards breaking through the central ridge before turning south along a course that is parallel to the ridge. Its major tributary is the Kudeti which drains the eastern part of the ridge. The western part of the city which consists of more recent residential and other developments is drained by the river Ona and its numerous tributaries that include Alalubosa, Oshun and Yemoja streams (Ayeni, 1994).

Climate

Oyo State exhibits the typical West African monsoon climate marked by distinct seasonal shift in wind patterns. Between March and October, the city is under the influence of moist maritime south-west monsoon winds which blow inland from the Atlantic Ocean. This is the rainy season. The dry season occurs from November to February when the dry dust–laden winds blow from the Sahara desert. The area experiences high relative humidity and generally two rainfall maxima regimes during the rainfall period of March to October. The mean temperatures are highest at the end of the Harmattan (averaging 28°C), that is from the middle of January to the onset of the rains in the middle of March. Even during the rainfall months, average temperatures are between 24°C and 25°C, while annual range of temperature is about 6°C.

Geology

Ibadan is covered with basement complex rock, which are mainly metamorphic rocks and of Precambrian age with few intrusion of granites and porphyries of Jurassic age. These rocks can be grouped into major and minor rock types. The major types are quartzites of the meta-sedimentary series and migmatites. The minor rock types include pegmatites, quartz, aplite, diorites, amphibolites and xenoliths.

Soil

The soils of Ibadan region were formed from rocks of the Precambrian basement formation, especially granite gneisses, quartz schist, biotite gneisses and schist. They were formed under moist semi-deciduous forest cover and belong to the major soil group called ferruginous soils (D'Hoore, 1964; Hopkins, 1965). Aweto (1994) identified four main soil associations in Ibadan region on the basis of soil parent materials. They are Iwo, Okemesi, Egbeda and Mamu soil associations. The soils of Iwo associations were formed from coarse-grained granites and gneisses and those of Okemesi from quartz gneisses, schist and quartzites. Those of Egbeda and Mamu were formed from fine-grained biotite and schist, and from sericite schists respectively.

Vegetation

The southern parts of the city are covered by the rain forest and derived savannah. Much of Akinyele, Oluyole and Lagelu local government areas were covered by the rain forest. The composition is basically the large tall crowned trees, mixed with thick undergrowth. However, anthropogenic activities have led to an acute reduction in forest coverage.

Methodology

The survey research design was used for the study. The study covers 11 local government areas (LGAs) affected by the flood. The study utilized both primary and secondary data. The primary data were obtained by physical observation and in-depth interview of affected households in the 11 LGAs. Twenty-five affected households were randomly selected in each of the 11 affected LGAs. In all, a total of 275 households were interviewed. In-depth interview was also carried out with key officials of State Ministry of Environment and Housing. The study also relies on the data from the Oyo State Government Task Force on Flood Prevention and Management report. Data collected were analyzed using simple descriptive statistics such as frequency tables, pie charts and percentages.

FLOODING IMPACTS IN THE CITY OF IBADAN

The impacts of the 26 August 2011 flood disaster in the city of Ibadan were obtained from respondents who are mainly households in the affected LGAs. The responses of the respondents on the impacts of the flood are summarized in Table 1.

Table 1 indicates that the major impact of the flood was destruction of household property. About 44 per cent of the respondents in the affected neighbourhoods identified the destruction of property as the major impact. According to the Oyo State Government Task Force on Flood Prevention and Management report, property worth over 100 billion naira was destroyed. Also, 30.9 per cent of the respondents identified damage to city infrastructure. Most cities in Nigeria are characterized by weak infrastructural facilities. The destruction of the few available infrastructures in

Table 1. Flooding Impacts in the City of Ibadan.

Impacts	No. of Respondents	Percentage of the Respondents
Loss of lives	50	9.1
Destruction of property	167	44.0
Damage to city infrastructure	87	30.9
Loss of livelihoods	34	12.4
Social dislocation	10	3.6
Total	275	100.0

Source: Fieldwork, 2011.

the city such as roads, culverts, bridges and electricity will in the long run increases the vulnerability of the people to disasters. For instance, the Eleyele Waterworks was destroyed with most of the treatment plants covered by flood. The bridge of Ona River at Apete was also destroyed, cutting off the Apete community from the rest of the city for 4 days. Plates 1–4 shows property and infrastructure destroyed during the flood. Some

Plate 1. Damaged Apete Bridge.

Plate 2. Apete Bridge Under Reconstruction with Bamboo Sticks.

Plate 3. Flooded Buildings.

Plate 4. Damaged Culverts and Electric Poles.

residents were trapped by floods and could not leave their houses for several days until the flood water receded. As indicated, by about 9.1 and 12.4 per cent of the respondents, many residents lost their loved ones and their means of livelihoods to the flooding.

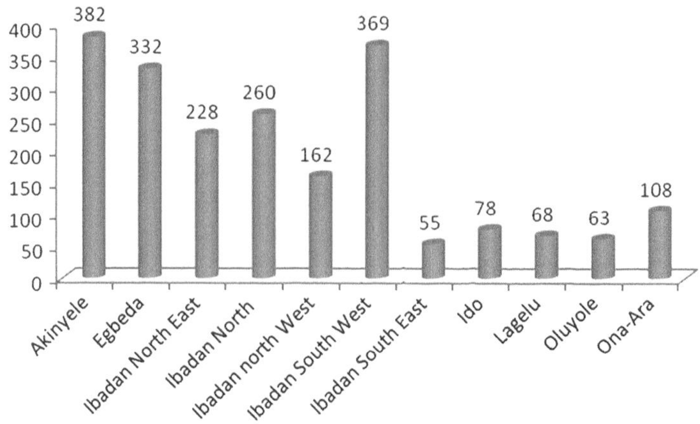

Fig. 1. Number of Buildings Flooded During the 26 August 2011 Flood Events in Ibadan. *Source:* Report of the Task Force on Flood Prevention and Management (2011).

Fig. 1 shows the number of buildings flooded during the 26 August 2011 flood events in the city of Ibadan.

Data from the report of the Task Force on Flood Prevention and Management indicates that 2,105 houses were flooded during the flood event. A break down of the figures shows that Akinyele LGA has the highest number of flooded buildings in the 11 affected LGAs. This is closely followed by Ibadan South West and Egbeda LGAs with 369 and 332 flooded buildings respectively. Flooding also precipitates environmental health hazards such as the outbreak of diseases, arising from drinking surface water and well water which have been polluted as a result of flooding (Report of the Task Force on Flood Prevention and Management, 2011). The flooding also created serious disruption to traffic flow on the motorable roads which were not affected by the flooding that led to emergency diversion of traffic to hitherto neglected and almost impassable roads as alternative routes. Transport fares also went up astronomically for commuters in the affected areas causing a lot of delays and hardship for the commuters.

VULNERABILITY FACTORS IN THE CITY OF IBADAN

Ibadan has for a long time suffered varying degrees of flooding. For instance, there were flooding in the watersheds of Ogunpa and Kudeti streams (the two major streams in Ibadan) in 1955, 1960, 1961, 1963, 1969, 1978 and 1980. The study sought to determine from the respondents the immediate vulnerability factors that contribute to flooding in the city of Ibadan. These are summarized in Table 2.

The respondents identified construction of buildings on flood plains as one of the major factors contributing to the vulnerability of the city to flooding. About 34 per cent of the respondents identified this factor. According to Akintola (1981), by the early 1960s, the flood plain settlement along the Ogunpa River valley was almost completed, such that the mean distance of buildings along the Ogunpa River valley is a mere 11 metres to the river banks, whereas the mean distance of floodable land is 90 metres. This practice contributes to flood by narrowing the river channel and also increasing the volume of water that flows into the river.

Also revealed from Table 2 is the indiscriminate dumping of refuse on the streets and other open spaces. The indiscriminate dumping of solid wastes especially of non-biodegradable polythene materials like pure water sachets and other similar packaging materials is a common practice in Ibadan metropolis. In addition, whenever it rains, residents especially in the core areas and unplanned suburbs dispose their solid wastes directly onto streets and gutters without any consideration of its consequences. This practice leads to the blocking of drainage channels. Also listed by the respondents is the high rate of urbanization in the city. Specifically, about 17 per cent of the respondents identified this factor. Until 1970, Ibadan was the largest city in sub-Saharan Africa (Lloyd, Mabogunje, & Awe, 1967). In 1952, it was estimated that the total area of the city was approximately 103.8 square kilometers (Areola, 1994). In terms of demographic growth, Ibadan experienced geometrical increase between 1851 and 1921 (Mabogunje,

Table 2. Vulnerability Factors of Flooding in Ibadan Metropolis.

Vulnerability Factors	No. of Respondents	Percentage of the Respondents
Construction on flood plains	95	34.5
Deforestation	45	16.4
Indiscriminate dumping of refuse	57	20.7
Excessive rainfall	20	7.3
Poor infrastructure	11	4.0
High rate of urbanization	47	17.0
Total	275	100.0

Source: Fieldwork, 2011.

1962). The 1991 census in Nigeria put the population at 1,222,570 (Ayeni, 1994) with a density of 475.11 person per square kilometre. Its population is estimated to be about 2,550,593 according to the 2006 estimates by the National Population Commission. Its projected population by 2010 using 3.2 per cent growth rate is about 2,893,137.

Deforestation has also been identified as contributing factor to flood problem in Ibadan. Some areas such as the Agala and Igbo Agala forests were deliberately preserved in Ibadan under teak and cassia forest during the colonial period. These were mainly the hills in and around Ibadan. The flood plain around the Eleyele River was equally consreved for the same reason. The destruction of these forests especially following the 1993 national election crisis has aided flooding in Ibadan metropolis due to the reduction in the infiltration capacity of these areas. This was confirmed by Akintola (1994) in a study of infiltration process in Ibadan city, which indicated varying rates or capacities for different types of urban land use surfaces. The changes in imperviousness of the metropolis between 1965 and 1994 are shown in Table 3.

In general, open surfaces heavily trampled upon by vehicles, human beings and animals have lower infiltration rates than grass surfaces, and thus generate higher excess water during rainfall. By implication, the large network of paved/tarred roads and several cemented surfaces in Ibadan has contributed to the loss of infiltration capacity. The corollary of this is that almost all the rainwater is released into river channels at the same time, thus leading to floods.

Another factor identified by the respondents is excessive rainfall. About 7.3 per cent of the respondents identified this factor. The International Institute of Tropical Agriculture (IITA) rain gauge recorded an all time high

Table 3. Percentage of City Surface Impervious to Water Infiltration.

Selections of City	Percentage in 1965	Percentage in 1994	Percentage Increase	Rate of Increase
Traditional core	15.4	42.5	27.1	1.80
Modern low density	4.3	17.3	13.0	0.98
Modern high density	11.2	30.2	19.0	1.20
Utilities and reservations	3.6	17.5	13.9	1.93
Mean values	9.5	28.4	18.9	1.31

Source: Akintola (1994).

of 187.5 millimetre rainfall on 26 August 2011 accompanied by high wind speed as high as 65 kilometre per hour (Report of the task Force on Flood Prevention and Management, 2011).

DISASTER RISK MANAGEMENT IN THE CITY OF IBADAN

The major aim of disaster risk management is to minimize human loss and economic damages, while making use of the natural resources for the benefit and well-being of the people. Disaster risks cannot be eradicated completely, but they can be managed. Consequently, disaster risk management does not strive to eliminate flood risks but to mitigate them. This may be achieved either by reducing flood risks to an acceptable level or by retaining, sharing or transferring flood risks through respective measures (World Meteorological Station/Global Water Partnership, 2008).

In this context, the study sought to determine the disaster risk management strategy in the city of Ibadan. The in-depth interview conducted among the key officials of the State Ministry of Environment and Housing indicates there is no state agency responsible for disaster management. According to the respondents, there is only one national agency and that is the National Emergency Management Agency (NEMA) charged with disaster management in Nigeria. NEMA was established via Act 12 as amended by Act 50 of 1999 to manage disasters in Nigeria. The primary objective of NEMA is to coordinate the provision of immediate and timely succour to disaster victims. This involves providing relief materials, establishing camps (where necessary) for internally displaced persons and ensuring proper and efficient management of the camps.

Disaster risk management is designed to protect livelihoods and the assets of communities and individuals from the impact of hazards by:

1. *Mitigation*: reducing the frequency, scale, intensity and impact of hazards through the provision of infrastructure (construction of earth bunds, gabion cages, contour planting, check dams, strengthened dwellings and public buildings, raised river banks, re-forestation and storm drains) and other non-infrastructure measures (public health campaigns, vaccination programmes (both for livestock and humans), introducing new agricultural practices such as short maturation or drought resistant varieties of cereal crops, promoting dialogue between communities in conflict, relocation of settlements, and awareness and education programmes);

2. *Preparedness*: strengthening the capacity of communities to withstand, respond to and recover from hazards, and of government, implementing partners and concern to establish speedy and appropriate interventions when the communities' capacities are overwhelmed; and
3. *Advocacy*: favourably influencing the social, political, economic and environmental issues that contribute to the causes and magnitude of impact of hazards.

The strategy of NEMA for disaster management primarily focused on response *after* disasters. This includes the provision of humanitarian aspects of relief, such as providing medical care, food and water, search and rescue, and containing the secondary disasters after the disaster had occurred. This approach undermined the three cardinal goals of disaster risk management as identified above. Also revealed by the respondents is the fact that there is no collaboration between the State Ministry of Environment and Housing and the NEMA. This has made it difficult in terms of building of synergy in disaster risk reduction in the Ibadan metropolis. They noted that the presence of NEMA is only felt in the event of natural disasters such as flooding.

The respondents also noted that there is no short-, medium- and long-term plan for disaster management in the city. With respect to flooding, the respondents observed that there is no weather information at the city level, thereby making predicting and forecasting of the occurrence of flooding difficult. The interviewees noted that their ministry has put in place adequate measures to ensure that individual developers do not construct their buildings at the flood plains. However, the rapid and uncontrolled expansion of the city has made it impossible to enforce such regulations. Some of the officials of the ministry also observed that the city lacked a master plan and all efforts to develop one have continuously proved abortive.

All these, they noted, have perpetually frustrated the effort of the ministry in disaster risk management. The officials noted that they have not at any time carried out hazard and vulnerability assessment of neighbourhoods in the city despite the long history of flood disasters in the city. One obvious fact gleaned from the in-depth interview is the fact that the officials lacked the capacity to carry out any proper risk management in the city. Also, the approach of the city agency in the management of disasters has remained ad hoc and contingency in nature and this was demonstrated by the state government which set up a task force on flood prevention and management in the eve of the 26 August 2011 flood disaster in the city.

CONCLUSION

Climate change has been predicted to increase the incidence of flooding across the globe and the developing countries have been identified to be most vulnerable. The 26 August 2011 flood disaster in Ibadan metropolis can be seen as a tip of the iceberg. It is therefore necessary that appropriate strategies, most especially disaster risk management framework, be instituted in the metropolis. In the city of Ibadan, it is obvious that flooding will continue to occur and the high vulnerability factors as itemized in this chapter make for appropriate framework for disaster risk management in the city a necessity. Consequently, the chapter provides the following policy recommendations:

1. The various vulnerability factors such as construction on flood plains and the indiscriminate dumping of refuse on side drains should be redressed by relevant agency.
2. The vulnerability of the city to flooding makes case for the establishment of emergency response agency at the state level, which should be trained on contemporary techniques for disaster risk management (hazard assessment and vulnerability assessment).
3. There is the need for synergy between the State Ministry of Environment and Housing as well as the NEMA.
4. It is important to involve all stakeholders and other residents of neighbourhoods susceptible to flooding, especially in the design of programmes towards disaster risk management.
5. There is need for strategy change by the agency charged with disaster management. The after response strategy should be changed to that of disaster mitigation, preparedness and advocacy.
6. The state should embark sensitization programmes using various media to inform residents of the dangers of bad environmental behaviour such as construction on flood plains and the indiscriminate dumping of refuse.

REFERENCES

Adeoye, N. O., Ayanlade, A., & Babatimehin, O. (2009). Climate change and menace of floods in Nigerian cities: Socio-economic implications. *Advances in Natural and Applied Sciences*, *3*(3), 369–377.

Akintola, E. O. (1981). The hydrological consequences of urbanization: A case study of Ibadan city. In P. O. Sada & J. S. Oguntoyinbo (Eds.), *Urbanization processes and problems in Nigeria*. Ibadan University Press.

Akintola, E. O. (1994). Flooding phenomenon. In M. O. Filani, F. O. Akintola & C. O. Ikporukpo (Eds.), *Ibadan region*. Ibadan: Heinemann.

Areola, O. (1994). The spatial growth of Ibadan city and its impact on the rural hinterland. In M. O. Filani, F. O. Akintola & C. O. Ikporukpo (Eds.), *Ibadan region*. Ibadan: Rex Charles Publication.

Askew, A. J. (1999). Water in the international decade for natural disaster reduction. In Leavesley, et al. (Eds.), *Destructive water: Water-caused natural disasters, their abatement and control. IAHS Publication* No. 239.

Aweto, A. O. (1994). Soils. In M. O. Filani, F. O. Akintola & C. O. Ikporukpo (Eds.), *Ibadan region*. Ibadan: Heinemann.

Ayeni, B. (1994). The metropolitan area of Ibadan: Its growth and structure. In M. O. Filani, F. O. Akintola & C. O. Ikporukpo (Eds.), *Ibadan region*. Ibadan: Department of Geography, University of Ibadan.

Clark, M. J., Priest, S. J., Treby, E. J., & Crichton, D. (2002). *Insurance and UK floods a strategic reassessment*. Report of the Tsunami Project. Universities of Southampton, Bournemouth and Middlesex.

Crichton, D. (1999). The risk triangle. In J. Ingleton (Ed.), *Natural disaster management* (pp. 102–103). Leicester: Tudor Rose.

D' Hoore, L. J. (1964). *Soil map of Africa*. Lagos: CCTA.

Fadare, W., & Oduwaye, L. (2009). Rebranding Lagos through regeneration, *Proceedings of the real corp 2009: Cities 3.0 – smart, sustainable, integrative; strategies, concepts and technologies for planning the urban future*, Tagunsband Sites, 22–25 April 2009.

Fleming, G. (2001). *Learning to live with rivers*. London: The Institution of Civil Engineers.

Hopkins, B. (1965). *Forest and Savannah*. Ibadan: Heinemann.

International Strategy for Disaster Reduction (ISDR). (2004). *Living with isk: A global review of disaster reduction initiatives*. Geneva: International Strategy for Disaster Reduction, United Nations.

Jha, J. L., Bloch, R., Bhattacharya, N., Lopez, A., Papachristodoulou, N., Bird, A., ... Barker, R. (2011). *Five feet high and rising: Cities and flooding in the 21st century*. Policy Research Working Paper No. 5648. The World Bank, Washington, DC.

Johnstone, W. M., & Lence, B. J. (2009). Assessing the value of mitigation strategies in reducing the impacts of rapid-onset, catastrophic floods. *Journal of Flood Risk Management*, *2*, 209–221.

Lloyd, P. C., Mabogunje, A. L., & Awe, B. (1967). *The city of Ibadan*. Cambridge: Cambridge University Press.

Mabogunje, AL. (1962). The growth of residential districts in Ibadan. *Geographical Review*, *1*, 56–77.

Obot, I. D. (1986). Rural development programme of the Directorate for Food, Roads and Rural Infrastructure in Cross River State: A perspective view. *Journal of the Nigerian Institute of Town Planners*, *VIII & XI*, 92.

Okpala, D. C. I. (1986). *Institutional problems in the management of Nigerian urban environment*. Ibadan: NISER.

Oyo State Report on Flood Prevention and Management. (2011). Assessment of the 26 August 2011 flood disaster in Ibadan Metropolis. Ministry of Environment, Ibadan, Oyo State.

Smith, K. (1996). *Environmental hazards*. London: Routledge.

Thywissen, K. (2006). *Components of risk: A comparative glossary*. Bonn: United Nations University, Institute for Environment and Human Security (UNU-EHS).

Tomori, M. A. (n.d.). *Ibadan metropolitan area and the challenges to sustainable Development*. Availalbe at https://webmail.iu.edu/horde/services/go.php?url=http%3A%2F%2FFmacosconsultancy.com%2FIbadan%2520metropolitan.html. Accessed on February 13, 2012.

UN. (2010). *World urbanization prospects: The 2009 revision*, CD-ROM edition, data in digital form (POP/DB/WUP/Rev.2009), Department of Economic and Social Affairs, Population Division, the United Nations, New York, NY.

Wirth, L. (1938). Urbanism as a way of life. *American Journal of Sociology*, *XLIV*(1).

World Bank. (2010a). *Cities and climate change. An urgent agenda*. The International Bank for Reconstruction and Development.

World Bank. (2010b). *Development and climate change: A strategic framework for the World Bank group, interim progress report*. Washington, DC: The World Bank.

World Meteorological Station/Global Water Partnership. (2008). *Urban flood risk management. A tool for integrated flood management*. APFM Technical Document No. 11, Flood Management Tool Series, WMO/GWP, World Meteorological Station/Global Water Partnership, Geneva.

CHAPTER 12

GLOBAL ENVIRONMENTAL CHANGES AND IMPACTS ON FISHING ACTIVITIES IN THE NORTHERN COAST OF SÃO PAULO, BRAZIL

Sônia Regina da Cal Seixas, Michelle Renk, João Luiz de Moraes Hoeffel, André Luiz da Conceição and Gabriela Farias Asmus

ABSTRACT

Purpose – *Scientific studies have shown that the coastal zone is one of the regions showing great vulnerability to the impacts of global environmental change. For the region, impacts that may directly affect the economy and daily life of the communities of coastal municipalities were predicted. It occurs from phenomena, e.g. temperature rise, sea level rise, salinity, acidification of the seas and extreme events.*

Methodology/approach – *The mariculture labourers and artisan fishers from the Cocanha beach (Caraguatatuba city, north coast of the state of*

Urban Areas and Global Climate Change
Research in Urban Sociology, Volume 12, 301–319

ISSN: 1047-0042/doi:10.1108/S1047-0042(2012)0000012015

São Paulo), a lookout group, are the first persons to perceive the environmental change impacts in their daily contact with natural resources. Thus, the aim herein was to (a) verify their perception on the amount and quality of mussels and fish catch and (b) if such changes could be related to global environmental changes. In order to do so, semi-structured interviews with this social group were conducted in November 2009–February 2011 and codified by using the NVIVO8 software.

Findings – *The results indicated a decrease in fish stocks and, according to interviewees, such decrease was tied in with changes in fishing, in climate, and in beach landscape. Moreover, the category related to increased water temperature was highly significant, since the fishing and mariculture activities are directly influenced by this factor.*

Research limitations – *The perception evaluation through interviews with artisanal fishermen and shellfishermen implies that many subjective aspects are present, as well as the role of the media that has published significant information on climate change in the contemporary world.*

Originality/value of paper – *There are few studies that present these perspectives; however, the authors believe that it is possible, while recognising the possible limitations of the method, to recognise something that is already perceived by the community and may help to evaluate the reality they face and to contribute actually to the construction of future public policies.*

Keywords: Global environmental change; fishing; mariculture; perception; north coast of the state of São Paulo

INTRODUCTION

Today's severe and intensely increasing global environmental changes are considered to be a profound crisis in the 21st century. This crisis involves changes to climate and ecological systems that affect biogeochemical cycles and urban infrastructure. These changes also harm economic and social activities and human health and profoundly affect poor populations living in the tropical regions of the world (Giddens, 2010; Kotir, 2010; Min, Zhang, Zwiers, & Hegerl, 2011; Patz, Gibbs, Foley, Rogers, & Smith, 2007).

According to Marengo and Valverde (2007), there is evidence that extreme events (droughts, floods, heat and cold waves, hurricanes, storms)

have affected the planet in a different way that has caused a rise in sea level and impacts on biodiversity, human health, economic activities, e.g. agriculture and the energy sector.

Considering the size of its population, economy, technology and science, São Paulo should have a prominent position in the national debate on global environmental change, especially in regard to climate change. A large scientific investment has been made to assess adaptation and mitigate the anticipated effects on agricultural production in coastal areas, biodiversity, urban infrastructure, public health, and economic development (Confalonieri, 2005; Hogan, 2008; Marengo, 2006; Seixas, Hoeffel, Vieira, Mello, & Vianna, 2010). Because it is one of the coastal regions which will suffer great impacts from environmental changes, the coast of São Paulo was chosen for a study. Some of these impacts will directly affect the economy and daily life of communities in coastal municipalities, and according to Bindoff et al. (2007), phenomena such as temperature and sea level rise, salinity, acidification of the seas and extreme events will have a direct effect on these regions.

Among the expected consequences for the coastal region, Solomon et al. (2007) highlight the loss of river valley areas, coastal mangrove swamps, and increased damages that stem from coastal flooding, and the combination of sea level rise with uncontrolled human land occupation.

Considering the characteristics of the coast of São Paulo, e.g. high population, urban areas, economic concentration, we expect that the occurrence and intensification of climate change will directly and indirectly impact regional economic activities, especially those related to fishing and aquaculture.

Artisan fishers and mariculture labourers and their families represent communities that are closely related to marine resources. Their daily contact with these resources provides knowledge about water and tides, biodiversity, and economic activities on site. These traditional communities may be considered a lookout group for environmental changes to the coast because they are among the first to perceive the impact of changes on natural resources (Hogan, 2008).

Thus, this study aimed (a) to verify the perception of mariculture labourers and fishers in the Cocanha beach (Caraguatatuba city, northern coast of the state of São Paulo) of the amount and quality of mussels and fish catch and (b) if any changes could be related to global environmental changes. Simultaneously, this work will contribute to ongoing projects in the region, e.g. *Urban growth, vulnerability and adaptation: Social and ecological dimensions of climate change on the coast of São Paulo* (process no.

2008/58159-7, Programme of Global Climate Change Research – PFPMCG, Foundation for Research Support of the State of São Paulo – FAPESP) and *Global environmental changes, vulnerability, risk, subjectivity: A study on the north coast of São Paulo* (process no. 2010/20811-5).

METHODOLOGY

To verify the perception of fishers and mariculture labourers from Cocanha beach concerning global environmental changes, 13 semi-structured interviews in two distinct periods were carried out: November 2009; February 2011.

The questionnaires were composed of open questions that attempted to lead interviewees to talk on changes perceived in the art of fishing along with their experience as fishers of changes in water, changes in quality or in fish catch quantity (or both), changes in fishing practice (e.g. changes in the equipment used), and changes in everyday life in general. After this interviewees were asked to suggest explanations for the perceived changes so that we could assess the connection between global environmental change and fishing activity. Other aspects such as income sources, time spent when fishing and family data were also researched during the interview.

The transcribed interviews were codified from the NVIVO8 software. At first, we carried out a query to identify the most frequent terms in the reports. These terms have formed the subcategories in which the reports were coded.

The subcategories created (free nodes) were organised into general categories (tree nodes) and represented the central idea of the conversations (Fig. 1).

STUDY PLACE: COCANHA BEACH, CARAGUATATUBA

Cocanha beach is located in the northern portion of Caraguatatuba city and bound on the north–south axis by Mococa beach and Massaguaçu beach (Fig. 2).

Because it is located in the city, it is considered to be the 'gateway to the north coast', the Cocanha beach and other beaches in the region, were strongly influenced by tourism and a growing process of land occupation

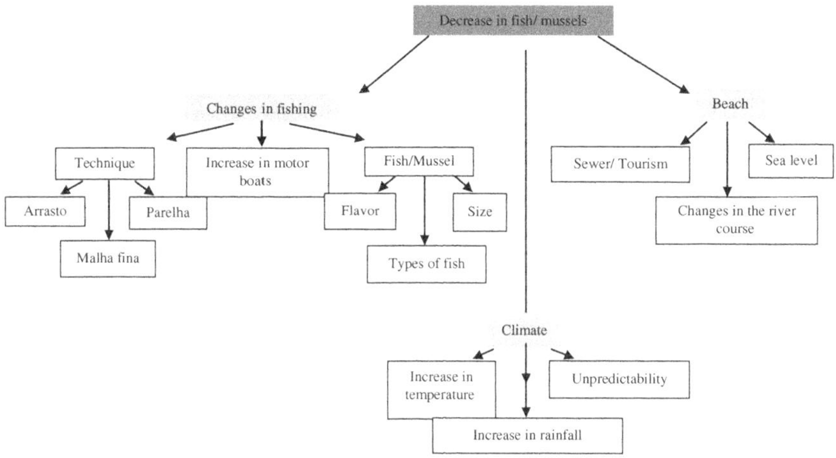

Fig. 1. Layout of Categories and Subcategories Found in the Reports of Fishers and Mariculture Labourers from Cocanha Beach (Caraguatatuba City). *Source*: Own elaboration, 2011.

Fig. 2. Aerial View of Cocanha Beach in Caraguatatuba City (São Paulo, Brazil). *Source*: Google Earth, 2011 (adapted).

that began in the mid 1950s and was a result of the opening of the north coast highway that connected it to the Paraíba valley (later known as *Rodovia dos Tamoios* [Tamoios Highway], SP-099) and the improvements in regional infrastructure that resulted from the work to expand the Port of São Sebastião (Souza, 2009).

Cocanha beach is important to the region economically both for tourism and for being the largest producer of brown mussels, i.e. *Perna perna* (Linnaeus, 1758), in the state of São Paulo, with nearly 30 tons/year (Marques, Bordon, Alves, & Medeiros, 2008).

Mussel Culture in Cocanha Beach

In 2004, the shellfish production of Cocanha Beach presented 4.8% of the Brazilian production; from such total, 79.5% is represented by cultivation of the species *Perna perna* followed by oysters (20.5%) and generated US$ 9.3 million in revenues that year (Boscardin, 2008).

According to Boscardin (2008), in 2000 mussel cultivation showed a production of 11,760 tons and that in 2004 production dropped to 10,370 tons. Another important factor is that the state of Santa Catarina is responsible for 94.5% of the yearly production, while the remainder (5.5%) is distributed among the states of Espírito Santo, São Paulo and Rio de Janeiro (Boscardin, 2008).

Cultivation of mussels on Cocanha beach began in 1988–1989 through the *Projeto Martim Pescador* [Halcyon Project].[1] This project aimed at empowering the community through mussel culture and focused on fixing fishers to their native communities (Brasil, 2011; Marques et al., 2008).

According to Cocanha mariculture labourers, mussel cultivation is accomplished through long lines with 50 m in length each. These structures are attached to buoys which guarantee the fluctuation of the system. The cords containing seed mussels have on average 2 m in length and are arranged in long parallel lines, with an approximate range of 0.5 m.

The aquaculture park in *Ilha da Cocanha* [Cocanha island] occupies an area of approximately 62,000 m^2, 5–8 m deep; there were 9 crops in operation in 2009, and the total capacity is 11 crops (Silvestre, 2009). Cultivation is divided into cycles lasting 8–9 months, the long-line average production is 2 tons; such production intensified from November to January due to seasonal demand (Silvestre, 2009).

In the state of São Paulo, mussel cultivation is carried out by hand by traditional families and fishing communities and employs family labour

(Marques et al., 2008; Ostrensky, Boeger, & Chammas, 2008). For Fagundes, Gelli, Otani, Vicente and Fredo (2004), mussel culture is as an alternative income for fishing communities that depend on fishing and allows these communities to remain in their places of origin after the decline of fish stocks from overfishing and pollution.

The interview showed that (Table 1) most fishers and mariculture labourers were born and had resided in the region for an average of 19.5 years and are from 34 to 54 years of age.

We observed in the field activity that the entire family participated in the process and that it is mostly wives and siblings who clean the mussels. In this regard, Pestana, Pie, and Pilchowski (2008) point out that aquaculture in general presents great potential for labour and income generation and for job generation. This was observed in Cocanha considering the large number of people who see mussel cultivation as supplementary income.

According to Fagundes et al. (2004), mussel culture performed in family systems proves to be economically viable because demands for investments and operating costs are relatively low. However, authors such as Ostrensky and Boeger (2008) and Ostrensky et al. (2008) identified some problems for Brazilian mariculture labourers and mariculture that include challenges of inserting new technologies and markets and of harmonising industrial and family productions.

The issue of sustainability should be considered important in mariculture because seed mussels are frequently collected from natural deposits and this impacts the reproduction of the species in their natural environment. Manzoni (2004) stresses that production of seeds in the laboratory may be a viable solution to reduce such impact. This solution, however, can generate a financial expense that was not previously recorded because this activity is presently extractive.

With regard to seed production, the use of artificial collectors that was observed on Cocanha beach was highly efficient. This technique is discussed in the study of Marques et al. (2008) which details the production of seed in the region studied. According to the authors, this technique provides satisfactory results and environmentally friendly products and represents an alternative way of collecting the seed crop on a farm instead of from the environment.

The use of the collectors can generate self-sufficiency in seed production, reduce extraction from the environment, enhance environmental sustainability and marine cultures and contribute to reducing production costs (Marques et al., 2008) and the cost increases that result when it is necessary to buy larvae produced in laboratories.

Table 1. Profile of Mariculture Labourers and Fishers Interviewed.

	Place of Birth	Place of Residence	Activity Period	Age (years)	Year of Interview
Interviewee 1	Atibaia (SP)	Cocanha	22 years	–	2009
Interviewee 2	Curitiba (SP)	Cocanha	7 years	34 years	2009
*Interviewee 3**	Caraguatatuba (SP)	Cocanha	20 years	39 years	2009
Interviewee 4	Capricórnio (Caraguatatuba)	Capricórnio (Caraguatatuba)	17 years	38 years	2009
*Interviewee 5**	Caraguatatuba (SP)	Cocanha	27 years	37 years	2009
*Interviewee 6**	Caraguatatuba (SP)	Cocanha	20 years	39 years	2011
Interviewee 7	Caraguatatuba (SP)	Caraguatatuba (SP)	Over 30 years	57 years	2011
Interviewee 8	Jacareí	Cocanha	15 years	–	2011
Interviewee 9	Caraguatatuba (SP)	Cocanha	3 years	–	2011
*Interviewee 10**	Caraguatatuba (SP)	Cocanha	13 years	40 years	2011
Interviewee 11	Caraguatatuba (SP)	Cocanha	Over 30 years	55 years	2011
Interviewee 12	Caraguatatuba (SP)	Cocanha	30 years	46 years	2011
Interviewee 13	Caraguatatuba (SP)	Cocanha	Over 20 years	38 years	2011

*Interviewees in the periods of 2009 and 2011.
SP, São Paulo.
Source: Own elaboration, 2011.

Artificial seed collectors are a great environmental alternative. Concerning production, however, according to the accounts made by the region's mariculture labourers, a reduction in area of cultivation occurred. In Cocanha, each mariculture labourer has a grant for an area of 2,000 m^2 of water layer for cultivation. The collectors are positioned in this same area and because these areas cannot be expanded the productive area is reduced. Although this issue does not make the use of artificial collection impracticable, it draws attention to the regulation of areas Ostrensky and Boeger (2008) as a major challenge to the industry.

On the north coast of São Paulo, a survey conducted by the *Instituto de Pesca* [IP – Fisheries Institute] has identified 4,100 acres suitable for mussel cultivation. This is an area with potential to produce roughly 200,000 tons (Brasil, 2006). This data shows that productivity studied in the region also depends on the regulation and expansion of cultivation areas.

In addition to the cultivation farms, there was also the existence of structures both at sea and on the sand where mussels are cleaned. According to interviewees, this phase of mussel processing is extremely laborious and performed mostly by hand.

Cleaning is composed of the removal of barnacles and other organisms that are fixed to the mussel shell. This process is usually carried out mussel by mussel on benches with the aid of a machete. At this moment the seeds that return to the farm on new strands are selected, and according to interviewees within a maximum of 8 hours, in order to avoid compromising the seeds.

With regard to this procedure, some mariculture labourers interviewed commented on the need for mechanisation of the process of waste removal and underlined the difficulty of establishing this without funding. This question reflects the need to introduce new technologies in mussel culture and planning to harmonise industrial production with family production because in family production the process is segmented, done by hand and requires more labour.

Sale is done mostly to direct by local buyers; tourists (for consumption on the beach) and the reports showed there was the beginning of a market to supply restaurants. Currently, however, the demand outweighs production. Interviewees attribute this growth in local market to the mussel festival held with the incentive of the local authorities. This festival achieved its ninth edition in 2010 with 8,000 participants, and 4,000 kg of mussels were consumed (Almeida, 2010).

Due to the acceptance and enhancement of the activity, the *Associação dos Pescadores e Maricultores da Praia da Cocanha – MAPEC* [Association of Fishers and Mariculture Labourers from Cocanha Beach] was founded in 1998–2000 as a technical support for the activity and the mariculture labourers from Cocanha beach. The association was guided by the Fishing Institute and the *Associação dos Maricultores do Estado de São Paulo – AMESP* [Association of Mariculture Labourers from the State of São Paulo] (Marques et al., 2008; Prado, 2001).

RESULTS AND DISCUSSION

The fishers and mariculture labourers from the study region organise in MAPEC in a structure near the beach which has 22 boxes and a central warehouse.

The interviewees said there were a total of 40 associates in 2009, and 35 in 2011. At first 37.5% of the interviewees had been engaged in mussel culture. The percentage of mariculture labourers rose roughly 5% in 2011 due to decreased fishing partners.

The interviewee 3 indicated the existence of 40 members, i.e. 15 mariculture labourers and from them only 10 cultivate effectively. The interviewee 10 however reported a total of 35 members and only one avid fisher. There is thus a disparity in the interviewees reports and tendency of a decrease in the number of fishers and also of mariculture labourers in the group studied.

In conjunction with these figures, the interviews often bring the term 'hobby' and 'sport' linked to the activity. This is legitimate in all cases because of the difficulty of surviving from income brought in exclusively from fishing. Data on the income sources of the interviewees (Table 2) shows that only one fisher (interviewee 6) has fishing as the sole source of his income and three mariculture labourers show mussel culture as their only activity. Thus, 69% of interviewees have additional income to both fishing practice and mariculture in other areas such as general labour services.

During field research the brief contact with the interviewees' family members showed that fisher activity that once passed from generation to generation today arouses little interest among offspring who are opting for other activities practised on the beach, e.g. fireman, lifeguard.

Another point highlighted in Table 2 is that fishing was mentioned by 71% of interviewees as a complementary activity (considering only those who develop a second source of income), compared with 28.5%, allocated for the mariculture developed two decades ago on the site.

Table 2. Income Source Profile of the Interviewees.

	Main Income Source	Supplementary Income
Interviewee 1	Fishing and mariculture	None
Interviewee 2	Mariculture	Fishery
*Interviewee 3**	Mariculture	None
Interviewee 4	Tourism	Mariculture
*Interviewee 5**	Fishery	Mariculture
*Interviewee 6**	Fishery	None
Interviewee 7	Local market	Fishery
Interviewee 8	Retired	Farm manager and fisher
Interviewee 9	Mariculture	None
*Interviewee 10**	Mariculture	None
Interviewee 11	Trade	Fishery
Interviewee 12	Sailor	Fishery
Interviewee 13	Fishing and mariculture	None

*Interviewees in the two field research.
Source: Own elaboration, 2011.

The issues mentioned above, e.g. the low number of workers actively involved, the lack of incentive of the younger generation and fishing perceived as a complementary income may suggest that fishing is no longer seen as a profitable economic activity that is able to support the families who depend on it.

In seeking possible causes of this we found (from the conversation) that it is neither the economic value assigned to the fish catch (market consumer) nor the will to perform the activity but mainly the decline of fish stocks that is causing this breakdown.

This same downward trend is highlighted by the interviewees and can be seen in the assessment of marine and estuarine fish production in the state of São Paulo for the last 40 years (Fig. 3).

According to the *Programa de Monitoramento da Atividade Pesqueira – PMAP* [Fishing Activity Programme Monitoring Activity] (IP, 2011), preliminary data from the production of marine and estuarine fishing in the state of São Paulo suggests that the year 2010 stands out as having the lowest production on record with a total of 22094.6 tons discharged on the coast of São Paulo.

Thus, given the apparent decline in fish production, the mussel culture which initially presented itself as a complement or alternative is now considered a central activity in the community studied. It is important to

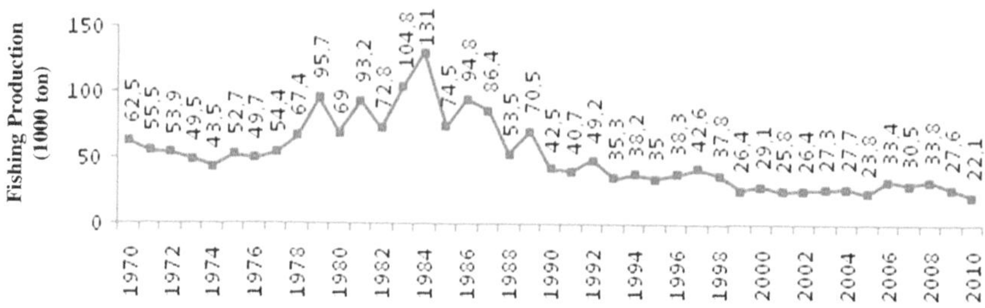

Fig. 3. Marine and Estuarine Fish Production in the State of São Paulo (1970–2010). *Source*: Fishing Institute, 2011.

note that the issue of reduced production is also seen in the interviews of mariculture labourers.

Even though the opinions of fishers and mariculture labourers differ as to the origins and causes of the decrease because of the peculiarities of each activity, the drop in fish kilograms per box runs parallel to losses in kilogram per mussel rope. The decline in fish stocks and losses in the cultivation of mussels can be identified as core items in the interviews and main factors behind the changes occurring to fishing.

Considering the production drop as a starting point, the interview assessment emphasises three categories – changes to fishing, climate and beach landscape, and consists of those that appear most frequently in the reports. Interviews also assess changes occurring in recent decades in the labour activity of fishers and mariculture labourers in the region studied.

Changes in Fishing

According to interviewees, changes in fishing involve three key factors: (a) equipment used for fishing (b) increased use of motorboats and (c) changes in fish catch. Among the fishers approached, we observed that more than 85% attribute the fish catch drop mainly to the use of bull trawl technique.

Tuna boats which approach the coast to collect bait were also cited; the trawl fishery, however, encompasses 69% of the reports. A large percentage could be related to the production of Atlantic seabob (*Xiphopenaeus kroyeri*) in Caraguatatuba, which according to IP (2011) represents 46% of total fishery production of the municipality (Fig. 4).

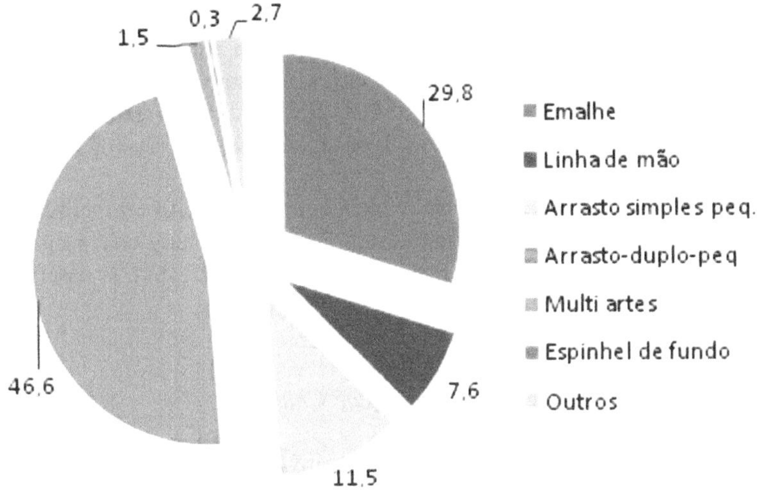

Fig. 4. Fishing Equipment Used in Caraguatatuba in 2010 (as percentage). *Source*: Own elaboration using data from IP (2011).

Another matter when considering the impact on the number of fish is the number of nets used. According to fishers the most appropriate tools used in the region of Cocanha is the gillnet. Although the interviewees declare that there has not been a change in their method of fishing, they highlight the growing number of motorboats, both for fishing purposes and speedboats and other boats for tourism.

Changes in Climate

The second factor identified in the interviews related to fish production drop were changes to climate. Interviewees speak of rises in temperature, increased rainfall and the unpredictability of weather conditions. These items are perceived as major interferences in both fishing activities and mariculture.

Significantly, the temperature rise is cited by 97% of interviewees and reflects the sharp interference this poses to the fishing industry and to mussel culture.

Importantly, among mariculture labourers, the temperature rise forms one of the most striking environmental changes to the mussel culture because the ideal temperature limits for mussel development ranges from 22 to 26 degrees. This limitation is the great reason for a higher volume of production in colder regions, e.g. the state of Santa Catarina.

With higher temperatures there is a proliferation of barnacles which compete with mussels for food and fixation substrate increasing losses even more.

The interviewees (54%) also cite rainfall combined with temperature rise. Of these, 42% represent interviews from 2009. Something else pointed out by fishers is the difficulty of predicting the weather that is essential to determine when to go out to sea.

Beach Landscape Changes

The third category indicated in the reports involves changes which occurred in the Cocanha beach landscape. Such changes were largely caused by the expansion of tourism in the local areas and by the modification of sediment deposits along the coast.

According to interviewees, there were marked changes to the beach. The reports pointed out that where the structure is seen today there was previously a mangrove swamp and that the sand strip in the estuary area was narrow.

The morphological changes emphasised by the interviewees, e.g. changes in sea level and flooding in the structure area, can be correlated to the diversion of the river and to the level of the mangrove area with earth cited by interviewee 13.

In the interviews, 76% of fishers and mariculture labourers attributed the changes to the growth of tourism in the region. Tourism is also connected to the diversion of the river and also appears as the cause of issues related to sewage, waste production and increased motorboats.

In regards to sewage, two aspects were observed: (a) production, i.e. kiosks and product trade on the beach and (b) coverage of sewage treatment in the municipalities. According to technicians of the sewage treatment company of the state of São Paulo – SABESP, 86% of the municipalities' population is being supplied with water along with a worrying 45% of collection, removal and sewage treatment (Renk, 2010). There are also reports of a lack of collection and treatment of sewage from the beach.

Coupled with this reality is the fact that during the holiday period the population triples and this overloads infrastructure, damages water quality and causes health problems. There is also the problem of contact with sewage-contaminated water, exposing bathers to viruses, bacteria and protozoa (CETESB, 2011; Caraguatatuba Prefecture, 2010). Fig. 5 illustrates such point.

CETESB has a basic indicator to classify the density of faecal coliforms on beaches (CETESB, 2011). According to Fig. 5, 15 local examined beaches (e.g. Tabatinga, Cocanha, Lagoa Azul, Prainha, Indaiá, Porto Novo) are classified as inappropriate most of the time during holidays especially in the summer. Concern about this is seen in the interviews of 86% of fishers and 100% of mariculture labourers. Because mussels are biological filter feeders, water contamination by faecal coliform directly affects product quality.

When looking at interviews from different periods such as reports from 2009, it is seen that in 2011, 73% call attention to waste production and 85% highlighted the sewage issue. It is important to note that the reports show that along with these negative aspects there is also an economic dependence on existing tourism resources in the region.

Field research showed that beach tourism is increasingly attracting the interest of the local community. Besides the mussel festival supported by the authorities, one can use boats (banana boat, speedboats) to visit the cultivation area and the islands in the region, i.e. island and islets of Cocanha.

On Cocanha island it was possible to follow the significant growth of visitation. In 2009, the island had as tourist attractions a trail to its top, a mansion and another long beach. Right at the entrance to this beach there is a makeshift restaurant that serves mussels. In 2011, this restaurant presented infrastructure and a team of roughly 30 people catering to people visiting the site on the weekends.

One can see that the same individuals who need a consumer market for their product watch and suffer the impacts from these activities because

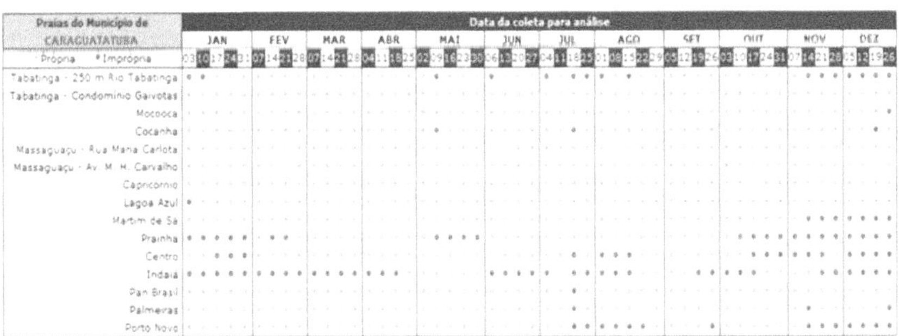

Fig. 5. Weekly qualification of the Caraguatatuba beaches during 2010. *Source*: CETESB, 2011.

they are usually carried out without planning, and this lack of planning overloads local infrastructure (sanitation). The subtle differentiation between the positive and negative aspects of tourism in the region became explicit in the interviews.

Environmental Changes and their Relationship to the Fishing Industry and to Mussel Cultivation on Cocanha Beach

In general, according to the interviewees the changes experienced in recent decades to fishing activity, climate, and beaches may be contributing to both (a) fish stock and number of fishers who can just live off fishing (as already mentioned) and (b) type and characteristics of fish and water quality.

According to IP (2011), low production in 2010 is mainly due to the discharge drop of Brazilian sardinella, this production level had returned to normal from 1999 to 2005, and to corvine discharge drop which has declined over the past 7 years. These are the species that most stand out in Fig. 6.

Other interferences to species diversity and affecting the quantity of fish catch are changes brought about by increased use of motorboats, both for fishing and for tourism.

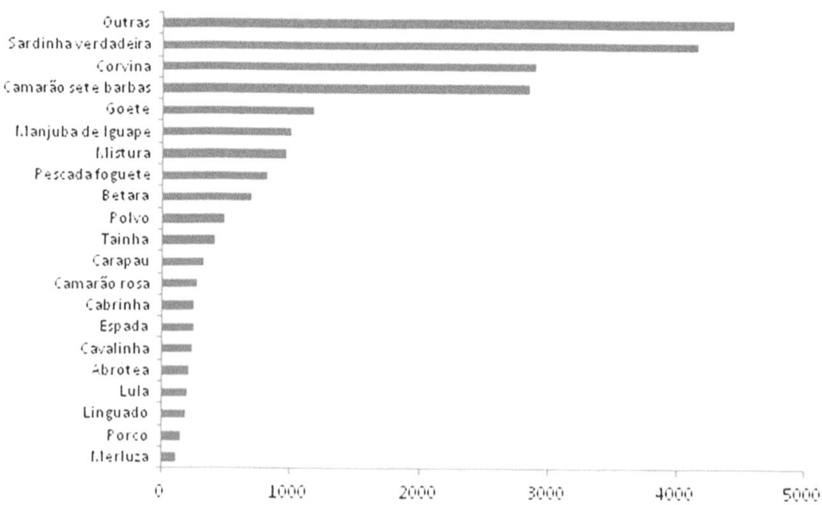

Fig. 6. Variety of fish landings in São Paulo in 2010. *Source*: Own elaboration using data from the IP (2011).

As to fish catch, changes can be seen in availability, size and flavour. According to the interviewees, this may be directly linked to changes in technique, in water contamination (mainly diesel and sewage from boats) and in the environment (climate, beach).

FINAL CONSIDERATIONS

According to the perception of fishers and mariculture labourers from Cocanha beach, global environmental changes showed three main categories of change: fishing, climate and landscape.

In the opinion of interviewees, anthropogenic activities are the main cause of these changes. Using diesel oil in vessels and inadequate drainage of sewage affects water quality and hence the quantity of fish catch and of mussel production.

Among the categories listed that relate to climate change, temperature rise was seen as the most expressive. Thus, one can observe that fishers and mariculture labourers have great insight about environmental changes especially those changes that are related to water temperature because this interferes directly with the fishery resources on which the community depends. It would be wise to deepen these studies because this social group could greatly contribute to both identifying the problems and mitigating them. Furthermore the relationship between this issue and nature is deep and intimate and it should be valued.

NOTE

1. Developed by the Department of Environment of São Paulo, with support from the Fishing Institute and FAO (Brasil, 2011).

ACKNOWLEDGEMENTS

We thank the São Paulo State Foundation for Research Support (FAPESP) for the support by using two projects in progress: 2008/58159-7 and 2010/20811-5.

REFERENCES

Almeida, S. (2010, November 09). Caraguatatuba-9° Festival do Mexilhão atraiu cerca de 8 mil pessoas Foram consumidos quatro mil quilos. *Notícias de Caraguatatuba*, Retrieved from http://www.onoticiado.com.br/noticias/caragua/4775-caraguatatuba-9o-festival-do-mexilhao-atraiu-cerca-de-8-mil-pessoas-foram-consumidos-quatro-mil-quilos.html

Bindoff, N. L. J., Willebrand, V., Artale, A., Cazenave, J., Gregory, S., Gulev, K., ... Unnikrishnan, A. S. (2007). Observations: Oceanic climate change and sea level. In S. Solomon, D. Qin, M. Manning, Z. Chen, M. Marquis, K. B. Averyt, ... H. L. Miller (Eds), *Climate change 2007: The physical science basis. Contribution of Working Group I to the fourth assessment report of the Intergovernmental Panel on Climate Change.* New York, NY: Cambridge University Press. Retrieved from http://www.ipcc.ch/publications_and_data/ar4/wg1/en/ch5.html

Boscardin, N. R. (2008). A produção Aquícola Brasileira. In A. Ostrenky, J. R. Borghetti & D. Soto (Orgs.) *Aquicultura no Brasil: o desafio é crescer* (pp. 27–72). Brasília: SEAP/PR.

Brasil. (2006). *Secretaria Especial de Aquicultura e Pesca. Balanço das Águas* (52 pp.). Brasília: SEAP/PR.

Brasil. (2011). *Secretaria de Educação Profissional e Tecnológica. Instituto Federal Fluminense. Colônia de Pescadores Z-8 Benjamin Constant.* Retrieved from http://pesca.iff.edu.br/nucleos/sudeste-02/Colonia_dos_Pescadores.pdf

Companhia Ambiental do Estado de São Paulo (CETESB). (2011). *Qualificação Semanal das Praias.* Retrieved from http://www.cetesb.sp.gov.br

Confalonieri, U. E. C. (Org.) (2005). *Análise da Vulnerabilidade da população brasileira aos impactos sanitários das mudanças climáticas.* Relatório Final de projeto de pesquisa. Rio de Janeiro: Ministério da Ciência e Tecnologia.

Fagundes, L., Gelli, V. C., Otani, M. N., Vicente, M. C. M., & Fredo, C. F. (2004). Perfil sócio-econômico dos mitilicultores do litoral paulista. *Informações Econômicas, 34*(5), 47–59.

Giddens, A. (2010). *A Politica Da Mudança Climática.* Rio de Janeiro: ZAHAR.

Hogan, D. J. (Coord.) (2008). Urban growth, vulnerability and adaptation: Social and ecological dimensions of climate change on the coast of São Paulo. Programa FAPESP de Pesquisa sobre Mudanças Globais – PFPMCG (processo n. 2008/58159-7), Núcleo de Pesquisas Ambientais/Universidade Estadual de Campinas.

Instituto de Pesca (IP). (2011, January). *Informe da Produção Pesqueira Marinha e Estuarina do Estado de São Paulo* (p. 5). São Paulo: Instituto de Pesca.

Kotir, J. H. (2010). Climate change and variability in sub-Saharan Africa: A review of current and future trends and impacts on agriculture and food security. *Environment, Development and Sustainability, 13*, 587–605.

Linnaeus, C. (1758). Tomus I. Systema naturae per regna tria naturae, secundum classes, ordines, genera, species, cum characteribus, differentiis, synonymis, locis. Editio decima, reformata. Holmiae. (Laurentii Salvii): [1–4], 1–824. Retrieved from http://taxonomicon.taxonomy.nl/Reference.aspx

Manzoni, G. (2004). Mitilicultura: A produção de sementes em laboratório pode ser uma soluç ão viável. *Revista Panorama da Aquicultura, 14*(83), 31–33.

Marengo, J. A. (2006). Mudanças Climáticas Globais e seus Efeitos sobre a Biodiversidade: Caracterização do Clima Atual e Definição das Alterações Climáticas para o Território Brasileiro ao Longo do Século XXI. Brasília: Ministério do Meio Ambiente e Secretaria de Biodiversidade e Florestas.

Marengo, J. A., & Valverde, M. C. (2007). Caracterização do clima no Século XX e Cenário de Mudanças de clima para o Brasil no Século XXI usando os modelos do IPCC-AR4. *Revista Multiciência*, *8*, 5–28.

Marques, H. L. A., Bordon, I. C. A. C., Alves, J. L., & Medeiros, A. M. Z. (2008, December). Produção de mexilhões jovens (sementes) por maricultores da Praia da Cocanha, Caraguatatuba (SP). *Revista Tecnologia & Inovação Agropecuária. Agricultura Familiar-Série Técnica Apta*. *1*(2), 87–93.

Min, S. K., Zhang, X., Zwiers, F. W., & Hegerl, G. (2011). Human contribution to more-intense precipitation extremes. *Nature*, *470*, 378–381. doi: 10.1038/nature09763

Ostrensky, A., & Boeger, W. A. (2008). Principais problemas enfrentados atualmente pela aqüicultura brasileira. In A. Ostrenky, J. R. Borghetti & D. Soto (Orgs.) *Aquicultura no Brasil: o desafio é crescer* (pp. 135–158). Brasília: SEAP/PR.

Ostrensky, A, Boeger, W. A., & Chammas, M. A. (2008). Potencial para o Desenvolvimento da Aqüicultura no Brasil. In A. Ostrenky, J. R. Borghetti & D. Soto (Orgs.) *Aquicultura no Brasil: o desafio é crescer* (pp. 159–182). Brasília: SEAP/PR.

Patz, A. J., Gibbs, H. K., Foley, J. A., Rogers, J. V. E., & Smith, K. R. (2007). Climate change and global health: Quantifying a growing ethical crisis. *EcoHealth*, *4*(4), 397–405.

Pestana, D., Pie, M. R., Pilchowski, R. W. (2008). Organização e Administração do Setor para o Desenvolvimento da Aqüicultura. In A. Ostrenky, J. R. Borghetti & D. Soto (Orgs.) *Aquicultura no Brasil: o desafio é crescer* (pp. 115–134). Brasília: SEAP/PR.

Prado, L. R. T. (2001). *Pescadores de Caraguatatuba*: Histórias e Estórias. Caraguatatuba: Fundação Educacional e Cultural de Caraguatatuba – FUNDACC.

Prefeitura Municipal da Estância Balneária de Caraguatatuba (Caraguatatuba Prefecture). (2010). Dados Gerais. Retrieved from http://www.caraguatatuba.sp.gov.br/

Renk, M. (2010). *Percepção de riscos no setor energético: um estudo de caso envolvendo o Gasoduto de Mexilhão/Petrobras em Caraguatatuba, Litoral Norte Paulista*. Master's thesis. Retrieved from Biblioteca digital da UNICAMP (http://www.bibliotecadigital.unicamp.br/document/?code = 000802812).

Seixas, S. R. C., Hoeffel, J. L. M., Vieira, S. A., Mello, L. F., & Vianna, P. V. C. (2010). Mudanças ambientais globais, vulnerabilidade, risco e subjetividade: um estudo sobre o Litoral Norte Paulista (processo n. 2010/20811-5). São Paulo: FAPESP.

Silvestre, F. (2009). *Determinação da produção de resíduos sólidos provenientes do cultivo de mexilhões na região da Ilha da Cocanha-Caraguatatuba, SP*. Master's thesis. Retrieved from http://www.teses.usp.br/teses/disponiveis/21/.../Dissertacao_Fausto.pdf

Solomon, S., Qin, D., Manning, M., Chen, Z., Marquis, M., Averyt, K. B., ... Miller, H. L. (Eds). (2007). *Contribution of Working Group I to the fourth assessment report of the Intergovernmental Panel on Climate Change*. New York: Cambridge University Press.

Souza, S. C. (2009, September). Elites Políticas em Caraguatatuba. *Seminário Nacional Sociologia & Política UFPR*. 1st. Retrieved from http://www.humanas.ufpr.br/site/evento/SociologiaPolitica/GTs-ONLINE/GT2/EixoIV/elites-politicas-SamuelSouza.pdf

CHAPTER 13

CLIMATE CHANGE AND THE NORTH COAST OF JAKARTA: ENVIRONMENTAL JUSTICE AND THE SOCIAL CONSTRUCTION OF SPACE IN URBAN POOR COMMUNITIES

Rita Padawangi

ABSTRACT

Purpose – *Many cities are located in coastal areas and many of them are identified as prone to climate change impacts, especially sea level rise and floods. Master plans of cities can feature responses to these challenges, as in the case of Jakarta's master plan 2010–2030. However, as this chapter will argue, the top-down nature of planning would likely produce, reproduce, or reaffirm unjust urban geographies in the name of climate change adaptation. North Jakarta and its coastal area, which were prone to climate change risks, has been home for more than 40,000 poor households, most of which live in houses less than 50 m² in informal settlements with lack of basic needs infrastructures. This chapter addresses the question, "How are poor communities in the north coast*

Urban Areas and Global Climate Change
Research in Urban Sociology, Volume 12, 321–339

ISSN: 1047-0042/doi:10.1108/S1047-0042(2012)0000012016

of Jakarta affected by extreme weather events, and how are their everyday experiences addressed in master plan Jakarta 2010–2030?"

Methodology/approach – *Analysis is based on community profiles, census information, and a workshop with representatives of these communities. This chapter will also analyze relevant parts of Jakarta's 2010–2030 master plan. The discussion covers the following: (1) the making of place-based communities of the urban poor in the north coast of Jakarta compared to the master plan, and (2) the impact of climate change on the urban poor's livelihoods in the north coast.*

Findings – *The current master plan 2010–2030 features plans to mitigate climate change and environmental risks for the coastal area, especially sea level rise, land subsidence, and pollution. The study reveals that North Jakarta communities were unaware of what the city planners have drafted, but most of them realized climate challenges based on their everyday experience. They aspired to be involved in the planning process, but their informal status hampered their opportunity to be heard.*

Originality/value of chapter – *Rather than looking at how Jakarta as a city is affected by climate change, this chapter focuses on specific communities in North Jakarta that are prone to climate change-induced risks. Climate change impacts are spatially unequal, and even in the same region that theoretically bears the same risks, the impact distribution of climate change can be unequal for different social groups. The chapter also questions the ability of urban planning to respond to these challenges when planning practice itself has not yet taken into account citizens' social awareness and participation meaningfully.*

Keywords: Environmental justice; community participation; urban poor; production of space; Jakarta

INTRODUCTION

Over the years, the face of Jakarta's north coast has changed. Historically, the Indonesian capital's coast had been a busy place for trade, commerce, and settlements. However, environmental challenges have crept in along with the speedy growth of the city over postindependence economic development. From an estimated population of 1.17 million in 1948, it has grown to over 9.6 million in 2010 (Badan Pusat Statistik, 2010; Silver, 2008).

Limited consideration to environmental management has brought environmental hazards to the city, which include floods, land subsidence, and salt water intrusion. Jakarta's north coast has become the most vulnerable area to these environmental challenges. Land subsidence in North Jakarta reaches 8 mm per year and sea level rise is projected at 3–5 mm per year (Manik & Marasabessy, 2010).

Besides being regularly hit by these environmental challenges, North Jakarta also hosts considerably more poor households compared to the other four regions of Jakarta. In 2008, there were 43,480 poor households in North Jakarta, that is, approximately 150,000 inhabitants (Badan Pusat Statistik, 2010). Most of these urban poor households live along riverbanks and dams (UrbanLAB, 2011a). Due to pollution of most rivers in Jakarta, the land on which these poor communities settle is often polluted as well.

This chapter highlights the importance of environmental justice in climate change mitigation planning, featuring a case study of urban poor communities in North Jakarta. In doing so, this chapter analyzes how their everyday experiences are addressed in the recent master plan for Jakarta 2010–2030. There are two main issues in the discussion: First, this chapter looks at how the urban poor communities of the north coast of Jakarta construct their places of living in their everyday life, compared to the master plan. Second, there is a discussion on how climate change challenges are addressed in the master plan and how they would impact the urban poor's livelihoods in the north coast. The analysis takes into account community profiles based on census and on-the-ground mapping. Most current views and experiences are captured through observation and a workshop with community representatives in March 2011.

ENVIRONMENTAL JUSTICE

The living conditions of Jakarta's poor communities are in stark contrast with the image of Jakarta's development as a megaurban region in Indonesia. Postindependence urban projects were marked by the industrialization of the city and the growth of business districts spurred by the opening of the economy to global capital flow. More recent urban projects, especially after the 1997 Southeast Asian economic crisis, were dominated by consumption-driven lifestyle and leisure, showed by the increasing presence of global franchises in the city (Douglass, 2010; Rimmer & Dick, 2009).

The concept of justice in urban studies had been discussed from various angles: Social justice, spatial justice, and environmental justice are among them. Undeniably, the spatial dimension of justice and the equitable distribution of spatial resources are relevant in the case of North Jakarta, for the concentration of urban poor population in the area and the mounting environmental challenges they face (Dikeç, 2001; Soja, 2010). The term environmental justice reflects a focus on environmental issues, but also includes the notion of space and geographical location, since environmental issues are pinned down on specific locations, areas, and communities. It is important, therefore, to look at the context of urban space and spatial justice in discussing environmental justice.

Spatial justice and environmental justice are essential to analyze how climate change affects communities. Instead of treating climate change as a blanket phenomenon that would affect everyone, the environmental justice perspective looks at the differential impact of climate change on different social groups. Climate change impacts are spatially unequal; some areas are more prone than others, and the north coast of Jakarta is more vulnerable to sea level rise and floods. Even in the same region that theoretically bears the same risks, the impact distribution of climate change can be unequal for different social groups, as will be demonstrated in the case of Jakarta's north coast. The notion of justice, therefore, has to be seen further down to the ground – in this case, urban communities – rather than just on a general map. Social awareness and people's participation and activism are crucial parts in environmental justice and spatial justice, and are essential to avoid the narrow definition of justice as distributional justice (Barnett & Scott, 2007; Soja, 2010; Young, 1990). Consequently, an analysis of spatial-environmental justice and climate change impacts should discuss people's awareness on how climate change has or would affect them. The analysis should also include mobilization efforts to address community disadvantages due to climate change impacts.

As previously discussed, Jakarta's growth has repeated the trend of environmental degradation for the sake of economic growth, due to lack of environmental law enforcement. Rather than valuing space for its natural qualities, capitalism dominates nature to produce abstract spaces (Lefebvre, 1991 [1974]; Smith, 2008). Lefebvre's theory on the production of space was of central importance in various writings on social justice and urban spatial justice. The poor do not have adequate financial resources to secure their rights. Nature is only valued as leisure to be consumed, but in the production of urban spaces nature is defeated. The poor often have to live in environmentally challenged areas with higher climate risks. Nevertheless, wealth is not the only factor that affects accessibility to relatively more secure homes with respect to environmental challenges. Racial characteristics, political affiliation, gender, and professions are also among the reasons of social disadvantages that potentially become barriers to participatory modes of urban development, which in turn affect the inclusiveness of planning process for environmental justice (Harvey, 1973, 2006; Lefebvre, 1991 [1974]; Soja, 2010; Young, 1990).

Some scholars cite social capital as an important factor in social mobilization for environmental justice. Communities need political and technical skills as their local adaptive and mitigative capacity. They would also need to be included in participatory processes to address social, political, and economic inequality issues and to build social capital (Aylett, 2010). Carpenter, Daniere, and Takahashi (2004) went further to call for more attention to gender in studies linking social capital and environmental or infrastructure management in urban contexts. However, existing skills and networks that were usually listed as social capital had to be complemented by sufficient information to build a momentum for social mobilization. The example of Jakarta's urban poor communities along the north coast will show that awareness of existing conditions is important, but information on future urban plans is necessary for social mobilization and participatory processes to materialize.

CLIMATE CHANGE, SEA LEVEL RISE, AND FLOODS

The Intergovernmental Panel on Climate Change (IPCC) through their Second Assessment Report of Working Group II identified that climate change induced by greenhouse gases could significantly affect coastal populations and ecosystems (Bijlsma et al., 1996; Nicholls, Hoozemans, & Marchand, 1999). Meanwhile, most of the world's urban landscapes are still located near the sea, along the river bank, or in deltas (UN-HABITAT, 2008). People in coastal flood plains in the developing world are more prone to regular flooding "with consequent disruption and economic loss, and at the extreme, severe loss of life" due to considerably less protection by structural measures that are usually found in the developed world. This would be worsened by land subsidence, which usually occurs in cities with excessive groundwater extraction (Nicholls et al., 1999). Meehl et al. (2005) projected that sea level would continue to rise "unabated with proportionately much greater increases compared to temperature." They argued that although greenhouse gases concentrations could be stabilized, global warming is already significant enough so that sea level would still rise.

The immediate risk that urban areas in coastal flood plains experience from climate change is the floods related to sea level rise. In the case of Jakarta, the city has also endured decades of groundwater overextraction, which led to land subsidence, especially in the areas toward the north coast. In addition to high tide floods that became more severe in recent years, Jakarta also faced severe floods, notably in 2002 and 2007, which were attributed to heavy rainfall. In 2007, more than 70% of the city area was flooded. Besides falling victim to floods, communities in the north coast were also prone to evictions in the name of climate change mitigation.

UNEQUAL LANDSCAPES OF JAKARTA'S NORTH COAST

The allocation of urban spaces in Jakarta for industrial growth from late 1960s to mid-1980s mirrored the emphasis of providing land as spaces of production (Douglass, 2010). Industries that were located in watersheds often dumped their wastes directly into the rivers, and environmental law enforcements remained weak. Various studies have shown that industrial waste was one of the primary causes of pollution in rivers, especially Citarum and Ciliwung that are connected to Jakarta (The Water Dialogues, 2009). New towns and large housing estates dominated Jakarta's urban projects from mid-1980s to 1997, but even these were mostly geared toward middle class and above (Douglass, 2010; Rimmer & Dick, 2009). Rivers and waterways were sacrificed for economic growth; plantations and farms were cleared to make way for housing. In the meantime, pockets of urban poor communities persisted and grew in various areas in Jakarta – notably along polluted riverbanks and waterfronts that become dumpsites of wastes.

Wastes in Jakarta's waterways would eventually accumulate in the north coast, where all the rivers end. Despite the existence of a busy trade port and industrial zones, urban poor communities in Jakarta are visible along the north coast. At the same time, there is a growing trend of urban renewal projects for the upper-middle class. There were at least five "Integrated Development Zones" proposed in North Jakarta in 2005 and 2006 alone, totaling more than 4 million square meters (Herlambang, 2010). These "Integrated Development Zones" typically include residences and shopping and leisure areas. For example, the Pantai Mutiara Integrated Development Zone features high-end apartments, waterfront houses with individual boats, and a recreation park Taman Rekreasi Pantai Marina.

These megaprojects are usually spared from existing environmental challenges experienced by poor communities from day to day. For example, water supply infrastructure in North Jakarta is inadequate: In some communities with water connection, there had been almost no water in the pipes for many years (Haryanto, 2010). Many poor communities also have to endure living among wastes and polluted sludge from the rivers (Fig. 1). In contrast, the megaprojects usually feature waterfront living as a desirable lifestyle that has a selling point, and these developments would secure adequate infrastructures and services to be marketable. The planning processes that result in these megaprojects rarely, if ever, involve the affected communities, even if it would lead to their eviction. As Lefebvre (1991 [1974]) has indicated, nature has become an image that was sold as lifestyle, but in reality the capitalist production of space has resulted in the defeat of nature – the environmental degradation. However, in the north coast of Jakarta, both cases stood nearby: waterfront as lifestyle for one class was full of environmental challenges for another.

Although economic class is a factor that determines the living conditions in the north coast of Jakarta, it would be an oversimplification to limit environmental justice issue in this area by class alone. The north coast of Jakarta is home to an ethnically diverse population with a variety of political associations and professions. In Cilincing and Kali Baru area, for

Fig. 1. Houses over Garbage and Polluted Waters in Muara Baru, North Jakarta, in 2010. *Source*: Author.

example, most of the household heads are fishermen, mostly from Javanese and Sundanese ethnic background. There are also some Maduranese,[1] Bugis,[2] and local Betawi ethnicities. The fishermen migrated to Jakarta's north coast several decades back by their boats from other areas along the northern coast of Java island.

THE NORTH COAST AND JAKARTA MASTER PLAN 2010–2030

In 2010, the government of Jakarta Special Capital Zone (*Daerah Khusus Ibukota* Jakarta) issued Jakarta's master plan 2010–2030. The plan showed the vision of North Jakarta as a reclaimed coast line and further development of trade ports, which included Tanjung Priok, Marunda, Sunda Kelapa, Muara Baru, Muara Angke, and Kepulauan Seribu (Kali Adem). The master plan was a detailed version of the regional plan that covered Jakarta, Bogor, Depok, Tangerang, Bekasi, Puncak, and Cianjur[3] and that was released in 2008 as a presidential decree (54/2008), which indicated Jakarta's north coast as a prime area for infrastructure, development, and environmental qualities. The area was planned as residential, commercial, light industries, and "selected prime economic activities," which was far from the image of the current poor communities in the area.

The 2008 regional plan had included the reclamation of Jakarta's north coast for the areas marked by "P3" in the map (Fig. 2). This was followed by Jakarta's master plan 2010–2030 that showed the reclamation down to the land use plan, which was mostly for industrial areas in the east,

Fig. 2. Regional Plan for Jakarta's North Coast (*Source*: Jakarta Special Capital Zone Provincial Government, 2010).

commercial areas in the center, and residential areas in the west. These residential areas were marked as "horizontal" and "vertical" residences, which indicated that many homes would be converted into high-rise buildings.

The reclamation plan for Jakarta's north coast was accompanied by two climate change mitigation plans. The first was JEDI/JUFMP (Jakarta Emerging Dredging Initiative/Jakarta Urgent Flood Mitigation Project), a US$ 150 million project funded through World Bank's soft loan. The second was Jakarta Coastal Defense Strategy, which was planned by the Central Government and Jakarta Provincial Government, assisted by the Government of the Netherlands, Rotterdam City, and Deltares consulting (UrbanLAB, 2011a). The World Bank-funded plan was still waiting for approval from the Ministry of Finance, and the Jakarta Coastal Defense Strategy was still undergoing research. The Jakarta Coastal Defense Strategy initially proposed to build a large sea wall to protect Jakarta Bay from sea level rise and to dam the flow from Jakarta's 13 rivers. The sea wall may be built within 8–16 km from the coast line.

It was obvious that the regional plan, the master plan, and the climate change mitigation plans were made with Jakarta's environmental challenges in mind. The plan also had a "sociological basis" of "spatial planning by the Government and the local government with respect to people's rights," in line with the national law on spatial planning (Law 26/2007).[4] Nevertheless, it remained unclear how the current community spaces would fit in the new plans. The reclamation of North Jakarta was an abandoned vision in the past when it was deemed environmentally harmful by the Ministry of the Environment in 2003. After the Supreme Court overruled the decision in April 2011, the reclamation plan was revived and the permits for reclamation that were issued in 1992 could proceed.[5] These plans differ from the current everyday life of the communities; for example, the fishermen community in Cilincing and Kali Baru were rezoned as light industries.

Although a sea wall would protect the coast from sea level rise, its impact on Jakarta Bay fisheries has not been sufficiently addressed in the plan. With the current condition of high pollutants in Jakarta Bay, the fisheries are already affected. Fishermen in North Jakarta have experienced repeated times when they could not work for several months due to overly high pollutant level. High level of pollution was also cited by Jakarta's sea and farm bureau as the reason to stop oyster farming in Jakarta north coast in 2010. The bureau chief, as quoted in several national news agencies in October 2011, warned of the danger of continuous consumption of green oysters from Jakarta Bay. While green oysters were mentioned as useful to help clean the pollutants, the level of harmful chemicals and metal in the sea water could cause cancer, organ failures, and even death.[6]

SOCIAL CONSTRUCTION OF EVERYDAY SPACES AND ENVIRONMENTAL CHALLENGES

Despite the environmental challenges that poor communities had to face at their doorsteps, many members have stayed for a few decades. The history of communities in Muara Baru area at the west side of the north coast could be traced back to early 1949, when fishermen from Bugis, Makasar, and east-coastal Java landed in Jakarta Bay. They began their economic activities there as fish traders at the port and the fish market. Later on, when the authorities established a marketplace for fisheries products, more migrants came from the west coast of Java – Banten and Serang. These migrants settled on the coast land and also used the land for plantation. Sugar cane was one of the main farm products in the past; although the area had stopped producing sugar cane now, the term was used to name an area formerly used for sugar cane plantation (UrbanLAB, 2011b).

The fish marketplace has been the workplace for 20,000–25,000 people, including small fishermen, since its establishment in 1996. Over 50% of poor households in Muara Baru depend on the fish marketplace, usually as laborers. Other work options include working as laborers in nearby industries, working as fishermen, being informal vendors, scavengers, or pedicab drivers. Business in the fish market is as high as 4 billion rupiah (close to US$450,000) per day, since the place involves thousands of laborers and many kinds of services such as boat services, fish processing, and fish storage.

Cilincing and Kali Baru, the east part of Jakarta's north coast, also have migrant communities from Bugis and Java's east coast. In these areas, many have continued their profession as fishermen or work in fishery-related activities, such as keeping fish ponds to breed fish and prawns. Most of the current fishermen originated from Indramayu in west Java, and many of these fishermen from the poor communities were still using traditional fishing nets. Community members are also involved in fishery-related businesses, such as becoming oyster peelers, laborers of fishery products, or workers in the production of salted fish. Being laborers in nearby industries is also another option. The women usually double-shift as housewives and laborers: either as laborers in salted fish industries or as oyster peelers. Professions such as fishermen and oyster peelers are also found in poor communities in Kali Adem – also on the east part of Jakarta's north coast.

The housewives also double-shift by either doing laundry for nearby upper-middle class residences or working at the fish market. The women's earnings have been useful to support the family, especially when the fishermen could not go out to the sea for climatic or environmental reasons. Oyster farming and peeling still persists several months after the authorities issued a ban on green oyster farming in Jakarta Bay.[7] Although the ban is to protect the public from food poisoning, the persistence of oyster farming in Jakarta's north coast shows that it is more than just a profession. Green oyster business has played a role in household and community dynamics, by being a part of how the community constructed their spaces and identities (Fig. 3).

These poor communities were continuously challenged by the accumulation of garbage from the polluted rivers and by health problems. However, they were also reluctant to move to other places. Many families have experienced displacement when their homes – which were considered illegal – were torn down by the authorities. In a workshop discussion, representatives of Cilincing fishermen indicated that they could not remember how many times their homes were torn down: "at least three or four times." They always came back to the area and rebuild their homes because it was "the place that we knew. We are fishermen; fishing is the life that we know." Some of them have moved several times within the area because their homes were repeatedly torn down. About 200 fishermen moved from Bintang Mas

Fig. 3. Oyster Peelers in Kali Adem. *Source*: UrbanLAB, Tarumanagara University.

district, which was rebuilt as Marina Ancol, and resettled at Muara Karang Dam. Later on, they moved again because the area was allotted to build Pluit Mega Mall.

There were also instances when the Jakarta provincial government resettled these communities. For example, on October 22, 2003 – around the Eid holiday – the authorities tore down homes in Kali Adem without prior notice. They cited the 2002 flood as the reason and that the area was part of the river dredging. After the incident, Kali Adem communities who had Jakarta identity cards were placed at the Cinta Kasih Muara Angke Apartment by Buddha Tzu Chi Foundation, while those without Jakarta identity cards would get financial assistance to come back to their villages, mostly Indramayu. They also received approximately US$55 (Rp. 500,000) from the authorities and aid from the World Bank. However, in less than six months, some members of the community who were placed in the apartment came back to their location at the north coast, mostly because they could not afford the rent and because they could not find a place to tug their boats and to place their fishes. Simone (2009) noted that urban poor and working class always have to "re-align what is paid attention to and considered urgent, valuable, and important." This was the working logic in these communities; their work that they were familiar with became a priority to fulfill daily needs. Their coming back to places with many environmental risks did not mean they preferred to be in such a poor estate, but it was due to urgent needs that were necessary.

Despite their poor condition, many of these communities had structured programs that increased their sense of belonging. In Muara Baru, there were several programs facilitated by various nongovernmental organizations to mobilize community members. For example, Kalyanamitra, a women's NGO, worked in Muara Baru since 2008 to empower women by providing sewing and handicraft classes. LBH APIK (*Lembaga Bantuan Hukum* APIK/APIK Legal Aid Institute) has worked with the community since 2004 by providing legal assistance, especially in cases of domestic abuse, while the Urban Poor Consortium has been active in Muara Baru to focus on community empowerment in terms of water services, displacement, and birth certificate. The Urban Poor Consortium also promoted savings and children's learning group. Other poor communities in Jakarta's north coast also received similar programs from various NGOs (Table 1). The fact that they have stayed in the area for many years also supported them to get to know their neighbors.

At first sight, outsiders who visit these communities would typically find the area messy. Narrow alleys separate the houses. In Muara Baru, some

Table 1. Community Programs Among North Jakarta's Urban Poor.

Community	Community Group Activities
Kali Adem	1. Buddha Tzu Chi Foundation Provides free medications to communities in Muara Angke and Kali Adem. The foundation also worked with the North Jakarta government to build Cinta Kasih Apartment 2 to cater for those whose homes were torn down in 2003. 2. Mustika A business group, previously known as KEMILAM, led by Mr. Naling. The group bridges its members to other parties outside the communities to strengthen them to fight displacement. The members pay fees to the group, but the fees can be used to help those in need. Mustika can also facilitate complaints to the National Human Rights Commission. 3. Community Legal Aid Body Provides assistance in overcoming community problems, especially eviction. 4. Urban Poor Consortium Provides advocacy to organize communities to fight evictions.
Muara Baru	1. Kalyanamitra Provides trainings to women, especially in sewing and handicraft. 2. LBH APIK (APIK Legal Aid Institute) Provides legal assistance, especially for domestic abuse cases. 3. Urban Poor Consortium (UPC) Provides community empowerment and advocacy for water services, eviction, and birth certificate. UPC also facilitates savings and children's learning group. 4. Trisakti University community program Teaches English to primary school children.
Cilincing/Kali Baru	1. Cilincing Community Program Supported by an NGO, Wahana Visi, this program provides economic empowerment through training programs to become mechanics and drivers, children's learning groups, and community health services. 2. JITC Provides family planning program, tutorials for students, basic food assistance for the elderly, free medications, education for adults. 3. Buddha Tzu Chi Foundation Provides free medications and scholarships. With Indonesia National Bank and Carrefour, this foundation launches home makeover programs. 4. Indramayu community organization 5. Kali Baru Association for the Deceased Assists funeral services for deceased community members. Members of the association pay approximately US\$0.25 each (Rp. 2,000). The heir of the deceased member would get US\$110 (Rp. 1,000,000).

Source: UrbanLAB (2011b).

of the houses nearer to the coast were constructed on bamboo poles that lifted them over the water filled with garbage. The alleys were made of wood and garbage underneath were visible from the gaps in-between.

Interestingly, these urban communities remained close-knit as they mostly knew each other. It was common for people to know the whereabouts of their neighbors. Observations in the area showed that it was usual practice to ask around when looking for a particular community member. In a few occasions, some would even volunteer to look for the persons or to take people to where they were. While the place might be confusing for outsiders, the alleys and houses were obviously legible for the communities. They have spent most of their lives there and could notice newcomers and were able to differentiate visitors at first sight. Although they were poor, sharing food was common among neighbors and guests.[8] In Jakarta where the production of urban spaces is driven by the logic of capital accumulation, however, it remains ambiguous how these interactions are to be priced (Simone, 2009). It is difficult, if at all possible, to capture these experiences in a master plan, especially when the planning process is top-down. Environmental, spatial, and social justice issues require inclusionary planning process, otherwise the distribution of environmental risks would sustain the ongoing dynamics of domination and oppression in the city (Young, 1990, 2002).

The urban poor community is aware of the environmental challenges they face on a daily basis. In a survey done in 2011, many of them cited garbage, high tide flood, sanitation, and poor planning as their environmental challenges. Sea level rise has increased high tide floods on the north coast of Jakarta, and this phenomenon is worsened by Jakarta's land subsidence. In Muara Baru, the place that gets the worst impact of high tide flood is Kebun Tebu (Sugar Cane Plantation) Block H, in which the flood could reach 0.5 m high. In Kali Adem, the community even preferred to live in their fishing boats rather than in their homes to avoid floods that could come from high tide as well as heavy rainfall.

PLANNED SOLUTIONS TO CLIMATE CHANGE CHALLENGES

Community members did not deny their aspiration to live in an environmentally better living condition. However, they were skeptical of resettlement plans that would relocate them into high-rise apartments. The case of Kali Adem community's resistance toward apartment living arrangement has shown that urban renewal projects rarely, if any, integrated the communities' social spaces into the new settlements. While these projects have identified the environmental problems that plague the communities, they only exchanged the solution to environmental problems with new socioeconomic problems, in which the people experienced difficulties in settling in the apartments because they had to change jobs and social interaction with neighbors.

With a new master plan that addresses the impact of climate change on the city of Jakarta, these urban poor communities are subjected to urban renewal, especially with the reclamation plans of Jakarta's north coast. The sea wall plan to protect Jakarta from sea level rise is also another physical intervention in the name of climate change impact mitigation that would alter the lives of these communities. Nevertheless, the urban poor in North Jakarta were not aware of these plans. When informed of the land use plans, the reclamation, and the sea wall plan, they expressed concerns about the future of their communities and professions, especially for the fishermen.

Jakarta Emerging Dredging Initiative (JEDI) or Jakarta Urgent Flood Mitigation Project (JUFMP) is now on the map for the Kebun Tebu area in Muara Baru. Although the project faced funding arrangement delays, the Jakarta provincial government announced that they would start in March 2012.[9] Discussions between the World Bank and the authorities have taken place since 2008, but until March 2011 community representatives were unaware of the program and of possible evictions that might affect them. The program will dredge 13 rivers and build 4 dams to tackle flood problems of Jakarta, evicting 10,000 homes along Pluit Dam and 750 in parts of Kali Adem River along its way (Setiawati, 2010). Households with Jakarta identity cards will be resettled in rental apartments, while those without Jakarta identity cards will be sent back to their villages of origin. Muara Baru community members, however, expressed concerns that the resettlement plan would be problematic, because the resettlement housing plan would move Kebun Tebu residents away from the fish market, where most of them work. A Tarumanagara University faculty member, Darrundono, echoed this concern by suggesting that it would be likely to find people going back to the area to be nearer to their workplace (Setiawati, 2010). Moreover, the housing provided to them would only be rental apartments that they would not be able to own. This would potentially lead to another case of Kali Adem in which the people could not afford the rent.

In a research workshop facilitated by Tarumanagara University in March 2011, community representatives were introduced to the master plan and the sea wall plan and were asked of their response to possible resettlement options. In small group discussions, their first response was a rejection to high-rise living. However, when the facilitator explained that high-rise apartments were unavoidable due to the increasing density of Jakarta as a whole, they said, "As long as we can own the apartment and we can still work as usual, we should be able to cope with it." Their reaction showed that communities would have higher appreciation when they were involved in a discussion.

Further small group discussions revealed that urban poor communities in the north coast of Jakarta have experienced climate change impacts on their everyday activities. Inarguably, they have been adapting to the life with environmental challenges and they remained most prone to climate change impacts. Nevertheless, when it comes to planning at the city level to mitigate climate change, their spaces were the first ones to be sacrificed. Their status as "informal" or "illegal" settlers had put them in a disadvantaged position. Rental apartments were considered as a good solution in the eyes of the planners, since these communities did not have land titles to start with. Property ownership was valued as the sole measure of entitlement to urban space, and there was no place for socially constructed spaces in the climate change mitigation plan. Resettling them into organized high-rise rental apartments was seemingly an environmentally better living arrangement for these communities.

The master plan's contradiction to the everyday realities of Jakarta's fishermen also indicates that the fishermen do not have space in the new master plan. This solution saw environmental justice only from a distributional perspective, in which poor communities are taken off their garbage-filled environment, polluted water, and the threats of high tide floods. Yet, the ignorance toward their socially constructed spaces and the lack of voice they have in the planning process signify that even the resettled communities would not form a just environment. Ilyas Saad, the Deputy of Environmental Planning in the State Ministry of the Environment, had mentioned in 2010 that reclamation would potentially affect fishermen's livelihoods. The nod toward reclamation, in spite of its incomplete environmental impact assessment, carried with it the interests of big developers that were engaged in the projects.

Meanwhile, urban poor communities on the north coast of Jakarta have had experiences in organizing themselves to address their problems, but were still left behind in the planning process. Their participation in planning usually featured conflicts with the authorities rather than consensus.[10] The

Muara Baru community had years of experience in mobilizing to demand for reliable water supply services and most communities have mobilized themselves to fight eviction threats that came repeatedly. The exclusion of community voices shows that climate change mitigation, which is supposed to be beneficial for the most prone communities, has become another tool to displace urban poor communities, giving way to formal and commodified spaces that would result from the North Jakarta reclamation plan, following the city development trends as a whole.

CONCLUSION: CLIMATE CHANGE MITIGATION AND ENVIRONMENTAL JUSTICE

Environmental challenges and the mismatch between developments – mostly urban – and environmental sustainability have garnered much attention since climate change debates gained momentum. As a city that faces many classic environmental problems caused by economic development and industrialization, it is conceptually promising that Jakarta has a 20-year master plan that takes into account climate change mitigation and environmental rehabilitation efforts. However, urban plans that address climate change challenges need to recognize that inequality exists. The impact of climate change would be different among various groups. In a city with existing problems like Jakarta, a master plan that addresses environmental problems and mitigates climate change impacts needs to integrate environmental justice concept. This means that there needs to be recognition of community voices in the mitigation efforts.

A top-down climate change mitigation plan and the lack of community voices in the new master plan of Jakarta indicate that the solutions have not addressed the very core of the environmental problems, which is about the unequal positions between polluters and those who suffer from environmental degradation. There needs to be recognition of the communities that are most prone to climate change impacts and they need to be involved in the planning process. Otherwise, the master plan would sustain these inequalities and injustices among different groups in the city, which in turn would worsen much of the existing social problems. The lack of attention toward the communities' socially constructed spaces would not result in an environmentally just city that is required to comprehensively deal with climate change challenges.

NOTES

1. The geographical origin of the Maduranese is Madura island, a smaller island in the east of Java.

2. Bugis ethnic group originated from Southern Sulawesi, a considerably large island in the center of the Indonesian archipelago.

3. Bogor, Depok, Tangerang, Bekasi, Cianjur are neighboring cities around Jakarta. Puncak highland is a popular weekend destination for the upper-middle class Jakarta inhabitants.

4. Quoted from the academic script of detailed spatial plan of Jakarta 2010–2030 (Laporan Akhir Penyusunan Naskah Akademis dan Raperda Rencana Detail Tata Ruang DKI Jakarta).

5. WALHI, one of the NGOs that rejected the reclamation plan, claimed that the April 2011 Supreme Court decision was much influenced by developers who had interests in reclamation-related developments.

6. "Kerang Hijau Tak Boleh Diternakkan di Teluk Jakarta," *Kompas*, October 25, 2011; "Jakarta Akan Bebas Budi Daya Kerang Hijau," BeritaJakarta.com, October 25, 2011; "Awas! Warga Jakarta Diimbau Jangan Makan Kerang Hijau," *Republika*, November 6, 2011.

7. As previously mentioned, in 2010 the authorities decided to stop green oyster farming in Jakarta Bay. This decision would affect approximately 5,000 oyster farmers in the area and the women who worked as oyster peelers. Jakarta's Sea and Farm Bureau Chief claimed that these farmers would be retrained as catfish farmers to substitute their work.

8. During a field visit in March 2011, a middle-aged woman whom we visited gladly offered the fruit salad she was making to us and her neighbors who passed by.

9. "JEDI Project to begin in March Next Year," *The Jakarta Post*, November 24, 2011.

10. Aylett (2010) discussed two types of participation: Habermasian (consensus) and Foucauldian (conflict). He argued that usually both types exist at the same time. In the case of North Jakarta, however, consensus-based participation has yet to materialize, due to the lack of available participatory processes.

REFERENCES

Aylett, A. (2010). Participatory planning, justice, and climate change in Durban, South Africa. *Environment and Planning A*, *42*, 99–115.

Badan Pusat Statistik. (2010). *Sensus Penduduk 2010 – Provinsi DKI Jakarta*. Retrieved from http://sp2010.bps.go.id/index.php/site?id=31&wilayah=DKI-Jakarta. Accessed on August 9, 2012.

Barnett, C., & Scott, D. (2007). Spaces of opposition: Activism and deliberation in post-apartheid environmental politics. *Environment and Planning A*, *39*, 2612–2631.

Bijlsma, L., Ehler, C. N., Klein, R. J. T., Kulshrestha, S. M., McLean, R. F., Mimura, N., … Warrick, R. A. (1996). Coastal zones and small Islands. In R. T. Watson, M. C. Zinyowera & R. H. Moss (Eds.), *Impacts, adaptations and mitigation of climate change: Scientific-technical analyses* (pp. 289–324). Cambridge: Cambridge University Press.

Carpenter, J. P., Daniere, A. G., & Takahashi, L. M. (2004). Social capital and trust in Southeast Asian cities. *Urban Studies*, *41*(4), 853–874.

Dikeç, M. (2001). Justice and the spatial imagination. *Environment and Planning A*, *33*, 1785–1805.

Douglass, M. (2010). Globalization, mega-projects and the environment: Urban form and water in Jakarta. *Environment and Urbanization Asia*, *1*(1), 45–65.

Harvey, D. (1973). *Social justice and the city*. Oxford: Blackwell.

Harvey, D. (2006). *Spaces of global capitalism*. London: Verso.

Haryanto, U. (2010). Indonesian water warriors flood streets of Jakarta. *Jakarta Globe*, *23*, 2.

Herlambang, S. (2010). On City commercialization: A preliminary research. *International Colloquium Series 2010: The strategy and concept for sustainable future of Jakarta*. Jakarta, Indonesia: Tarumanagara University.

Lefebvre, H. (1991 [1974]). *The Production of Space*. Oxford: Blackwell.

Manik, J. M., & Marasabessy, M. D. (2010). Tenggelamnya Jakarta dalam Hubungan dengan Konstruksi Bangunan Beban Megacity. *Makara Sains*, *14*(1), 69–74.

Meehl, G. A., Washington, W. M., Collins, W. D., Arblaster, J. M., Hu, A., Buja, L. E., … Teng, H. (2005). How much more global warming and sea level rise? *Science*, *307*, 1769–1772.

Nicholls, R. J., Hoozemans, F. M. J., & Marchand, M. (1999). Increasing flood risk and wetland losses due to global sea-level rise: Regional and global analyses. *Global Environmental Change*, *9*, S69–S87.

Rimmer, P. J., & Dick, H. (2009). *The city in Southeast Asia: Patterns, processes and policy*. Singapore: NUS Press.

Setiawati, I. (2010). Relocation of 5,000 families in JEDI Project Confirmed. *The Jakarta Post*, June 22, 2010.

Silver, C. (2008). *Planning the megacity: Jakarta in the twentieth century*. Oxford, UK: Routledge.

Simone, A. (2009). *On Provisional Publics and Intersections: Remaking District Life in North Jakarta*. Retrieved from http://www.scribd.com/doc/17815978/Simone2UrbanGeo. Accessed on January 2, 2012.

Smith, N. (2008). *Uneven development: nature, capital, and the production of space* (3rd ed.). London: The University of Georgia Press.

Soja, E. W. (2010). *Seeking spatial justice*. Minneapolis: University of Minnesota Press.

The Water Dialogues. (2009). *Indonesia: Contextual analysis in water supply and sanitation sector*. Retrieved from http://www.waterdialogues.org/documents/8.6ContextualAnalysis.pdf. Accessed on November 8, 2011.

UN-HABITAT. (2008). *State of the world's cities 2008/2009*. London: Earthscan.

UrbanLAB. (2011a). *Preliminary report: North Coast of Jakarta*. Jakarta, Indonesia: Tarumanagara University.

UrbanLAB. (2011b). *Laporan penelitian Jakarta utara: Kali Adem, Cilincing, dan Muara Baru*. Jakarta: Tarumanagara University.

Young, I. M. (1990). *Justice and the politics of difference*. Princeton, NJ: Princeton University Press.

Young, I. M. (2002). *Inclusion and democracy*. Oxford, UK: Oxford University Press.

CHAPTER 14

THE GENDER DIMENSIONS OF CLIMATIC IMPACTS IN URBAN AREAS: EVIDENCE AND LESSONS FROM KAMPALA CITY, UGANDA

Buyana Kareem

ABSTRACT

Purpose – *The chapter explores the gender dimensions of climatic impacts in urban areas and draws lessons for cities in Africa.*

Methodology – *The data presented was generated through focus group discussions with female and male residents of Kasubi-Kawaala neighborhood in Kampala city as well as extensive review of relevant literature.*

Findings – *Climatic impacts in Kasubi-Kawaala and Kampala city at large, mainly include prolonged dry spells, erratic heavy rains, and seasonal floods, which destroy physical infrastructure, expose households to environmental health hazards, contaminate air and water sources, and lead to unprecedented spread of cholera and malaria. These climatic impacts on one hand do worsen gender inequalities across different urban sectors, while on the other such gender inequalities contribute to the intensity of climatic impacts. These are the gender dimensions of climatic*

Urban Areas and Global Climate Change
Research in Urban Sociology, Volume 12, 341–356

ISSN: 1047-0042/doi:10.1108/S1047-0042(2012)0000012017

impacts in urban areas that require deep examination while planning to adapt or reduce emissions.

Research limitation – *The methods used to collect data were qualitative in nature and therefore no statistical data was obtained on gender inequalities and climatic impacts. But the review of different literature did enable the study gain relevant descriptive statistics on the effects of climatic change in Kampala city.*

Value of the chapter – *There have been studies on gender and climate change in Africa, but many of these have focused on rural settings and women in particular. This chapter provides a relational understanding on women's interface relative to men's interface with climatic impacts in Kampala with the aim of drawing lessons that can be applied to local circumstances in different African cities.*

Keywords: Gender inequalities; climatic impacts; urban areas

INTRODUCTION

Like gender, climate change is both a global and sector-wide concern. The risks associated with high temperatures and heavy or no rainfall have and may continue to hamper the delivery of quality services in sectors like transport, housing, and health (Blakely, 2007; Berrang-Ford, Ford, & Paterson, 2011; Bulkeley et al., 2009; Füssel, 2007; Lwasa, 2010; Tompkins & Adger, 2004). Although the survival of women and men in urban areas depends on such climate-sensitive sectors, much of the knowledge generated so far on gender and climate change in Africa is largely rural-centered. Many a study have concluded that women in rural areas are more vulnerable than men due to their reliance on rain-fed agriculture and natural resources for a living (DFID, 2008; Economic Commission for Africa, 2009; Oxfam, 2008; UNDP, 2007; UNIFEM, 2009). Even though such a conclusion makes useful linkages between gender and climate change, it is drawn within a context that does not consider the gender dimensions of climatic impacts in urban areas. As a response, the chapter explores two inversely connected urban gender issues: (i) how do climatic impacts worsen gender inequalities across different urban sectors; and (ii) how do urban gender inequalities contribute to the intensity of climatic impacts? The chapter also draws lessons on adaptation and mitigation, using qualitative data from Kampala city, Uganda's capital.

METHODOLOGY

The study was undertaken in the neighborhood of Kasubi-Kawaala, located in the northwestern part of Kampala city. This neighborhood is known to be a host of unplanned settlements; a pattern that aggravates climatic hazards, especially flooding that makes residential dwellings inhabitable, contaminates air and water sources, restricts mobility due to impassable commuter routes, and reduces the longevity of physical infrastructure. It is these social-environmental occurrences that made the neighborhood suitable for holding discussions on urban gender inequalities and climatic impacts with three (3) focus groups including: a women-only, men-only, and mixed-sex group. The sex-based segregation of groups aimed at acquiring gender-disaggregated information and enabled the immediate grouping of responses to map the relationship between gender inequalities and climatic impacts. Each of the groups comprised eight residents that were purposively selected from the neighborhood to conduct a gender-sensitive mapping of the gender inequalities across urban sector and how these are worsened by and contribute to the intensity of climatic impacts in the city. Focus group discussions were enriched by the review of a UNHABITAT assessment report on cities and climate change in Kampala (2009) as well as other pieces of relevant literature. This further enabled the study to obtain descriptive statistics on the effects of climatic changes in the city.

Climatic Impacts in Kampala City

Kampala is an inland city managed by Kampala Capital City Authority (KCCA), whose records in regards to service delivery indicated a number of development projects that have been undertaken to enhance both women's and men's capacity to deal with environmental challenges. Examples of such projects include the Ecological Sanitation Project (ECOSAN), the Kampala Integrated Environmental Management Project (KIEMP), the Sustainable Neighborhoods in Focus Project (SNF), and the Kyanja Edible Landscape Project (KELP) among others (Kampala city development plan, 2010–2011). However, population pressure on city services coupled with the destruction of the natural environment by a rapidly expanding built-up city has constrained the desired outcomes of these projects amid changes in temperature and precipitation. According to a UNHABITAT report (2009), climate changes in Kampala entail an increase/decrease in precipitation and rising temperatures. Changes in precipitation mean an increase in water that usually turns into runoff due to the low levels of utilization by households and commercial entities in Kampala city.

The impact of increased runoff has already been felt in the form of severe flash floods that usually destroy backyard gardens, make roads impassable, accelerate contamination of air and water sources, and contribute to the intensive spread of diseases like cholera and malaria, which overstretch the already resource-constrained public health care system in the city. A study undertaken by Lwasa (2010) revealed that poor households commit an average of 15.5% of their annual total income on preventive and curative measures for environmentally related diseases. In cases of extreme weather conditions, there is a likelihood that such costs increase and the most affected will be women, who always take care of the sick at home and shoulder the household economic burden in case the main income generator is ill.

Flooding has also put the city's roads, culverts, and drainage system at risk. The UNHABITAT report (2009) noted that the extended rains and floods of December 2006 to February 2007 caused serious damage to housing, schools, and businesses within and outside Kampala city. The report further documented that an estimated 1,600 km of urban road segments are highly vulnerable to flooding and extreme weather events. These climatic impacts have also contributed to the continued prevalence of malaria, dysentery, and cholera epidemics. Cholera cases and frequency has increased in the last 8–10 years due to erratic rainfall contributing to slow onset floods. This reinforces the point that climate change is a sector-wide issue because of the hindrances it creates within the functioning and use of physical infrastructure, health systems, urban transport, housing, schools, and businesses.

In Kampala, these sector-wide impacts are likely to be aggravated by the current absence of a sustainable urban development agenda that has a gender-responsive adaptation plan. A review of planning documents and reports at KCCA revealed that Kampala has no adaptation or mitigation plan in place and all the initiatives related to the generation of climate change information are externally driven by UNHABITAT and other development partners. The available information is also not used innovatively by departments that require it to come up with a strategic and well-coordinated response to climate change while building on the knowledge and experiences of women relative to men at neighborhood level.

The Inverse Connection between Gender Inequalities and Climatic Impacts across Urban Sector

From the study, the key signifier of gender inequality is imbalances in access and control over resources and opportunities with urban transport, health, energy, housing, urban agriculture, and waste management, which for instance results in the exclusion of women from decisions on how the household income should be spent to meet living, travel, child rearing, and leisure needs. The other inequality discovered by the study is that the mobility of girls and women in the neighborhood and to the city center is under men's control. This results in not only less access to opportunities for paid work and exclusion from business networks within the city, but also lower participation in community-level projects on environmental issues like climate change. This section focuses on how these inequalities contribute to the intensity of climatic impacts and how the impacts of climate change do worsen gender inequalities across different urban sectors.

Transport

Transportation services in the neighborhood of Kasubi-Kawaala are characterized by omnibuses that take the biggest share of what both women and men use to reach out to the city center. This is followed by motor cycles, bicycles, and privately operated buses for travel to the country side. The women's group said that they mainly use walking and omnibuses to balance child rearing and commercial responsibilities, although a few own private cars, especially among higher income groups. Men, on the other hand, use motor cycles, omnibuses, and private cars to search for economic opportunities around the city. Respondents in all the group discussions agreed that private cars, omnibuses, and motorcycles are the key emission-intensive travel modes in the transportation sector due to their dependence on petrol and diesel for movement as well as maintenance. This resonated with the city-level assessment made by UNHABITAT in 2009. The assessment report indicated that vehicle emissions are the major source of pollution in Uganda, with 80% of the diesel and 100% of the petrol contributing to emissions within the transportation sector. The same report documented that on average one omnibus produces 18,536 kg of carbon annually, while one big bus produces 61,847 kg of carbon a year, which means that if women's and men's travel is increasingly dependent on omnibuses, the level of carbon emissions from this sector is likely to increase.

Urban transport is also a major source of employment for men, as owners and operators of buses, private cars for hire, and motorcycles. To women, this meant that men's employment patterns as well as travel behavior contribute more to the burning of fossil fuels, especially petroleum and diesel, on which the movement of vehicles and motorcycles depends. The respondents further noted that men are not only unaware of the damage caused by their travel activities to the atmosphere but also unwilling to change because of the higher socioeconomic standing that comes with accessing or owning a car in the household. A few of the women did disagree by arguing that they usually travel by walking despite the multiplicity of daily trips for maintaining the household and their businesses, a phenomenon that men can emulate. Further discussions, however, revealed that women do prefer walking not because of they are aware of the climatic impacts of motorized transport but rather the inability to pay the fares that arise out of multiple trips. This is reinforced by a GTZ study in 2007, which revealed that women's lesser socioeconomic position is largely responsible for the lower incidence of vehicle use and a higher incidence of walking in urban areas.

Conversely, men's preference for motorized transport is partly driven by road safety issues. Men argued that although deaths and injuries are a problem associated with motor car accidents, the victims are usually pedestrians and cyclists. To the men, this makes nonmotorized transport quiet unsafe for people who frequently move to, within, and from the city center despite its environmental friendliness. The differential impact of road accidents on pedestrians and cyclists was further confirmed by records obtained from the Uganda Police Headquarters in Kampala (2011). The records indicated that 43% of the accidents recorded at Kiira police station in Kampala city affected pedestrians, whereas 21% affected cyclists. A similar pattern was recorded at Kawempe police station, where 32% of the accidents recorded had pedestrians as the victims and 25% were cyclists. This coupled with the absence of an adequate pedestrian footpath network in the city downplays the use of nonmotorized modes of transport as an emission-reduction measure.

From the women's point of view, crime and fear of violence lowers the use of nonmotorized transport. According to female respondents in the study, travel by walking at night requires not only aggressive policing but also adequate street lighting which is not readily available at neighborhood level in Kampala city. Respondents further reported that the presence of police does not effectively reduce crime and the perception of risk among pedestrians and cyclists. Men added that so many of their colleagues who earn a living from cycling have been mugged, beaten, and killed at night even when the police are within the neighborhood. The respondents concluded that the safest way to travel at night is by use of a car. These gendered experiences imply that the reduction of emissions through nonmotorized means of travel cannot succeed unless the personal security of pedestrians and cyclists is assured at neighborhood to city level. Therefore, the gendered nature of road safety and personal security ought to be considered if carbon control within the transport sector is to take route. Depending on the methods used in a study, it is also possible to use gender as a lens for analyzing human activities that contribute to green house gas emissions (GHGs) in the transport sector.

Health

The key health-related climatic impact discovered in the neighborhood is that of flooding, which contributes to the spread of waterborne diseases, especially cholera and malaria. Women said that children and the elderly members of the household are usually the most affected by these sanitation- as well as flood-related diseases. The respondents added that since men are often in charge of spending the household income, women are sometimes unable to purchase medicines required to cure the diseases. The study further revealed that men's control over financial resources in the household is based on a social order that gives household heads (usually males) power over others even in circumstances where the women may be contributing considerably to the survival of the home. These gender power relations in regards to control over financial resources mean that women have to seek curative or preventive health care services under the direct permission of men.

In situations of waterborne epidemics due to flooding, men as the controllers of finance in the home usually prioritize and offset their health care costs more easily than women. According to the responses acquired, women are expected to invest labor and time in caring for cholera-stricken children, maids, and other household members. This health care burden limits their capacity to cope because the additional caring roles mean absence from their paid work, yet flexible working schedules are hard-to-find in the informal and semiformal sectors on which women rely for employment. Men's control over women's and girls' mobility around the neighborhood further lowers their capacity to search for alternative sources of health care support such as borrowing drugs from local clinics and retail shops. Discussions with some of the women revealed that girls often miss school because they are either sick or have to attend to ill household members as their mothers opt to work during day and perform the care roles at night. On-time access to health centers is additionally constrained by flooded community access roads, which together with the damage made on home-based businesses and the residential environment makes women experience anxiety and depression. Flooded homes also make children and disabled persons susceptible to physical injury, as they try to have their way to water sources, shops, markets, and other places around the neighborhood. These physiological and socioeconomic health care burdens are experienced in a city with an under-resourced health care system, which aggravates the problem.

The other health impact is indoor air pollution, an issue that was raised by both female and male respondents. It was cited as a common phenomenon among households that most times use biomass fuels to cook food and boil water in inadequately ventilated living spaces. While the magnitude and frequency of women's exposure relative to men's exposure to such pollution could not be measured by the study, respondents agreed that women and girls are the key victims because they are expected to spend much of their day cooking indoors as mothers, domestic workers, and employees in local restaurants. According to the UNHABITAT report (2009), indoor air pollution has several health risks that arise from inefficient combustion and unstable mixture of particulate matter, carbon monoxide, hydrocarbons, nitrogen oxides, formaldehyde, and benzene. The risks include middle ear infection, tuberculosis, perinatal mortality (still births and death in the first week of life), low birth weight, eye irritation, cataract, asthma, and oral cancer. If vulnerability to these health dangers is partly differentiated by the distinctive roles that women as compared to men take on socially, then it is rational to investigate the gender-specific health impacts of climate change and further assess the readiness of urban health systems in enabling vulnerable groups cope with the resultant illnesses.

Urban Agriculture

Agriculture in the area of study was characterized by growing of food crops, rearing of poultry and animals. The main food crops include banana, cassava, maize, and vegetables. Rearing of poultry, pigs, goats, and cows is also a common phenomenon in the area. According to an agricultural census held in Kampala city between 2008 and 2009, more than one-third of the households in the city practice some kind of agriculture and women comprise the largest number of producers as men concentrate more on marketing (Uganda Bureau of Statistics, 2010). The discussions held revealed that most of the food is grown by women to feed their families and the surplus is sold to generate money that enables women supplement the incomes needed to feed and educate children. The rearing of poultry and animals, however, is done by both women and men for sale to market places within and outside the neighborhood.

Women noted that they have access but less control over arable spaces including backyards, rooftops, and pieces of land owned either privately or by institutions. This gender inequality limits women's capacity to adopt environmentally sustainable farming practices, such as drought-resistant cropping or mixed farming that allows the planting of trees to control erosion in the gardens and compounds. The other inequality is that women usually depend on word-of-mouth to access to climate information due to restricted mobility and time constraints associated with domestic responsibilities. This makes their investments in urban agriculture more vulnerable to weather-related hazards due to insufficient knowledge about seasonal variations and the likely impacts on productivity. The study also found out that women and children provide most of the labor, usually as part of their daily domestic routines and sometimes as paid labor. This poses a disproportionate burden on women and children because they have to spend long hours under the scorching sun, gardening, rearing, and marketing agricultural products, which exposes them to inhalation of dust from contaminated soil or gaseous emissions from cars and motorcycles passing by.

Although the study could not statistically measure the impacts of these health-related occurrences, the hazards associated with urban agriculture have a link with changes in climatic conditions, and women are the most affected. Therefore in seeking to develop more commercial forms of urban agriculture for poverty reduction, attention should be paid to gender-responsive climate-resilience strategies such as the promotion of green enterprises – a concept that encompasses (i) nutrient reuse for soil conditioning and emission reduction; (ii) urban greening for recreation, natural resource conservation, and reduction of flood intensity; (iii) vertical urban use of urban spaces for large-scale production; (iv) institutional linkages through the promotion of microscale school, hospital, and home gardens; and (v) creation of jobs through producing food and nonfood products for the market. A gendered approach to these strategies is more efficient because women constitute the largest number of urban farmers.

Waste Management

The structuring of household gender roles determines the distribution of waste management activities. From the study it was revealed that gender roles account for the differentiations in paid and unpaid waste management work. With regard to unpaid waste management work, the responses given showed that collection of wastes in containers, sacks, and polythene bags is mainly done by women. This pattern is premised on socially shared expectations of gender appropriate behavior inside the household:

> [... it is women who cook and clean the house ... the items we use are packed in polythene bags, boxes or paper bags that can be kept for future use or thrown away ... and after cooking we have to collect and put the banana leaves, ash and dirty water in the dust bin to make sure that your kitchen is clean because as women the kitchen is our office so it our role to ensure that it is clean ... you cannot cook from a place that is dirty because your food can get contaminated which may harm the family members and taking them to the health center is very expensive ...] reasoned one female respondent in Kasubi-Kawaala.

> [... I wake up everyday at 6.00 am to prepare my children for school, cook for them and clean around the house ... before I go to Kasubi Market I have to make sure that all the rubbish is collected and thrown in the rubbish pit behind our house for drying ... at the end of the week I always burn the dried rubbish because I do not see why I should be charged money for rubbish that I can deal with in my own home ...] argued another female respondent in Kasubi-Kawaala.

From the evidence provided above, women handle the discarded materials from other household members: it is part of the definition of who they are in society and what they are "supposed" to do. Therefore, women's socially defined roles within the household and the socially constructed notions of cleanliness explain their presence in unpaid waste management work and the motivations underlying it.

Although women also participate in paid waste management work, the study found out that paid waste management activities outside the household are predominantly a male sphere. Men's traditional gender role of providing incomes for families and surviving in the city has partly accelerated the commercialization of waste management work in Kampala. According to the respondents, everybody selling goods in the markets pays a monthly charge to the people in charge (usually males) who are responsible for organizing and paying a group of laborers to collect the waste from each stall in the market; pack it in the sacks; and take it to the immediate collection center; where KCCA trucks load it using male labor; and take it to dumping sites in the peripheral parts of the city. Female respondents further acknowledged that ever since the collection of urban waste in the market became commercial, it has attracted a lot of men, especially migrants from other neighboring parishes like Bwaise, Namungona, Makerere, and Nansana. Male respondents too said that most of the people who collect waste on behalf of private companies are males from outside the community. The responses below also continue to reinforce the same points:

> [...I rent in Bwaise but I do most of my work in Kasubi Market .... I collect waste on behalf of the people and I receive a salary from the Director because I have to sort the waste ... pack it in the sacks ... and tell him to call the KCC people when the waste is ready for collection ... this is how I have been earning for two years now ...] said a male respondent from Kasubi Market.

> [... am making these solid waste stands (made of timber) at 30,000 shillings each for our Councilor ... he wants me to make for him 100 stands ... he is going to take them to the community ...] explained a male respondent at Kasubi Community Development Center (KACODA).

> [... you know cats and dogs have been tearing the sacks where people put the waste, even community members who deal in charcoal steal the sacks at night ... so I have invented these solid waste stands.... I will take them to the households I have already identified and each household will pay 500 shillings each time I collect the waste ... this will bring in income for my family and my organization ...] explained a male representative from KACODA.

Furthermore, women as compared to men differ in their priorities regarding household expenditure on urban waste. This accounts for the gender differentiations in the utilization of formal and informal urban waste management services. Men on one hand prefer using formal services and women on the other hand prefer informal services. According to male respondents, money to pay formal waste collectors is usually left at home with the spouses before going to work. But the spouses do not often use the money for the intended purpose. Women instead use the money for other household needs such as buying food supplements, buying water for home use, soap, sugar, and others save the money for future use. Female respondents said that women need to save part of the money to purchase other household items and this is why women prefer using children and male youth in the community to collect the waste at a cheaper price. Yet, the responses given by men revealed that such informal urban waste services rely on illegal waste dumps such as drainage channels, market places, and road sides for disposal. This kind of waste management phenomenon contributes to the blockage of drainage channels during heavy rains, which heightens the level of flooding in commuter routes and market places. The disposal methods used, both legal and illegal demarcation of dumping sites and provision of skips around the neighborhood, further engages both women and men in unsustainable waste management solutions. Emission of methane gas is characteristic of waste management sectors that rely on skips and landfills.

At community level, however, women and men engage in waste reuse and recycling activities. Some of the waste reuse activities include collection of plastic bottles, food stuffs, paper bags, and banana peelings for sale, drawing silt from drainage channels for construction of houses, water harvesting, selling chicken litter and saw dust. With regard to recycling, some of the activities include urine trials on home gardens, making charcoal briquettes, composting, making hand bags using polythene bags and drinking straws. Access to markets for reusable and recyclable materials is gender differentiated. According to the respondents the neighborhood-level market for products such as banana peelings is accessible and utilized by both males and females. But female respondents went ahead to argue that the city-level market is predominantly a male sphere. The respondents attributed this gender differentiation to women's limited access to information and socioeconomic ties, high transport charges, and having no control over public spaces which somebody can use as a collection center for large volumes of valuable waste.

Further enquiry was also made on how males compared to females access information on the recycling of wastes. This is because both female and male respondents agreed that engaging in the commercialization of waste materials most times requires either formal or informal training from inside and outside the community. Female respondents argued that most organizations which offer free formal training services usually choose male leaders. The respondents continued to explain that some women who wish to access the formal training are barred by domestic chores as well as lack of permission from the spouses. The male leaders were asked whether they organize return studies for their female and other male counterparts. The responses given revealed that the information got from formal training is usually shared informally, on an individual basis (man to man or woman to woman) and where both parties have a shared business interest. This implies that cleaner waste technology interventions that are targeted at community representatives may not at all times offer opportunities for equitable redistribution of the information gained, something that constrain the effectiveness of mitigation and adaption plans.

Energy

Female respondents in the study said that they are used to charcoal and firewood for cooking food and water for domestic consumption. Women also depend on the sale of charcoal and firewood to supplement their household incomes. Although these energy consumption patterns help women meet their domestic and commercial responsibilities, these activities are sustained by deforestation in the peripheral parts of the city, which constrains rain formation and makes erosion possible. Men on the other hand own motorcycles and a few do purchase used cars whose movement depends on petroleum and diesel products. In fact this has attracted motor garages in and around the two neighborhoods, where petroleum and diesel are used to render services to customers (usually male taxi and motorcycle drivers). These male-oriented energy consumption activities do result in burning of fossil fuels that emit carbon into the atmosphere. Therefore, the gendered nature of energy consumption is relevant in undertaking research on carbon footprints in urban areas, for purposes of targeting mitigation strategies at behavioral change rather than apportioning blame. If the transformation of energy systems takes off right from household to community and finally city level, there is need for gender-disaggregated information on the barriers that limit the uptake of renewable technology. Such information provides a basis for designing gender-responsive incentives that promote the installation of energy efficiency measures and renewable energy systems.

Housing

The sociocultural order that every responsible household head (usually male) should provide a habitable shelter for the family has contributed to less-vegetated surfaces in the neighborhood of Kasubi-Kawaala, which ultimately leads to a decrease in evaporative cooling. Settlement in informal and flood-prone areas was said to be driven by gendered stereotypes that a true family head should provide his spouse and children with shelter. This coupled with lack of knowledge on what it takes to settle in a wetland underlies men's involvement into a chain of environmentally insensitive housing activities. These include buying "cheaper" land in wetlands; metal stockpiling; cutting and transportation of timber; unregulated quarrying and transportation of stone, soil, and sand; construction of methane-emitting pit latrines and open sewage pits. This chain of housing activities ultimately leads to destruction of swamps, increases erosion, increases vulnerability to flooding of houses, and leads to burning of fossil fuels. Conversely, women have lesser bargaining power than men during the purchase of land for housing, relocation, or renovation of the residential dwelling. This division of gender roles and entitlements in the housing sector provides a strong entry point for promoting domestic energy efficiency and security through community-based, low-income, and green housing projects.

Lessons for Cities in Africa

Addressing the sector-wide impacts of climate change is a policy and urban management challenge for African cities. The challenge is even more complex since the cross-cutting issues of gender issues are usually overlooked while seeking to mitigate and adapt to the impacts of climate change. Therefore, one of the prospective approaches to climate change is gender-responsive planning. This kind of planning enables the understanding of vulnerability from a service end-user point of view by sex, location, and sector. The information can thereafter be used for evidence-based prioritization while making policies, plans, and budgets that can strategically and equitably enhance women's as well as men's capacity to cope at neighborhood to city and national level. In Kampala, national- and city-level assessments on climate change have been done including that on gender, with support from UNHABITAT. But capacity to utilize this information innovatively, in ways that can build a gendered response to climate change, was not visible at the time of the study.

Such a shortfall in institutional capacity does not necessarily mean climate change-focused training for planners and managers at city level, but rather improving the quality of urban governance and better management of urbanization through creation of climatic knowledge bases on the frequency and magnitude of climatic hazards, mapping gender differentials in vulnerability, opening sector-by-sector dialogue on how urban planning can be made climate compatible, and designing gender-responsive mitigation and adaptation plans. This cycle of planning activities provides inclusive and productive ways of addressing the challenges associated with low institutional capacity to enable women and men cope with climatic stressors from neighborhood to city and national level. It is also important to scale up the existing environmental projects in a given sector, like waste management, through mobilization from below by seeking greater involvement from NGOs, CBOs, and local political representatives who have a direct interaction with women and men at neighborhood level. Even with little or no knowledge of gender issues in climate change, the development of a city-wide agenda through meaningful participation of all stakeholders can bring different yet valuable experiences to the fore of planning and build an institutional culture where learning from and among each other is the avenue for consensus building on the future of a given city.

CONCLUSION

Gender inequalities can contribute to the intensity of climatic impacts and the impacts of climate change do worsen gender inequalities across different urban sectors. Dealing with this inverse connection means a shift in urban planning and mobilization of different stakeholders to build a gender-responsive adaptation plans. This is probably best achieved by considering gender when undertaking sector-by-sector climate change assessments and planning for adaptation.

ACKNOWLEDGMENTS

The information presented in this chapter would not be meaningful without support from the residents of Kasubi-Kawaala neighborhood in Kampala city. Thanks to UNHABITAT, Nairobi, Kenya and the Foundation for Urban and Regional Studies, Oxford, UK for funding the research. I am also grateful to the American Sociological Association Community and Urban Section for providing the opportunity to publish this material.

REFERENCES

Berrang-Ford, L., Ford, J. D., & Paterson, J. (2011). Are we adapting to climate change? *Glob Environ Change*, *21*, 25–33.

Blakely, E. J. (2007). *Urban planning for climate change*. Working Paper No. WP07EB1. Lincoln Institute of Land Policy.

Bulkeley, H., Schroeder, H., Janda, K., Zhao, J., Armstrong, A., Yi Chu, S., & Gosh, S. (2009). *Cities and climate change: The role of institutions, governance and urban planning*. Report prepared for the World Bank Urban Research Symposium on Climate Change, June 28–30, Marseille.

Department for International Development. (2008). *Gender and climate change: Mapping the linkages: A scoping study on knowledge and gaps*. BRIDGE: Institute of Development Studies, UK.

Economic Commission for Africa. (2009). *Gender and climate change: Women matter*. United Nations Economic Commission for Africa (UNECA).

Füssel, H. (2007). Adaptation planning for climate change: Concepts, assessment approaches and key lessons. *Sustainability Science*, *2*, 265–275.

GTZ. (2007). Gender and urban transport: Fashionable and affordable. A source book for policy-makers in developing cities. Kampala City Development Plan (2010–2012).

Kampala City Development Plan. Kampala (2010–2011). Kampala: Kampala City Council.

Lwasa, S. (2010). Adapting Urban areas in Africa to climate change: The case of Kampala. *Current Opinion in Environmental Sustainability*, *2*, 166–171.

Oxfam. (2008). Climate wrongs and human rights – Putting people at the heart of climate change policy.

Tompkins, E., & Adger, W. (2004). Does adaptive management of natural resources enhance resilience to climate change? *Ecology and Society*, *9*(2), 10.

Uganda Bureau of Statistics (2010). Urban agriculture census. Kampala: Uganda Bureau of Statistics (2008–2009).

United Nations Development Fund for Women. (2009). *Women, climate change and refugees*. New York, NY.

United Nations Development Programme. (2009). *Resource guide on gender and climate change*. New York, NY.

United Nations Human Settlements Programme. (2009). *Cities and climate change assessment report – Kampala Uganda*.